Modern Applied Regressions

Modern Applied Regressions creates an intricate and colorful mural with mosaics of categorical and limited response variable (CLRV) models using both Bayesian and Frequentist approaches. Written for graduate students, junior researchers, and quantitative analysts in behavioral, health, and social sciences, this text provides details for doing Bayesian and frequentist data analysis of CLRV models. Each chapter can be read and studied separately with R coding snippets and template interpretation for easy replication. Along with the doing part, the text provides basic and accessible statistical theories behind these models and uses a narrative style to recount their origins and evolution.

This book first scaffolds both Bayesian and frequentist paradigms for regression analysis and then moves onto different types of categorical and limited response variable models, including binary, ordered, multinomial, count, and survival regression. Each of the middle four chapters discusses a major type of CLRV regression that subsumes an array of important variants and extensions. The discussion of all major types usually begins with the history and evolution of the prototypical model, followed by the formulation of basic statistical properties and an elaboration on the doing part of the model and its extension. The doing part typically includes R codes, results, and their interpretation. The last chapter discusses advanced modeling and predictive techniques—multilevel modeling, causal inference and propensity score analysis, and machine learning—that are largely built with the toolkits designed for the CLRV models previously covered.

Chapman & Hall/CRC
Statistics in the Social and Behavioral Sciences Series

Series Editors
Jeff Gill, Steven Heeringa, Wim J. van der Linden, Tom Snijders

Recently Published Titles

Big Data and Social Science: Data Science Methods and Tools for Research and Practice, Second Edition
Ian Foster, Rayid Ghani, Ron S. Jarmin, Frauke Kreuter and Julia Lane

Understanding Elections through Statistics: Polling, Prediction, and Testing
Ole J. Forsberg

Analyzing Spatial Models of Choice and Judgment, Second Edition
David A. Armstrong II, Ryan Bakker, Royce Carroll, Christopher Hare, Keith T. Poole and Howard Rosenthal

Introduction to R for Social Scientists: A Tidy Programming Approach
Ryan Kennedy and Philip Waggoner

Linear Regression Models: Applications in R
John P. Hoffman

Mixed-Mode Surveys: Design and Analysis
Jan van den Brakel, Bart Buelens, Madelon Cremers, Annemieke Luiten, Vivian Meertens, Barry Schouten and Rachel Vis-Visschers

Applied Regularization Methods for the Social Sciences
Holmes Finch

An Introduction to the Rasch Model with Examples in R
Rudolf Debelak, Carolin Stobl and Matthew D. Zeigenfuse

Regression Analysis in R: A Comprehensive View for the Social Sciences
Jocelyn H. Bolin

Analysis of Intra-Individual Variation: Systems Approaches to Human Process Analysis
Kathleen M. Gates, Sy-Min Chow, and Peter C. M. Molenaar

Modern Applied Regressions: Bayesian and Frequentist Analysis of Categorical and Limited Response Variables with R and Stan
Jun Xu

For more information about this series, please visit: https://www.routledge.com/
Chapman--HallCRC-Statistics-in-the-Social-and-Behavioral-Sciences/book-series/
CHSTSOBESCI

Modern Applied Regressions

Bayesian and Frequentist Analysis of
Categorical and Limited Response Variables
with R and Stan

Jun Xu

CRC Press
Taylor & Francis Group
Boca Raton London New York

CRC Press is an imprint of the
Taylor & Francis Group, an **informa** business

A CHAPMAN & HALL BOOK

First edition published 2023
by CRC Press
6000 Broken Sound Parkway NW, Suite 300, Boca Raton, FL 33487-2742

and by CRC Press
4 Park Square, Milton Park, Abingdon, Oxon, OX14 4RN

CRC Press is an imprint of Taylor & Francis Group, LLC

ISBN: 9780367173876 (hbk)
ISBN: 9781032376745 (pbk)
ISBN: 9780429056468 (ebk)

DOI: 10.1201/9780429056468

Publisher's note: This book has been prepared from camera-ready copy provided by the authors.

In memory of my grandparents

Contents

Preface   xi

1   **Introduction**   1
    1.1   Categorical and Limited Response Variables   1
       1.1.1   A Brief History of CLRV Models   2
       1.1.2   Overview of CLRVs   3
    1.2   Approaches to Regression Analysis   4
       1.2.1   Frequentist Approach to Regression Modeling   4
       1.2.2   Bayesian Approach to Regression Modeling   4
          1.2.2.1   The Example of COVID-19   5
       1.2.3   Priors   6
          1.2.3.1   Conjugate Priors   6
          1.2.3.2   Informative, Non-informative, and Other Priors   7
       1.2.4   Markov Chain Monte Carlo (MCMC)   8
    1.3   Introduction to R   8
       1.3.1   RStudio   9
       1.3.2   Use R as Calculator   10
       1.3.3   Set Up Working Directory   11
       1.3.4   Open Log File   12
       1.3.5   Load Data   12
       1.3.6   Subset Data   13
       1.3.7   Examine Data   13
       1.3.8   Examine Individual Variables   14
       1.3.9   Save Graphs   15
       1.3.10   Add Comments   15
       1.3.11   Create Dummy Variables and Check Transformation   16
       1.3.12   Label Variables   17
       1.3.13   Label Values   17
       1.3.14   Create Ordinal Variables   18
       1.3.15   Check Transformation   18
       1.3.16   Drop Missing Cases   19
       1.3.17   Graph Matrix   19
       1.3.18   Save Data   20
       1.3.19   Close Log   20
       1.3.20   Source Codes   20
    1.4   Review of Linear Regression Models   21
       1.4.1   A Brief History of OLS Rregression   21
       1.4.2   Main Results of OLS Regression   22
          1.4.2.1   OLS Estimator and Variance-Covariance Matrix   23
       1.4.3   Major Assumptions of OLS Regression   23
          1.4.3.1   Zero Conditional Mean and Linearity   24
          1.4.3.2   Spherical Disturbance   24

		1.4.3.3	Identifiability	24
		1.4.3.4	Nonstochastic Covariates	25
		1.4.3.5	Normality	25
	1.4.4	Estimation and Interpretation	25
	1.4.5	A Brief Introduction to Stan and Other BUGS-like Software	31
	1.4.6	Bayesian Approach to Linear Regression	32

2 Binary Regression   **37**
	2.1	Introduction	37
		2.1.1	A Brief History of Binary Regression	37
		2.1.2	Linear Probability Regression	38
	2.2	Maximum Likelihood Estimation	40
		2.2.1	Simple MLE Examples	41
		2.2.2	MLE for Binary Regression	45
		2.2.3	Numerical Methods for MLE	46
		2.2.4	Normality, Consistency, and Efficiency	48
		2.2.5	Nonlinear Probability	49
	2.3	Hypothesis Testing and Model Comparisons	49
		2.3.1	Wald, Likelihood Ratio, and Score Tests	51
			2.3.1.1 Graphical Comparison of Wald, LR, and Score Tests	55
		2.3.2	Scalar Measures	56
		2.3.3	ROC Curve	59
		2.3.4	Goodness of Fit Measures: The Hosmer-Lemeshow Test	61
		2.3.5	Limitations of NHST	61
	2.4	Interpretation of Results	62
		2.4.1	Precision Estimates	63
			2.4.1.1 End-Point Transformation	64
			2.4.1.2 Delta Method	64
			2.4.1.3 Re-sampling Methods	64
		2.4.2	Interpretation Based on Predictions	65
		2.4.3	Interpretations Based on Effects	68
			2.4.3.1 Odds Ratios	68
			2.4.3.2 Discrete Rates of Change in Prediction	69
			2.4.3.3 Marginal Effects	71
		2.4.4	Group Comparisons	73
	2.5	Bayesian Binary Regression	74
		2.5.1	Priors for Binary Regression	75
		2.5.2	Bayesian Estimation of Binary Regression	78
		2.5.3	Bayesian Post-estimation Analysis	82
		2.5.4	Bayesian Assessment of Null Values	84

3 Polytomous Regression   **87**
	3.1	Ordered Regression	88
		3.1.1	Types of Ordinal Measures and Regression Models	88
		3.1.2	A Brief History of Ordered Regression Models . . .	89
		3.1.3	Cumulative Regression	89
			3.1.3.1 Model Setup and Estimation	89
			3.1.3.2 Hypothesis Testing and Model Comparison	93
			3.1.3.3 Interpretation	95
		3.1.4	Testing the Proportional Odds/Parallel Lines Assumption	98
		3.1.5	Partial, Proportional Constraint, and Non-parallel Models	101

      3.1.6   Continuation Ratio Regression 106

      3.1.7   Adjacent Category Regression 112

      3.1.8   Stereotype Logit 114

  3.2 Extentions to Classical Ordered Regression Models 116

      3.2.1   Inflated Ordered Regression 116

      3.2.2   Heterogeneous Choice Models 120

      3.2.3   General Guidelines for Model Selection 122

  3.3 Multinomial Regression 123

      3.3.1   Multinomial Logit Regression 123

      3.3.2   Multinomial Probit Regression 129

  3.4 Bayesian Polytomous Regression 130

      3.4.1   Bayesian Estimation 130

          3.4.1.1   Bayesian Parallel Cumulative Ordered Regression . . . 131

          3.4.1.2   Bayesian Non-Parallel Cumulative Ordered Regression . . . 132

          3.4.1.3   Bayesian Stereotype Logit Model 134

4 Count Regression   **137**

  4.1 Poisson Distribution 137

  4.2 Basic Count Regression Models 139

      4.2.1   Explore the Count Response Variable 140

      4.2.2   Plot Observed vs. Predicted Count Proportions 141

      4.2.3   Poisson Regression 142

      4.2.4   Contagion, Heterogeneity, and Over-Dispersion 150

      4.2.5   Quasi-Poisson Regression 152

      4.2.6   Negative Binomial Regression 152

  4.3 Zero-Modified Count Regression 157

      4.3.1   Zero-Truncated Models 158

      4.3.2   Hurdle Models 159

      4.3.3   Zero-Inflated Models 160

  4.4 Bayesian Estimation of Count Regression 162

      4.4.1   Bayesian Estimation of Negative Binomial Regression 162

      4.4.2   Bayesian Estimation of Zero-Inflated Poisson Regression 167

5 Survival Regression   **169**

  5.1 Introduction 169

      5.1.1   Censoring and Truncation 170

  5.2 Basic Concepts 172

      5.2.1   Time and Survival Function 172

      5.2.2   Hazard Function 173

  5.3 Descriptive Survival Analysis 174

      5.3.1   The Kaplan-Meier Estimator 174

      5.3.2   The Log-Rank Test 177

  5.4 Accelerated Failure Time Model 180

      5.4.1   Exponential AFT Regression 181

      5.4.2   Weibull AFT Regression 183

  5.5 Parametric Proportional Hazard Regression 186

      5.5.1   Exponential PH Regression 187

      5.5.2   Weibull PH Regression 188

  5.6 Cox Regression 190

  5.7 Testing the PH Assumption 192

  5.8 Bayesian Approaches to Survival Regression 194

5.8.1   Bayesian Estimation of Weibull PH Model Using `rstan`   195
5.8.2   Bayesian Estimation of Survival Models Using `spBayesSurv`   202

6   Extensions   **205**
6.1   Multilevel Regression   205
    6.1.1   Multilevel Logit Regression   206
    6.1.2   Multilevel Count Regression   211
    6.1.3   Bayesian Multilevel Regression   214
6.2   Causal Inference   216
    6.2.1   Average Treatment Effects   217
        6.2.1.1   Average Treatment Effects   217
        6.2.1.2   Average Treatment Effects for the Treated   218
        6.2.1.3   (Strong) Ignorability of Treatment Assumption   218
    6.2.2   Propensity Score Analysis   219
        6.2.2.1   Propensity Score Matching   219
        6.2.2.2   Mahalanobis Distance Matching   223
        6.2.2.3   Genetic Matching   224
        6.2.2.4   Coarsened (Exact) Matching   226
6.3   Machine Learning   228
    6.3.1   Basic Concepts   229
        6.3.1.1   Machine Learning and Statistical Learning   229
        6.3.1.2   Supervised Learning and Unsupervised Learning   230
        6.3.1.3   Regression and Classification   230
        6.3.1.4   Training, Validation, and Test   230
    6.3.2   Supervised Learning   230
        6.3.2.1   Regularization: Ridge and Lasso Regression   230
        6.3.2.2   Penalized Binary Logit   241
        6.3.2.3   Decision/Regression Trees   243
        6.3.2.4   Bagging and Random Forests   250
        6.3.2.5   Classification Trees   252

Bibliography   **257**

Index   **273**

Preface

No efforts are taken futile, and no
jades are made easy.

based on *Lotus Sutra* and *Book of Poetry*

I have always wanted to read a single-volume paperback of a statistics book that recounts historical anecdotes, traces evolutionary changes, provides intuitive conceptualization, formulates mathematical basics, illustrates the how-to with some open-source software, and showcases post-estimation analysis of the statistical modeling techniques under study. It turns out that the best solution is to write one.

There has been an explosive growth of statistical literature and applications of the models for categorical and limited response variables since the 1980s. Several classic texts already provide comprehensive coverage of the derivation, estimation, and inference of these models. This text differs from other books in four ways. First, it prioritizes the doing part of data analysis with an appropriate coverage of statistical theories and mathematical formulations. Readers can directly skip to the application part of each section and reproduce the results with ease without belaboring through theoretical derivation. This manuscript also provides numerous textual templates for post-estimation analysis and interpretation, wherever appropriate and possible, so that practitioners and students can easily replicate the results and extend to their own empirical problems. My years of experience in this area has convinced me to use this application–theory and then theory–application iterative learning cycle; that is, when one starts to learn a new statistical technique or mathematical concept, an easy empirical example and intuitive interpretation of this particular concept or technique—which should follow a very basic conceptual understanding of the topic to begin with—may have to precede intensive theoretical study of the topic, followed by a serious attempt to study the corresponding theory, back to application and then theory, and so on and so forth (learning spiral). Usually, one is advised to revisit the same topic multiple times by referencing to the same or different texts. But for most practitioners and students, this doing part may be the greatest hurdle to mastering a mathematical or statistical technique. And without it, many would simply surrender to frequently abstract and almost incomprehensible derivations without going much further. This is mostly an issue about confidence and perseverance.

A second important feature is the rich description of the origins and evolution of major types of models discussed in this text. Authenticity and realism from classics are frequently lost in rendition and re-interpretation. I have always found it fascinating and effective to learn about the evolution of these modeling techniques to better understand them. History itself is interesting enough. For example, Cox came up with the Cox proportional-hazards model after having several days of high fever, and it really prompted me to imagine what went through his mind in those days. Stories pique curiosity among both teachers and students. While finding out stories behind these models, I became more infatuated with these stories than the techniques themselves. Learning mathematics and statistics is supposed to be fun. And hopefully, this is the fun part for those avid learners.

I've also tried to add a commentary, critical, and sometimes narrative flavor to the writing to make it more enjoyable to read. This is often interwoven with the preceding feature. Based on my personal learning and teaching experience, I've come to realize that there can be a wide gap between theory and practice with some experiences and tricks to be acquired only through apprenticeship (implicit knowledge). This text mixes the two and attempts to fill in those missing gaps, including, for example, the criticism of NHST, discussions about group comparisons, and commentaries to essentialize causal inference and machine learning.

The last main feature of this book is its integration of the frequentist and Bayesian approaches so as to compare and contrast these two competing philosophical paradigms in statistical and data science. Over the past few decades, especially since the late 1980s and early 1990s, the Bayesian framework has metamorphosed from an obscure alternative to a serious challenger of the traditional Fisherian frequentist statistics, with the exponential growth in computing power and the "second-generation MCMC revolution." Doing Bayesian statistics follows the natural process of data analysis by empowering users with prior beliefs and updating their beliefs with new data. Bayesian inference can also estimate models that are otherwise too complex for the usual numerical algorithms to find answers (e.g., imposing the scale constraints in stereotype logit models). Thus, it is fruitful for contemporary data scientists and statisticians to become aware of its great potential. This book can be used both as a reference for applied researchers and a textbook for graduate students in behavioral, medical, and social sciences. It would be appropriate to assign this text in a graduate-level course on categorical data analysis or modern regression analysis, broadly defined. The following few sections elaborate on the main contents in each chapter by focusing on their unique features.

Organization of the Book

The first chapter introduces major types of categorical and limited response variables (CLRV) and general regression modeling strategies, both frequentist and Bayesian. It also provides an overview of the primary statistical environment used in this book—R and RStudio—and multiple Bayesian estimation software applications, with a focus on Stan. This introductory chapter concludes with a concise review of classical linear regression that serves as the basis for the non-continuous response variable models to be covered in later chapters. Both frequentist and Bayesian estimations of the OLS regression model and their associated results are presented and discussed in this section.

The second chapter of this text begins with a brief history and evolution of binary regression models, followed by a general discussion of maximum likelihood estimation (MLE). Then the discussion centers on null hypothesis significance testing (NHST) along with its limitations. This chapter also features various ways to interpret results, conduct group comparisons, and make Bayesian inferences about binary regression models. With a comprehensive guideline, this chapter illustrates different methods for interpretation with simple empirical examples. Instead of preferring one set of statistics, this chapter proposes using multiple quantities of interest, including both effects and predictions, global and local statistical measures, and differences and ratios, to provide a panoramic view of the results for substantive interpretation and triangulation. The Bayesian section of this chapter discusses how to run a binary logit/probit model using Stan along with the user-written R package, `rstan`, interpret results with credible intervals, compare and select models using the Watanabe-Akaike information criterion (WAIC) and the leave-one-out (LOO) statistic, and make Bayesian assessment of null values using regions of practical equivalent (ROPEs) and highest density intervals (HDIs).

Built on the preceding chapter on binary regression, the third chapter covers regression models with polytomous response variables. It sets out its discussion with the classic proportional odds model, including its model derivation, estimation, and post-estimation analysis. This chapter then proceeds with the parallel lines (PL) assumption and extensions to the PL model that relax this assumption along multiple dimensions, including the data-generation processes (cumulative, adjacent, sequential), how parameter constraints are imposed (parallel, partial, non-parallel, or partial with proportionality constraints), and choices of link functions (logit, probit, and cloglog). The discussion typically focuses on application and interpretation. This chapter also introduces other major types of extensions to the PL model by addressing issues such as heteroscedasticity (heterogeneous-choice ordered regression models) and distributional/process mixture (category-inflated polytomous models). The last section of this chapter covers Bayesian estimation and inference of polytomous regression models. Main (parallel cumulative logit) and special types (non-parallel cumulative logit and stereotype logit) of ordered regression models are sampled to illustrate how to use R and Stan to estimate these models, conduct Bayesian null hypothesis testing, and provide substantive interpretations.

The fourth chapter moves onto count regression models. This chapter opens with a review of the Poisson distribution, its assumptions and limitations, and how the Poisson distribution is linked to the binomial and normal distributions. Then the discussion focuses on the Poisson regression as a starting point for modeling count responses, and the problem and various origins of over-dispersion commonly seen in count response variables. The next few sections discuss popular remedies for over-dispersion and unobserved heterogeneity, including the negative binomial, hurdle, zero-inflated regression models, and count models with censoring and truncation. The last section of this chapter covers Bayesian estimation and inference of count regression models. Again, the focus is on application and interpretation.

Chapter five provides a gentle introduction to survival methods by focusing on survival data with right-censoring, including the non-parametric Kaplan Meier estimator, the parametric exponential and Weibull regression (both proportional hazard and accelerated failure time models), and the semi-parametric Cox regression. This chapter, though, does not cover advanced topics such as time-varying covariates or repeated time-to-events. The first section of this chapter elaborates on basic but important concepts and functions in survival analysis, such as time, probability density functions for the time-to-event variables, hazard function, survival function, cumulative hazard and cumulative survival functions, and the association among these functions. The following few sections discuss the workhorse in survival regression modeling, including exponential, Weibull, and Cox regressions. Special attention is paid to the interpretation of results. The last section covers the Bayesian estimation of survival models and their post-estimation analyses.

The last chapter of this manuscript briefly overviews advanced modern statistical models/techniques that add additional layers of complexity to or build on basic regression models for categorical and limited response variables along different dimensions. In particular, this chapter discusses multilevel analysis of categorical data, causal inference, and propensity score analysis, and introduces the relatively new framework of machine and statistical learning that serves as the basis for analyzing large dynamic data and AI-powered predictive analytics.

Data, Software, and Web Support

To effectively illustrate common challenges and issues in empirical data analysis, this manuscript uses only a few data files so that the same response variables, wherever appropriate, are analyzed using different models with their corresponding data management procedures. This manuscript uses R, the lingua franca in data science and statistics, throughout all chapters. Bayesian analyses are conducted primarily with Stan (a probabilistic programming language for Bayesian methods), `rstan` (a user-written R package that interfaces R with Stan), and `rstanarm` (an R wrapper package that integrates both Stan and `rstan` and nicely automates and largely simplifies the whole process). Many R coding snippets used in the book are adapted from examples from various R packages and related resources. Note that R is a member of GNU, a mass collaboration project for free and open-source software that Richard Stallman started in the early 1980s. So R and its user-contributed packages come with no cost. Stan is also an open-source software, free of charge for download, use, and extend. In the text, we use `typewriter` as the font style for R package and function names, R codes, and output outside code chunk boxes. While using functions from R packages, this text usually cites them once for the first application in corresponding sections or chapters. The online resources for this book, including R and Stan codes and supplementary notes, can be accessed at https://sites.google.com/site/socjunxu/home/statistics/modern-applied-regressions.

Acknowledgment

While growing from a graduate student who was drawn more into grand narratives than research methodology, I have learned statistics in a tortuous way. First, I would like to thank Dr. Charles Cappell for inviting me to trek on the less traveled terrain of social science and working in his quantitative research lab during my formative years at Northern Illinois. He is the one who first told me that I need to pick up calculus and linear algebra to have a real understanding of statistics. My gratitude also goes to the quantitative training that I received from the economics and sociology programs at IU-Bloomington. My collaborative work with Dr. J. Scott Long on the SPost 9 project and the learning of categorical data analysis under his tutelage helped me metamorphose from an analytical student to a statistical researcher. My coursework on econometrics with Drs. Konstantin Tyurin, Pravin Travedi and Tong Li taught me not only the fundamentals of statistical modeling but also how to ask the right statistical questions. Drs. Lahn Tran and Bruce Solomon at the Mathematics Department helped me lay a firm mathematical foundation for my quantitative works in later years.

I would also like to thank Dr. Andrew S. Fullerton at Oklahoma State and Dr. Jason Doll at Francis Marion. Andy's early email inquiries about sequential logit models initiated probably my most successful line of research on the ordered regression models, when I was in a period of self-doubt and disappointment. The encounter with Jason was accidental while he was still a graduate student in the Biology Department at Ball State. We quickly hit it off because of our shared passion for using Bayesian statistics and R through the *Applied Statistics and R Working Group*. His enthusiasm, intellect, and contentment, and his passion for fisheries is simply contagious. To him I owe considerably for helping me learn how to program Bayesian models with JAGS. Dr. Roger Wojtkiewicz, who took the risk to help recruit me to Ball State and later became my office neighbor, inspired me in many ways to aim high and achieve great through his own exemplary work and numerous small but stimulating talks. For that, I thank him very much. Additional thanks are owed to Jie Li at IU-Bloomington for reading an early draft of the section on machine learning. I am also grateful for all the resources provided by my department and Ball State for completing this manuscript, including a sabbatical semester slated to work on a different project.

In the writing process, I consulted authors of several R packages and new models for coding and other technical issues, and I can't thank them enough for their help. They include Drs. Mark Harris (for the ZiOPC model), John Kruschke (BEST), Thomas Yee (VGAM), Jas Sekhon (Match), Haiming Zhou (spBayesSurv), and the Stan development team. It took me almost four years to complete this manuscript, and I contacted many scholars for their expertise. I have to apologize if I miss anyone who is supposed to be on this list.

Not enough can I thank the two editors at Chapman & Hall/CRC, John Kimmel and Lara Spieker, and the copy editor at KnowledgeWorks Global, Shabir Sharif, without whose advice and support the completion of this manuscript is impossible. The comments and suggestions from anonymous reviewers for both my book proposal and specific chapters are also crucial for improving the manuscript in significant ways. The help and support from the editorial and typesetting team at Chapman & Hall/CRC is certainly critical in eliminating errors and producing a fine manuscript. Last but not least, I feel very much indebted to all my family members for proofreading this manuscript and their loving support. Much of the credit for completing this project should go to them.

1

Introduction

> The Way begets oneness, oneness engenders twoness, twoness breeds threeness, and threeness multiplies into all.
>
> in Chapter 42 of *Dao De Jing* by Laozi

1.1 Categorical and Limited Response Variables

Categorical and limited response variables cover a wide range, or even the majority, of variables. When it comes to measurement, all variables, to some extent, can be viewed as categorical and limited response variables, since very few or none would satisfy the standards of an idealized continuous variable, covering the full range of the real number line that can be infinitesimally divisible. A categorical variable is a variable that usually consists of, putatively, several mutually exclusive (see the discussion on fuzzy set for exceptions) categories that exhaust the measurement space. For example, a variable for measuring health insurance coverage can include private insurance, public assistance, and no coverage. A limited variable usually refers to a variable that is limited in some way while being compared with a theoretical continuous variable, and it straddles somewhere in between the two ideal types of categorical and continuous variables. For example, the number of days of having poor mental health in the past 30 days can only have integer values ranging from 0 to 30, with an upper truncation, or it can be the elapse of time, which is (artificially) unidirectional and can only be non-negative. Both variables are limited response variables. There is a large variety of categorical and limited response variables, and this manuscript primarily considers binary, ordinal, polytomous(multi-category), count, and survival time (time-to-event) response variables under the general regression framework, with introductory coverage of important extensions such as causal inference, multilevel modeling, and machine learning.

Early origins and ensuing developments of the analysis of categorical and limited response variables, such as those related to mortality data and risks (Graunt, 1662; Hald, 1990), in entomological studies (e.g., pest control) (Bliss, 1934), and in legal systems (e.g., the number of wrongful convictions) (Poisson, 1837), laid empirical and some theoretical foundation for the modern progress of such analyses. Karl Pearson and Udny Yule are usually considered to be the first two modern pioneers that provided effective conceptual and empirical tools in this area. Because of the complex nature of such variables, scholarly rifts existed even between Pearson and Yule in how to approach them. Pearson was a polymath, making substantial contributions to multiple important areas, and he was Yule's supervisor (and mentor-teacher) for a while (1893–1899) (Hald, 2000; Agresti, 2013). Pearson was known for his polemic, uncompromising, and vindictive dispositions, and he was also notorious for

DOI: 10.1201/9780429056468-1

his partly progressive and partly regressive thoughts. Albeit with exceptions, talents tend to have tempers.

The early divide between Pearson and Yule on the continuity vs. discreteness of nominal level variables also reflects fundamental issues of this type of variables that still haunt us today. Pearson, a proponent of the continuity thesis, contended that the correlation between two categorical variables in a contingency table reflects a relationship between two underlying continuous variables, and such correlation can be measured by the contingency coefficient, a correlation measure similar to the correlation coefficient in linear regression; Yule, however, posited that variables such as bacterial/viral infection and vaccination (vaccinated or unvaccinated) and survival (life/death) are discrete in nature, and cannot be approached with some underlying continuum or multivariate normal distribution (Agresti, 2013). From today's viewpoint, it would be imperious to solely argue for only one way or the other without invoking some contingency or qualification. For some nominal level categorical variables, such as attitudes and perceptions, they can be comfortably viewed as manifestation of some latent continuous variables, or multivariate normal hump in a multidimensional space. But for other variables/factors that have clear boundaries and intervals, such as birth/death and viral infections, a discrete approach appears more appropriate. Even so, a somewhat extrapolated argument can still be made that the birth or infection process follows a continuum, although there may exist some clear-cut thresholds that separate qualitatively different states. Most categorical and limited response variables can be argued both ways. The example that Pearson illustrated—the association between stature in father and son—to argue for an underlying continuum is actually somewhat tautological since the nominal/ordinal variable was created from a variable that can be viewed as a continuous one (Pearson, 1904). An interesting historical fact is that in the same article, Pearson, for the first time, used the term contingency—thus the origin of contingency table—to denote the total deviation from independence between two cross-classified variables (Pearson, 1904, p. 5). On the other hand, Yule was inclined to embracing the discrete nature of categorical/nominal variables due in part to his work on vaccination (Greenwood and Yule, 1915). Yule (1900) appeared to distance from the idea of latent continuity by proposing the coefficient of association, as opposed to the coefficient of correlation. In the same paper, Yule also popularized the idea of odds ratio $\kappa = \frac{\pi_{11}\pi_{22}}{\pi_{12}\pi_{21}}$ and proposed the normalized measure of odds ratio, Yule's $Q = \frac{\kappa-1}{\kappa+1}$, which is bound between -1 and $+1$.

A second important but much lesser discussed dyadic concept is fuzzy vs. crisp set. Usually, the categories or response levels of multi-category variables appear to have clear boundaries and are mutually exclusive, for example life and death, party membership, and voter turnout. There are, however, a large variety of categorical, nominal, limited response, and non-continuous variables—probably more so than those with crisp boundaries—that have uncertainty in measurement, and such uncertainty may come from multiple sources, including response bias, measurement coarsening, or a combination of both. The thresholds for hypertension, hyperglucemia, and poverty, for example, are fixed, at least over a certain time span and in some regions, but it would be hard to argue that the numbers in the vicinity of the thresholds necessarily belong to one or the other category. When this happens, which is not uncommon, we can argue that there is a fuzzy as opposed to crisp set. Unfortunately, the fuzzy set theory (Smithson and Verkuilen, 2006) and the fuzzy regression framework (Tanaka et al., 1982; Chukhrova and Johannssen, 2019) have not been fully developed and thus will not be formally treated in this text.

1.1.1   A Brief History of CLRV Models

Pearson and Yule are often viewed as the two preeminent statisticians that laid the foundational work for modern analysis of categorical and limited response variables, and they introduced such important statistical concepts as contingency table, χ^2 test, odds ratio, and correlation of association, just to name a few. Before the two, statistical works on such variables were largely descriptive (e.g., the mortality table by John Graunt in 1662), sporadic and tangential. The next important figure, arguably more so than Pearson and Yule and the lifetime foe of Pearson, is Ronald A. Fisher. Fisher introduced, formalized, and popularized in a series of works published from 1912 to 1922 the method of maximum likelihood (ML), which turns out to be one of the most important and widely used statistical techniques since the late 20th century until present (Aldrich, 1997a). Fisher also proposed the concept of degrees of freedom (df), and corrected the df for the χ^2 test of a two-way contingency table to be $(r - 1)(c - 1)$, where r and c denote the number of rows and columns, respectively. Arguably Fisher's statistical magnum opuses, *Statistical Methods for Research Workers* (Fisher, 1934) and *The Design of Experiments* (Fisher, 1935) devised and/or formalized many important statistical concepts and techniques useful for the analysis of categorical and limited response variables, including Fisher's exact test for a 2×2 table, Gosset's Student t distribution, and experimental design (e.g., the Latin Square and the factorial design). Fisher's innovative work coincided with some inceptive yet novel attempts at binary regression models, for example, the first application of the probit model by Bliss (1934) and Berkson (1944) for the logit model. The categorical and limited response variable models that ensued these early developments include the ordered regression variable model in the 1950s (Aitchison and Silvey, 1957), the multinomial logit model in the 1960s and 1970s (McFadden, 1973), the stereotype logit model in the 1980s (Anderson, 1984), the generalized linear model framework proposed in the late 1980s (McCullagh and Nelder, 1989), and the even more encompassing framework, the vector generalized linear and additive models proposed by Yee (2015). Coupled with the developments of the mainstream CLRV models are extensions and ingenious applications of these models in seemingly unrelated fields in statistics and data science, broadly defined, for example, Bayesian statistics, causal inference, and statistical as well as machine learning.

1.1.2 Overview of CLRVs

Prior scholarship has delineated distinctions among categorical and limited response variables along several sometimes overlapping dimensions (Agresti, 2013). First, these variables differ in their countability/uncountability. Binary, ordinal, and multinomial variables are all countable, but count, survival time, and continuous variables are uncountable since their numbers of categories are unlimited in theory. The second pair of antithetical concepts is continuity and discreteness. Binary, ordinal, and multinomial to count variables are discrete variables since only crisp points and no continuous value ranges/regions on the real line can be taken, whereas time and other variables that theoretically can take on values from a continuous range/region are continuous. Among the continuous variables, we can also have the interval and ratio distinction. Interval variables are those for which the difference in values has numerical meanings, for example, the temperature measured on the Fahrenheit scale. Such values, however, cannot be compared by taking ratios. So a four degrees Fahrenheit is not twice as warm/cold as a two degrees Fahrenheit. But two inches is the twice of the length/height of one inch, and thus length measured in inch is a ratio variable. Another quality that only time-to-event variables and some ordinal variables have is sequentially or temporally unidirectional; that is, greater values of these variables are grown/progress from smaller values. For example, a college education presupposes secondary and primary education, and a process has to go past five days to reach the sixth day. This text uses the general regression framework to cover models predicting these categorical and limited response variables. Note that this text does not explicitly consider censoring or truncation, except for time-to-event response variables (i.e., survival time analysis).

1.2   Approaches to Regression Analysis

While concentrating on the association between response variables and their covariates/predictors (these two terms are used interchangeably in this text), modern regression analysis has been advanced in several thrusts. The classical one still builds on the ordinary least squares (OLS) regression and its extensions by relaxing its various assumptions (e.g., general and generalized linear regression), by adding layers at different levels and directions (e.g., multilevel, spatial, and time series regression models), and concatenating cascades of multiple single-equation regressions (e.g., multilevel and structural equation models). In general, this thrust is grounded on the frequentist approach to statistics that views probability as estimable from frequencies of multiple mutually exclusive outcomes of an event (sometimes can be a combination of events), and this probability can be approximated or even reached through a long (actually an infinite) series of trials. The second thrust bases its framework on Bayesian statistics that takes a different philosophical approach to probability. Under the Bayesian framework, probability can be viewed subjectively, and it can be updated with new data. The third and probably currently most popular one is machine/statistical learning. This thrust evolves around the long-drawn infatuation with an intelligent system that can automate the decision-making process based on (usually a large amount of) data input. Machine/statistical learning is interested in devising computing algorithms that can improve themselves in prediction through systematically processing a large amount of experiential data. Although the last chapter briefly discusses some key concepts and reviews several major techniques in this area, the focus of this manuscript is still on the first two areas.

1.2.1   Frequentist Approach to Regression Modeling

The divide among the few alternative approaches to probability and statistical inference is not so much statistical and technical as philosophical and epistemological. The early discussion in this area was dominated by the frequentist approach, and it remains so in many disciplines outside computer science, data science, and statistics. The frequentist approach defines probability as the limit of the relative frequency of an infinite and usually independent series of trials of an experiment. Thus probability is undefined for any unrepeatable event. Under this framework, large sample size usually brings one closer to the truth—population parameters—which are usually viewed as fixed but unknown. Thus, concepts such as consistency and asymptotic properties belong to the frequentist approach. The rise of the frequentist approach is usually credited to Laplace for his foundational work on the normal distribution and the central limit theorem (Wilcox, 2003). Ronald. A. Fisher, along with Jerzy Neyman and Egon Pearson, established the frequentist approach on arguably firmer mathematical and statistical ground (Fisher, 1934, 1935; Neyman and Pearson, 1933). Arguably it is because this approach has not gone unchallenged, and sometimes even very seriously.

1.2.2   Bayesian Approach to Regression Modeling

Two new statistical approaches have shaken the world of statistics beginning from about three decades ago when they roughly reached the fruition of development. Machine (statistical) learning, a remedy for the then-dying-out artificial intelligence that focused on devising machinery/computer programs that would think like human brains, is to design computer algorithms that can achieve two routine tasks that human beings undertake on a daily basis—to classify and predict—but with much higher level of accuracy, consistency, and speed. Machine learning has revived artificial intelligence and is currently the workhorse of this broad scientific area. Machine learning entails a wide range of techniques and tools that originated in the fields of computer science and cognitive science and accidentally stepped onto statistics because of the many techniques and theoretical foundations that machine learning borrows from and builds on statistics.

In this manuscript, we focus on a second seemingly new and yet a parallel line of thoughts that has coexisted with the frequentist approach for centuries—Bayesian statistics. The origin of Bayesian statistics can be credited to Thomas Bayes—an English statistician, philosopher, and clergyman—in his paper, "An Essay towards solving a Problem in the Doctrine of Chances," published posthumously by his friend, Richard Price (Bayes and Price, 1763). In this splendid paper, Bayes (1763) did not yet come up with a mathematical formula for what is today known as the Bayes theorem; instead, he conducted a thought experiment to systematically determine the location of a ball thrown on a table or plane without observing the experiment outcome but being updated of the outcome in relation to the previous outcome, and he came up with this procedure that prior belief can be updated to arrive at posterior belief (Hald, 2000). In his 1774 memoir on the probability of the causes of events, Laplace ingeniously expressed the Bayes' theorem—albeit not in mathematical notation—that the probability of a cause (given the event) is proportional to the probability of the event (given the cause); or in Laplace's own words under the heading of Principle in the second section of this memoir that (in Stigler's English translation),

"If an event can be produced by a number n of different causes, the probabilities of these causes given the event are to each other as the probabilities of the event given the cause, and the probability of the existence of each of these is equal to the probability of the event given that cause, divided by the sum of all the probabilities of the event given each of these causes" (Stigler, 1986, pp. 364-365).

Using mathematical notation, the Bayes' theorem can be expressed as

$$P(C_k|E) = \frac{P(E|C_k)P(C_k)}{\sum_{i=1}^{n} P(E|C_i)P(C_i)} \tag{1.1}$$

where C_k denotes a possible cause and E for an event. With simpler Bayesian notation, this Bayes' formula can be written as $P(\theta|D) = \frac{P(D|\theta)P(\theta)}{P(D)}$, where D represents observed data and θ for hypothesis (about parameters or the distribution of parameters). Since D is usually viewed as fixed (relative to θ), we can rewrite the Bayes' theorem as $P(\theta|D) \propto P(D|\theta)P(\theta)$, or the posterior distribution is proportional to the product of prior $(P(\theta))$ and likelihood function $(P(D|\theta))$. $P(\theta)$ denotes our prior belief about the parameter distribution, and $P(\theta|D)$ is our posterior belief about θ updated with D.

1.2.2.1 The Example of COVID-19

Below, we use a simple example to illustrate the utility of the Bayes' theorem. Suppose an individual is running a fever and wants to figure out the probability of contracting COVID-19. So this person needs to compute the conditional probability of having COVID-19 given the symptom of fever. Based on the Bayes' theorem, one has $P(C|F) = \frac{P(F|C)P(C)}{P(F)}$, where C denotes COVID-19 and F for fever. We searched various sources for good estimates of the three probabilities on the right-hand side of the equation and used two weeks as the time frame for calculating incidence. To compute the marginal probability of getting coronavirus, we used a seven-day average estimate (for the week of September 13, 2021) from 1Point3Acres, a data clearinghouse of COVID-19 during the pandemic. In this case, $P(C) = 150,220/328,200,000 = 0.0004577087$. For the conditional probability $P(F|C)$, we consulted several online publications and selected a rough estimate of 0.8. The hard part is to find the probability of having fever in the general population, or $P(F)$. We were not able to find any exact statistics or estimates from reliable sources, probably because fever is usually classified into different kinds/causes. Here we use an estimate of $P(F) = 0.07$ based on a study about the prevalence of non-diarrhea or-cough-related fever among children in low-resource areas in a two-week period (Prasad et al., 2015). This appears to be an overestimate for individuals in the US, given the poor hygienic conditions in poor regions and the less-than-optimal immune system among the infant and children samples. Once the calculation is carried out using the Bayes' formula, we computed the final probability $(150220/328200000 \times 0.8)/0.07 = 0.005230957$; that is, one has a roughly up to 0.5% chance of getting coronavirus with the symptom of fever solely while this section is being written up. Note that this is a very crude estimate that uses an upper limit, and such estimate is not stratified by age or gender. So please be advised not to use it for one's own health assessment. But this example shows the power of the Bayes' theorem.

1.2.3   Priors

1.2.3.1   Conjugate Priors

Two major hurdles for using Bayesian statistics before it became popular, or even feasible, in the 1980s were first, its very high computational cost when simulation methods were run at a crawling speed, and second, to a lesser extent but equally challenging, its unorthodox philosophical approach to statistics and data analysis that were hardly embraced by mainstream statisticians. Let us focus on the first aspect. Although one can disregard the $P(D)$ term in the Bayes' formula $(P(\theta|D) \propto P(D|\theta)P(\theta))$, since it can be viewed as a normalizing factor, the product of the likelihood function $(P(D|\theta))$ and the prior $(P(\theta))$ can be either mathematically or computationally intractable. One workaround is to have conjugate priors. A conjugate prior is a prior probability distribution conjugate to the likelihood function; that is, once the prior is processed by (multiplied with) the likelihood distribution, the resultant posterior probability distribution and the prior are of the same distributional type. This means that one can have a closed-form solution for the posterior. For a detailed description about the match of conjugate priors and likelihood functions, please refer to Fink (1997).

The most commonly used example of such conjugacy is the binomial distribution (Lynch, 2010; Kruschke, 2013; Gelman et al., 2014). Below, we use some hypothetical examples from the ongoing COVID-19 pandemic. Assume that we have 200 COVID-19 patients age 65 years or older, and 45 died of complications in a six-week period or a mortality rate of 22.4%. Let us assume that the likelihood function follows a binomial distribution with the total number of trials set to $n = 200$ with 45 successes (deaths from COVID), and the probability of success is π in the population. Then the likelihood function can be written as $L(y|\pi) = \begin{pmatrix} 200 \\ 45 \end{pmatrix} \pi^{45}(1-\pi)^{(200-45)}$ or $L(y|\pi) \propto \pi^{45}(1-\pi)^{155}$. Next step is to select the prior for π. A natural candidate is the beta distribution that also takes a form similar to the likelihood, and once it is processed by the likelihood function, the resultant posterior still turns out to be a beta distribution. The probability density function (PDF) for the beta distribution is $f(\pi|\alpha,\beta) = \frac{\pi^{(\alpha-1)}(1-\pi)^{(\beta-1)}}{B(\alpha,\beta)}$, where $B(\alpha,\beta) = \frac{\Gamma(\alpha)\Gamma(\beta)}{\Gamma(\alpha+\beta)}$, $\Gamma(\theta)$ is the gamma function of a generic random variable θ, $\alpha, \beta > 0$ are shape parameters, and $\pi \in [0,1]$. Note that since $B(\alpha,\beta)$ does not involve the random variable π, we can rewrite the PDF as $f(\pi|\alpha,\beta) \propto \pi^{(\alpha-1)}(1-\pi)^{(\beta-1)}$. Here let us assume innocence about the disease, and a 60-40 percent chance would be a good place to start. For example, in a group of 150 elderly COVID patients, three-fifths (90) died of complications from COVID and the other two-fifths survived, and we want to use these statistics as our prior. Using a beta distribution

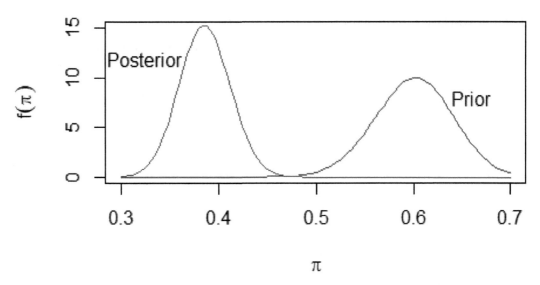

Figure 1.1
Prior and Posterior Distribution Plot

with $\alpha = 90$ and $\beta = 60$, the prior can be expressed as $p(\pi|\alpha, \beta) \propto \pi^{(90-1)}(1-\pi)^{60-1}$. Combining the two terms, $L(y|\pi)$ and $p(\pi|\alpha, \beta)$, we can easily derive our posterior to be $p(\pi|\alpha, \beta, y) \propto \pi^{45}(1-\pi)^{155}\pi^{(90-1)}(1-\pi)^{60-1} = \pi^{134}(1-\pi)^{214}$, which takes the form of a beta distribution with $\alpha = 135$ and $\beta = 215$. Please refer to Fig. 1.1 for graphing the beta prior and posterior distributions aforementioned.

1.2.3.2 Informative, Non-informative, and Other Priors

In addition to all the technical challenges, choosing an appropriate or fair prior constitutes the primary controversy for choosing the Bayesian method over its alternatives. On one side of this debate, including prominent figures in the development of Bayesian methods such as Bayes, Pascal, Laplace, Jaynes, and Jeffreys, all expressed preference of non-informative priors, priors that assign equal probability to all parameter values (Stone, 2013, pp. 91, 157), for example, uniform priors. Gelman (2002), on the other side, argues for the use of weakly informative priors and specific informative priors (Gelman on his and Stan's website). It is not that hard to understand that informative priors refer to prior parameter probability distributions that circumscribe the parameters to be (more likely) in certain area of the parameter space. For fairly well-established results in science, such as medical and physical research, Gelman (2002) suggests that one use highly informative prior distributions. For parameters that have more uncertainty in exiting literature, Gelman (2002) suggests the use of moderately informative priors. Towards the end, Gelman argues that most priors should be weakly informative since we have learned something about them, but not in a definitive way. Gelman (2019) provides several examples for such priors. For example, $N(0, 10)$ is a weakly informative prior, and $N(0, 1)$ is a generic weakly informative prior. Noninformative priors, such as flat priors (e.g., $U(0, 10)$) and near-flat priors (e.g., $N(0, 1e + 6)$), are not usually recommended, according to (Gelman, 2019).

1.2.4   Markov Chain Monte Carlo (MCMC)

When the posterior distribution follows some unknown form, one needs to use Markov chain Monte Carlo (MCMC) methods, a collection of random sampling methods that are similar in nature, to simulate the posterior distribution. The Markov chain part of the methods refers to a stochastic process, wherein current states only depend on the previous states, and accordingly memoryless (Lynch, 2010; Hayes, 2013; Kruschke, 2013; van Ravenzwaaij et al., 2018). This is ontological in the sense that once the information of current states is obtained, future and past states are independent of each other. Alternatively, if the current states are known, then one can sufficiently predict the future without referring to the past. The Monte Carlo part of the integrated sampling methods was devised and popularized by Enrico Fermi, Stanislaw Ulam, and John von Neumann around 1930-1940s (Robert and Casella, 2011). Monte Carlo methods use a large number of random experiments to find numerical solutions to problems that are often analytically invincible, for example, the calculation of an irregular shape without an existing formula or the simulation of a probability distribution with unknown forms/properties (e.g., the first or second moment). For layperson introduction to the MCMC methods, one can read chapters 9 and 11 in Hayes (2017); for accessible discussion, please refer to Lynch (2010) and Kruschke (2015); and for more technical details, please see Gelman et al. (2014). Below is the general procedure of a typical MCMC method (e.g., the Metropolis-Hasting algorithm).

- First, the algorithm randomly selects a starting value (e.g., first sample)

- Second, using some proposal probability distribution, the algorithm generates a proposal sample by adding some random noise to the first sample

- Third, the algorithm throws the two samples into the posterior and calculates their corresponding values up to a scale. If the posterior of the second sample is greater than the first, then accept it. If the second sample has a smaller posterior, the algorithm calculates the ratio of two posteriors, and usually uses a uniform distribution to randomly produce a number. If this number falls in the acceptance region proportional to the ratio, then the algorithm accepts this sample; otherwise, the second sample is rejected.

It can be seen that each sample is dependent on its previous one only; accordingly, it is a Markov chain. Despite such interdependence, it has been proven that this generic algorithm can find the target distribution very accurately in most cases. There are a variety of algorithms similar to the one aforementioned, including the Gibbs sampler, which is a special case of the Metropolis-Hasting algorithm, and the Hamiltonian Monte Carlo algorithm (a.k.a. hybrid Monte Carlo) used by Stan, a probabilistic programming language for Bayesian analysis (Carpenter et al., 2017; Stan Development Team, 2018, 2020).

1.3   Introduction to R

We use R as the statistical analysis environment exclusively for all the models discussed in this text, accompanied with discussions about other software applications for Bayesian analysis. R is an open-source free software for statistical analysis, and it is a member of the GNU Project that advocates users' freedom to create, extend, and use the software. Since its creation (based on the S language) by Ross Ihaka and Robert Gentleman in the 1990s, R has become the lingua franca among data scientists and statisticians across a wide range of disciplines and industries, and it has consistently remained at the top of several programming indices. The main advantage of R—like that in other modern programming languages—is that it is a primarily object-oriented language (it also supports procedural programming); that is, it can turn literally anything during the data analysis process–data, codes, intermediate output, and results–into an object with different class types and unique attributes that can be retrieved in later stages of analysis. R can be downloaded for free (as promised) from the R Project website at https://www.r-project.org/. Users can visit the free R manuals website (https://cran.r-project.org/manuals.html) to learn the full functionality of the R language. Below, we provide a brief introduction to R with an emphasis on functions from the `base` package that comes with R installation by default. There are several user-contributed R packages that have greatly enhanced basic functions in R, including the `tidyverse` package (data management and streamlined workflow for data science) (Wickham et al., 2019) and the `ggplot2` package (aesthetic plotting) (Wickham, 2016). But because this manuscript focuses on regression modeling and interpretation, we only have sporadic discussion of these packages to provide a quick taste of them. Interested users can read the corresponding manuals and visit the websites of the R packages aforementioned for a thorough study.

1.3.1 RStudio

RStudio is probably the best integrated development environment (IDE) for R out there, as far as ease and power are concerned. After users download, install, and open it, they should see four windows on the screen. Situated at the upper left corner is the syntax window, where one compiles R scripts (i.e., a syntax file to run commands in batch mode or selected parts). In that window, one can run a selected line of R codes by moving the cursor to that line and click the **run** button on the upper right corner of the syntax window. One can also run through the whole syntax file by clicking the **source** button. The bottom left window is the interactive command line or console, where one can interact with R in the background (i.e., type in commands line by line and hit enter to execute). The upper right window is the environment/history window where objects are listed, and bottom right window is usually where graphs and help files show up.

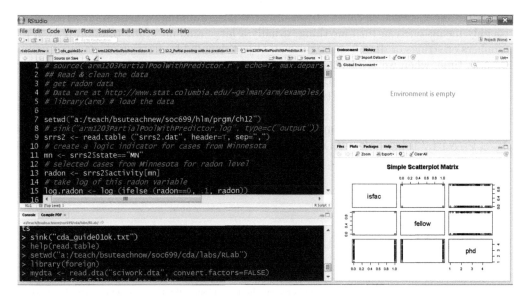

There are three ways whereby one interacts with R via RStudio. First, the interactive command line or the console in R/RStudio, which is at the bottom left window of the RStudio screen. One can type a command line and press enter to execute that command, which can sometimes be wrapped up and run through several lines. Second is to use a GUI (graphical user interface) to pull down dialog box and type in required information to communicate with R interactively. To work with a GUI, one can install the `Rcmdr` package in R first and then invoke that package in the R/RStudio environment by typing in "`library(Rcmdr)`" (Fox, 2017). Once the GUI is active, one can point and click and proceed from there. Third, one can type out commands in a text file in the upper left syntax window in RStudio and save it as an R script file with a name such as `cdaLinReg.R`. Then, in the console, type: `source("/thePathofRScriptFile/cdaLinReg.R")`. One can start learning R by issuing commands and examining results in the console window. Note that methods 1 and 2 are often useful for exploratory work, but working with R scripts (in the syntax window) should be one's primary working mode for serious work since it largely simplifies debugging and ensures reproducibility.

1.3.2   Use R as Calculator

Albeit like launching a missile to kill a fly, one can use R as a calculator. For example, with six reads of platelet count (k/ml) from six individuals, 100, 110, 120, 130, 140, 150, and 160, one can use R to calculate the average of these six numbers by first entering the data into the console,

```
# manually enter data
plateletCount <- c(100, 110, 120, 130, 140, 150, 160)
```

Here `c` is an R function to combine (hence "`c`") values into a vector or list of numbers and strings. The leftwards arrow sign ("`<-`") is an assignment sign used to assign what is after this assignment sign (e.g., data array/frame, function, matrices) to what (e.g., the object name) precedes this sign. It is also important to keep spaces around the `<-` sign to avoid confusion and unexpected errors. One purpose of this assignment procedure is to simplify the referencing process for future analysis.[1] In the platelet count example, we assign the name, `plateletCount`, to the list of numbers, 100, 110, 120, 130, 140, 150, and 160. This, in computer/data science, is called object-oriented programming, which, in layperson's terms, roughly means one can turn anything, usually a large amount of information (e.g., a list of numbers or strings or data frames), into an object denoted by a string of letters or numbers. So next time, when we invoke this name, the program knows we refer to the very structure of symbols or information, as how they are ordered. Objects usually have their attributes and behavior and can be structured for reusable code snippets and easy communication among themselves.

Below, for example, we compute the average of the six reads of platelet count ($\bar{x} = \sum_{i=1}^{N} x_i/N$) by using objects. One can apply the built-in R function, `mean`, to the data object, `plateletCount`,

```
# help(mean)
# RSiteSearch(mean)
mean(plateletCount)
```

```
## [1] 130
```

---

[1]In most cases, the assignment sign (or the leftwards arrow sign) serves the same function as the equal sign (=), but there is some slight difference between the two. Generally, the `<-` operator can be used at any level, whereas the = sign, as an assignment operator, can only be used at the top-level (for embedded R functions). For beginning and intermediate users, they can be used interchangeably, and usually there is no need to distinguish between the two.

If we need more details about an R function, we can use another built-in function, `help`, to find out. If we are interested in finding out more information about something (e.g., the `mean` function) on the web, we can use the `RSiteSearch` function. Note that in the R code and output block, the line beginning with the two-pound sign, "##," denotes output and is used to distinguish output from R codes.[2]

In an R script file, any line starting with a pound sign ("#") is to comment out anything following the sign; that is, R will take that line as text and print out that line as it is. Frequently, people use the pound sign to add comments (annotations). Below is another way to compute the same thing with a line of annotation above.

```
# how to do it without using the mean function
sum(plateletCount)/length(plateletCount)
```

```
## [1] 130
```

To get a measure of dispersion (how different numbers are using a representative one like mean as the benchmark), such as variance, $s^2 = \sum_{i=1}^{N} (x_i - \overline{x})^2 / (N - 1)$, one can use the `var` function,

```
# compute variance for a sample
var(plateletCount)
```

```
## [1] 466.6667
```

Below is a more tedious way of using low-level functions, including `sum`, `mean`, and `length`, to compute the same thing,

```
sum((plateletCount-mean(plateletCount))^2)/(length(plateletCount)-1)
```

```
## [1] 466.6667
```

After savoring the basics of R and RStudio, one is usually advised to learn its workflow of data analysis and graphics.

1.3.3 Set Up Working Directory

It is often a good habit to create a working directory for one's project, place all data and data-related documents (e.g., code books) in that directory, and perform analysis in and save results to the same directory since that will make the project clean and tidy. Doing so can also avoid all the hassles to specify the same directory path each time when one reads in or saves a file from/to the same folder. One can change directories with the `setwd` function.

Let us assume that our current working directory is `c:/temp` and we want to reset to `c:/stats/work4DrXu`. We can just type,

```
setwd("c:/stats/work4DrXu")
```

Note that one has to make sure that the working directory needs to have the data file to be read in and analyzed, or the system will produce error messages.

---

[2]This manuscript follows this delimiting practice throughout for the overwhelming majority of R code chunk and output, with a few exceptions (e.g., bootstrapping and Bayesian estimation) when computationally intensive analyses are executed outside dynamic reporting and results have to be copied and pasted.

1.3.4   Open Log File

Serious data analysis usually requires reliability and reproducibility; that is, one should get the same results with the same analytic procedures (e.g., same code chunks executed in the exact same order). Creating a log file, is an effective method to achieve such goals. Thus, before going into data management and analysis, one can open a log file, where both R codes (commands) and results will be saved. This can be accomplished by the `sink` function.

```
sink("cdaLabGuide01.log", split=TRUE)
```

If the `split` argument in the `sink` function is not set to `TRUE`, then the interactive console will display issued commands and the created log file will contain output only. If instead the `split` argument is set to `TRUE`, then the console will display both commands and associated output, and the log file again only contains the output without commands. The ideal scenario, for most quantitative scholars, is to have both commands and output displayed in the console and simultaneously saved to the log file. Then one has to combine the use of the `sink` with the `source` function, which will be covered at the end of this section. Alternatively, one can use simplified logging functions in user-contributed packages, such as `logr` (Bosak, 2022) and `log4r` (White and Jacobs, 2021).

1.3.5   Load Data

The next step is to read in the data, in this case, from an external source. In R, one needs to first load the `foreign` package to be able to read in data files of foreign/different formats than R data. An alternative is to use functions (e.g., `read_stata`) from the `haven` package (or the `tidyverse` package) to read in data in various formats (Wickham and Miller, 2021).

```
# load the foreign package
library(foreign)
mygss <- read.dta("data/gssCum7212Teach.dta", convert.factor=F)
```

Here the "<-" sign assigns a Stata data file read into the R environment with an object name "mydta" using the `read.dta` function from the `foreign` package. The `convert.factor` option is used to read in value labels for creating factor variables in R when the option is set to T (TRUE); otherwise, it is set to F.

In this case, we know that the data file `gssCum7212Teach.dta` is saved under the `c:/stats/work4DrXu` folder. If one does not have the file path information, the following line of codes can be used to find out the location of a file

```
shell('cd /d A: & dir gssCum7212Teach.dta /s')
```

```
Directory of c:\stats\work4DrXu
06/20/2022  08:46 PM        13,689,803 gssCum7212Teach.dta
               1 File(s)    13,689,803 bytes
```

One can see from the results that the `gssCum7212Teach.dta` Stata data file is saved under "c:\stats\work4DrXu" folder. The `shell` function is used to call the Windows CMD shell to communicate with the operating system. Every operating system has a shell programming system. In Windows, it is CMD, and in Linux (an operating system like Windows, but is free and open source), it is BASH. In the `shell` function above, one can see that we use two shell commands, including "`cd /d A:`", which means changing the drive letter to A, and "`dir gssCum7212Teach.dta /s`", which directs the terminal to search for a file named gssCum7212Teach.dta, and the "/s" option means including all subfolders on drive A in the search. The "&" sign is used to combine these two commands so that R can execute one command after another in a row. The single quotation marks tell R where is the beginning of the shell commands and where is the end.

1.3.6 Subset Data

After reading in a data file, one usually needs to manage and clean the data, for example, to keep some variables or observations that satisfy certain conditions especially when planning to work with a subset of that full data file. We can select cases using certain conditions with the `subset` function,

```
mydta <- subset(mygss, year==2000 & (age>=27 & age <= 65))
```

In this case, we select cases surveyed in year 2000 and aged between 27 and 65. We can also select variables as follows,

```
usevar <- c("health", "hlthc2", "age", "female", "educ", "impinc")
# [!usevar] to drop variables
usedta <- mydta[usevar]
```

Thereby, we keep only six variables, including `health`, `hlthc2`, `age`, `female`, `educ`, and `impinc` in this subset data object `usedta`. One can also use the `subset` function to select cases and variables simultaneously,

```
usedta <- subset(mygss, year == 2000 & (age >= 27 & age <= 65),
    select = c(health, hlthc2, age, female, educ, impinc))
```

When we have many variables in the data file and want to keep most of them and drop only a few, we can drop variables by first specifying the variable names to be dropped in a vector or list and second using the logical operator ! and the `%in%` operator to identify if an element belongs to a string of characters so as to drop these variables from the initial data file,

```
# drop variables
drops <- c("age", "female")
dropAgeFem <- usedta[ , !(names(usedta) %in% drops)]
```

One can use a similar method to drop cases. The following line of R codes drops cases aged between 18 and 25.

```
dropAge1825 <- usedta[!(usedta$age >=18 & usedta$age <= 25),]
# dropAge1825 <- usedta[which(usedta$age < 18 & usedta$age > 25),]
```

1.3.7 Examine Data

Naturally, the next step is to examine the data as a whole by using the `summary` and/or `str` function. From this point on, we will not get into details about R functions unless necessary, since one can easily get these details using the `help` function.

```
summary(usedta)
```

```
##       health           hlthc2            age            female
##  Min.    :1.000   Min.    :0.000   Min.    :27.00   Min.    :0.0000
##  1st Qu.:1.000   1st Qu.:0.000   1st Qu.:36.00   1st Qu.:0.0000
##  Median :2.000   Median :0.000   Median :43.00   Median :1.0000
##  Mean    :1.844   Mean    :0.175   Mean    :43.51   Mean    :0.5477
##  3rd Qu.:2.000   3rd Qu.:0.000   3rd Qu.:52.00   3rd Qu.:1.0000
##  Max.    :4.000   Max.    :1.000   Max.    :63.00   Max.    :1.0000
##  NA's    :39      NA's    :39
##       educ            impinc
##  Min.    : 0.00   Min.    : 5.808
##  1st Qu.:12.00   1st Qu.: 9.667
##  Median :13.00   Median :10.306
##  Mean    :13.65   Mean    :10.192
##  3rd Qu.:16.00   3rd Qu.:10.712
##  Max.    :20.00   Max.    :11.857
##
```

```
str(usedta, no.list = T, vec.len = 2)
```

```
## 'data.frame': 199 obs. of  6 variables:
##  $ health: int  2 1 2 2 2 ...
##  $ hlthc2: num  0 0 0 0 0 ...
##  $ age    : int  44 44 53 62 38 ...
##  $ female: num  1 1 0 0 1 ...
##  $ educ   : int  14 16 16 13 12 ...
##  $ impinc: num  9.56 11.86 ...
```

1.3.8   Examine Individual Variables

One can also look at individual variables using, for example, univariate frequency distributions.

```
summary(usedta$health)
```

```
##    Min. 1st Qu.  Median    Mean 3rd Qu.    Max.    NA's
##   1.000   1.000   2.000   1.844   2.000   4.000      39
```

```
# tabulate health
table(usedta$health, useNA =c("ifany"))
```

```
##
##    1    2    3    4 <NA>
##   58   74   23    5   39
```

```
# transpose the table by 90 degrees for an easy look
as.matrix(table(usedta$health, useNA ="ifany"))
```

```
##       [,1]
## 1       58
```

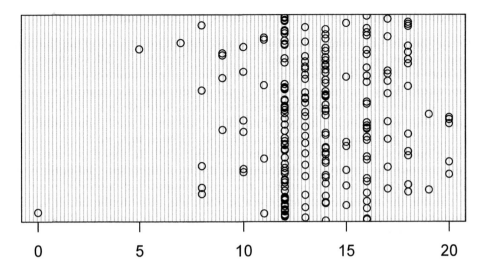

Figure 1.2
Dot Plot

```
## 2      74
## 3      23
## 4       5
## <NA>   39
```

Note that because we "de-factor" all categorical variables while reading in the Stata data file, all value labels are missing.

1.3.9   Save Graphs

We can also graph a simple dotplot using the `dotchart` function, such as the one in Fig. 1.2. The graph can be saved by preceding the `dotchart` function with the `png` function that specifies the graph file format (the portable network graphic format) and preferences for graphics (e.g., transparent background), and closing it with `dev.off()`.

```
#png(file="educ.png", bg="transparent")
dotchart(usedta$educ)


#dev.off()
```

1.3.10   Add Comments

In statistical programming, annotation is critical since one does not want to figure out those esoteric codes written long time ago. To add annotations/comments, one can start a line with the pound sign, "#".

```
# add comments after this pound sign in the same line, and nothing happens
```

To have a block comment (comments running for multiple lines), one could use the if structure "`if (FALSE) {anything}`". Note that the curly brackets can take multiple lines.

1.3.11   Create Dummy Variables and Check Transformation

Creating dummy variables is a common task in data analysis. Although much of the function of dummy variables can be achieved automatically by the `factor` and `ordered` functions along with factor variables in R, the needs to create explicit dummy variables can still arise in various occasions. When one has less than 12 years of education, for example, we may want to code that individual to be one, otherwise zero. For such recoding, it would help to generate a new variable with such dummy coding. Below, we create two dummy variables, `lths` and `lths2`, with the exact same coding schema but different methods. First, we use the `as.numeric` function. As long as the condition specified after the function is met, the cases will be set as one, and otherwise zero. Similarly, if the condition specified in the first argument after the `ifelse` function is satisfied, then its corresponding cases will be set to the value provided in the second argument; otherwise, cases will be set to the value given in the third argument. Note that unlike the commands in SAS or Stata, these two R functions usually leave missing data (i.e., `NA`) intact.

```
# This is method one
usedta$lths1 <- as.numeric(usedta$educ < 12)
# This is method two
usedta$lths2 <- ifelse(usedta$educ < 12, 1, 0)
# check using cross-tabulation
table(usedta$educ, usedta$lths1, useNA=c("ifany"))
```

```
##
##          0  1
##    0     0  1
##    5     0  1
##    7     0  1
##    8     0  5
##    9     0  4
##    10    0  6
##    11    0  5
##    12   54  0
##    13   23  0
##    14   36  0
##    15    7  0
##    16   26  0
##    17   10  0
##    18   13  0
##    19    2  0
##    20    5  0
```

```
table(usedta$educ, usedta$lths2, useNA=c("ifany"))
```

```
##
##          0  1
##    0     0  1
##    5     0  1
##    7     0  1
##    8     0  5
##    9     0  4
##    10    0  6
```

```
## 11  0  5
## 12 54  0
## 13 23  0
## 14 36  0
## 15  7  0
## 16 26  0
## 17 10  0
## 18 13  0
## 19  2  0
## 20  5  0
```

1.3.12 Label Variables

R does not have a built-in function to label variables, but one can install the `Hmisc` package and use its `label` function to label variable names (Harrell Jr et al., 2021).

```
# install.packages("Hmisc")
require(Hmisc)
label(usedta$lths1) <- "education < 12 years"
describe(usedta$lths1)
```

```
## usedta$lths1 : education < 12 years
##          n  missing distinct     Info      Sum     Mean      Gmd
##        199        0        2    0.307       23   0.1156   0.2055
```

One can also use functions from the `labelled` package (Larmarange, 2021) to label variables and value labels.

1.3.13 Label Values

Sometimes while examining the frequency distribution of variables of interest, especially those at the ordinal or nominal level of measurement, we want to know what a specific value stands for. In tabulating self-rated health, `health`, one may have problems to figure out whether the value range from one to four corresponds to health status from excellent to poor, or the other way around. Thus, we need to add value labels. To accomplish this, usually one turns to the `factor` function for nominal level variables and `ordered` function for ordinal variables.

```
table(usedta$health, useNA=c("ifany"))
```

```
##
##    1    2    3    4 <NA>
##   58   74   23    5   39
```

```
usedta$health <- factor(usedta$health,
       levels = c(1, 2, 3, 4, NA),
       labels = c("1exltHlth", "2gooddHlth", "3fairHlth", "4poorHlth"))
table(usedta$health, useNA =c("ifany"))
```

```
##
##  1exltHlth 2gooddHlth  3fairHlth  4poorHlth       <NA>
##         58         74         23          5         39
```

```
as.matrix(table(usedta$health, useNA ="ifany"))
```

```
##              [,1]
## 1exltHlth     58
## 2gooddHlth    74
## 3fairHlth     23
## 4poorHlth      5
## <NA>          39
```

Note that in the `labels` argument of the `factor` function, we include both values and labels for the value labels to ensure correct coding and easy reading.

1.3.14   Create Ordinal Variables

While examining the effects of education, one is often interested in how educational attainment/transition would affect various life outcomes. Below, we create such a variable, `edlv` (levels of education), using years of education, `educ`. If `educ` is below 12, equal to 12, between 13 and 16, and greater than 16, then `edlv` is coded as one, two, three, and four correspondingly.

```
table(usedta$educ, useNA=c("ifany"))
```

```
##
##  0  5  7  8  9 10 11 12 13 14 15 16 17 18 19 20
##  1  1  1  5  4  6  5 54 23 36  7 26 10 13  2  5
```

```
usedta$edlv[usedta$educ<12] <- 1
usedta$edlv[usedta$educ==12] <- 2
usedta$edlv[usedta$educ>12&usedta$educ<=16] <- 3
usedta$edlv[usedta$educ>16] <- 4
```

```
usedta$edlv <- ordered(usedta$edlv,
       levels = c(1, 2, 3, 4, NA),
       labels = c("1<hSchl", "2=hSchl", "3colg", "4grad"))
table(usedta$edlv, usedta$educ, useNA=c("ifany"))
```

```
##
##            0  5  7  8  9 10 11 12 13 14 15 16 17 18 19 20
##  1<hSchl   1  1  1  5  4  6  5  0  0  0  0  0  0  0  0  0
##  2=hSchl   0  0  0  0  0  0  0 54  0  0  0  0  0  0  0  0
##  3colg     0  0  0  0  0  0  0  0 23 36  7 26  0  0  0  0
##  4grad     0  0  0  0  0  0  0  0  0  0  0  0 10 13  2  5
```

1.3.15   Check Transformation

We can use the `plot` function to cross-examine our recoding in Fig. 1.3.

```
plot(usedta$educ,usedta$edlv, main="Education Recode Check",
       xlab="Years of Ed.", ylab="Levels of Attainment" )
```

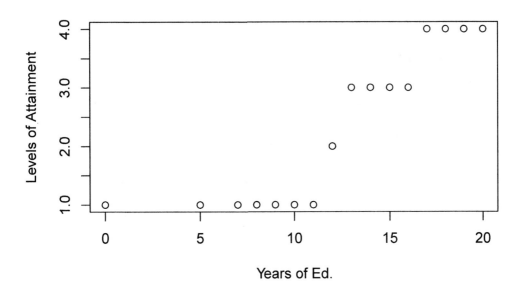

Figure 1.3
Recode Check Plot

1.3.16 Drop Missing Cases

Frequently, one needs to manage missing cases in their various forms. Below, we use the `complete.cases` function to drop missing cases with listwise deletion. Now the newly created `nomisdta` data frame contains cases without any missing value.

```
# use complete.cases to drop cases
dim(usedta)
```

```
## [1] 199   9
```

```
nomisdta <- usedta[complete.cases(usedta),]
# The following line does the same
#  usedta <- na.omit(usedta)
dim(nomisdta)
```

```
## [1] 160   9
```

From the change in the first dimension (rows) of the data frame, we can see how many cases are dropped due to missing.

1.3.17 Graph Matrix

One can use the `pairs` function to explore bivariate correlations using a pairwise matrix plot such as the one in Fig. 1.4.

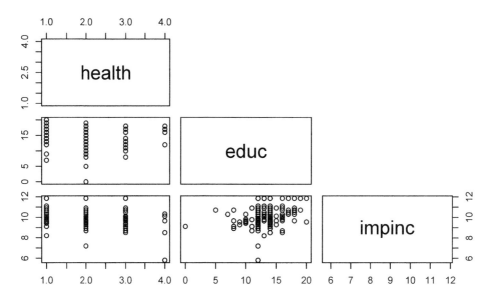

Figure 1.4
Bivariate Matrix Plot

```
pairs(~health+educ+impinc, data=nomisdta,
        main="Scatter Plot Matrix", upper.panel=NULL)
```

1.3.18   Save Data

After the preliminary exploration, we can save the data for other purposes.

```
save(nomisdta, file="hlthTeachCDA.Rdata")
# load(hlthTeachCDA.Rdata)
```

1.3.19   Close Log

The last step is to close the log file. We can simply type

```
sink()
```

1.3.20   Source Codes

After finalizing the R codes, one may want to run through the whole R script without interactively compiling the codes or exploring the data, or may want to reproduce results from that R script after making revisions. To accomplish that, one can use the **source** function in the console, which executes the R script specified in the quotation marks as follows,

```
source("yourOwnDirectoryWhereRScriptSaved/assignment01.r",
    echo = TRUE, max.deparse.length = 10000)
```

If the `echo` argument is set to `T` (i.e., `TRUE`) together with the `split` option set to `T` in the `sink` function, then the software will display both R codes and associated output in the console and save both in the log file where the `sink` function diverts to.

```
sink("cdaLabGuide01.log", split=TRUE)
```

This is very important: if one just runs the `sink` function without running `source` interactively in the console, R and RStudio will divert only the results, not the commands, to the log file. So to see both commands and results in the R console AND both commands and results in the log file, one needs to:

1. set the `split` option to `TRUE` in the first `sink` function at the beginning of the R script and

2. set the `echo` option to `TRUE` in the `source` function and

3. run the `source` function in the console or terminal.

So a well-structured R script file may look like the following

```
#source("yourOwnDirectoryWhereRScriptSaved/assignment01.r",echo=TRUE,max.
deparse.length=10000)
rm(list=ls(all=TRUE)) # remove all active objects in the memory
setwd("yourWorkingDirectory")
sink("cdaLabGuide01.log", split=TRUE)
...
...
YOUR R CODES IN THE MIDDLE
...
...
sink()
```

With the first `source` line silenced and beginning with the "#"sign, one can copy that line (without the pound sign), paste it into the console, and hit "Enter". One should see the codes run through and results produced in the console and saved to the log file simultaneously.

1.4 Review of Linear Regression Models

1.4.1 A Brief History of OLS Rregression

The ordinary least squares (OLS) regression is more commonly known as the linear regression model. The early development of the least squares methods and statistics, broadly defined, was closely associated with advances and exploration in the field of astronomy and geodesy. The first known precursor to the modern application of the linear regression model is Galileo in analyzing astronomical data of the new star of 1572. In his analysis, Galileo offered a

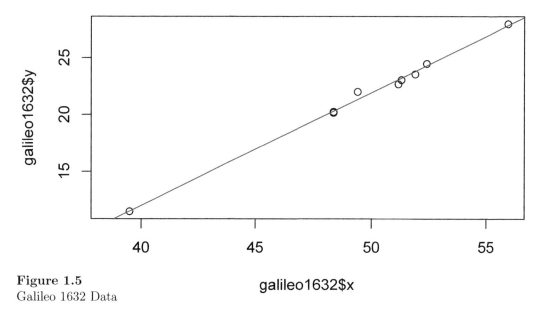

Figure 1.5
Galileo 1632 Data

surprisingly then-modern treatment of the data using what he calls "observational errors," for example, that the observations are distributed symmetrically around their true values (Hald, 2003, p. 150), although Galileo never explicitly used modern terms associated with least squares, nor did he come up with the least squares technique.

Fig. 1.5 is a graphical rendition and the least squares analysis of the data about the new star of 1572 (Galileo, 1632, pp. 294-295), where x corresponds to the altitude of the pole and y for the minimum altitude of the star. It can be shown that the (regression) line fits the data almost perfectly.

Today, the discovery of least squares is credited to both Adrien-Marie Legendre (1805) and Carl Friedrich Gauss (1809). With two names attached to the same discovery, as one would imagine, an anteriority dispute would usually occur. Legendre (1805, pp. 72-73) provided a first compendious exposition of the method of least squares in an appendix, titled "On the Method of Least Squares," in *Nouvelles Méthodes pour la Détermination des Orbites des Comètes* (*New Methods for the Determination of the Orbits of Comets*). Yet, Gauss somewhat disquietedly disputed that he already used this method—or what Gauss calls "...the principle that the sum of the squares of the differences between the observed and computed quantities must be a minimum..."—since 1795 (Gauss, 1809, p. 269). Since the discovery of the method of least squares is trivial compared with Gauss's other important work in mathematics and adherence to his personal motto—pauca sed matura (few, but ripe), Gauss's presumptive claim on the discovery of least squares is usually considered credible, along with his 1821 Latin paper on the Gauss-Markov theorem that shows that the OLS estimator has the minimum sampling variance of all linear unbiased estimators, hence best linear unbiased estimator (BLUE).

1.4.2   Main Results of OLS Regression

The ordinary least squares estimator is obtained by minimizing the sum of squared errors, $\sum_{i=1}^{N}(y_i - \widehat{y}_i)^2$, as the objective function, whereby y_i is the observed response variable, $\widehat{y}_i$ is its corresponding expected value, and we have the structural model, $y_i = \mathbf{x}_i\beta + \varepsilon_i$, where $\mathbf{x}_i$, $\boldsymbol{\beta}$, and ε_i correspond to the covariates matrix, the parameter vector, and the disturbance term (the disturbance term and the error term are interchangeable in this and following sections). Using matrix language, if e is define as a $N \times 1$ column matrix of $(y_1 - \widehat{y}_1, y_2 - \widehat{y}_2, ..., y_N - \widehat{y}_N)'$, then $\sum_{i=1}^{N}(y_i - \widehat{y}_i)^2$ can be re-expressed as $e'e$. Using vector calculus, the ordinary least squares estimator, $\widehat{\beta}_{\text{OLS}} = (\mathbf{X}'\mathbf{X})^{-1}\mathbf{X}'y$. If we assume $\text{Var}(\varepsilon) = \sigma^2\mathbf{I}$ (the spherical disturbance assumption as discussed below), then $\text{Var}(\widehat{\beta}_{\text{OLS}}) = \sigma^2\mathbf{I}(\mathbf{X}'\mathbf{X})^{-1}$.

1.4.2.1 OLS Estimator and Variance-Covariance Matrix

One can derive the ordinary least squares estimator in the following way,

$$
\begin{aligned}
e'e &= \left(y - \mathbf{X}\widehat{\beta}\right)' \left(y - \mathbf{X}\widehat{\beta}\right) \\
&= \left(y' - \widehat{\beta}'\mathbf{X}'\right)\left(y - \mathbf{X}\widehat{\beta}\right) \\
&= y'y - y'\mathbf{X}\widehat{\beta} - \widehat{\beta}'\mathbf{X}'y + \widehat{\beta}'\mathbf{X}'\mathbf{X}\widehat{\beta} \\
&= y'y - 2\widehat{\beta}'\mathbf{X}'y + \widehat{\beta}'\mathbf{X}'\mathbf{X}\widehat{\beta}
\end{aligned}
$$

To minimize this objective function with respect to $\widehat{\beta}$, we have $\frac{\partial e'e}{\partial \widehat{\beta}} = -2\mathbf{X}'y + 2\mathbf{X}'\mathbf{X}\widehat{\beta} = 0$. Thus, $\widehat{\beta}_{\text{OLS}} = (\mathbf{X}'\mathbf{X})^{-1}\mathbf{X}'y$. Based on this result, one can also derive the variance-covariance matrix of $\widehat{\beta}_{\text{OLS}}$ as follows,

$$
\begin{aligned}
V\left(\widehat{\beta}_{\text{OLS}}\right) &= \text{Var}\left((\mathbf{X}'\mathbf{X})^{-1}\mathbf{X}'y\right) \\
&= \text{Var}\left((\mathbf{X}'\mathbf{X})^{-1}\mathbf{X}'(\mathbf{X}\beta + \varepsilon)\right) \\
&= \text{Var}\left((\mathbf{X}'\mathbf{X})^{-1}\mathbf{X}'\mathbf{X}\beta + (\mathbf{X}'\mathbf{X})^{-1}\mathbf{X}'\varepsilon\right) \\
&= \mathbf{0} + \text{Var}\left((\mathbf{X}'\mathbf{X})^{-1}\mathbf{X}'\varepsilon\right) \\
&= (\mathbf{X}'\mathbf{X})^{-1}\mathbf{X}'\text{Var}(\varepsilon)\mathbf{X}(\mathbf{X}'\mathbf{X})^{-1} \\
&= \sigma^2\mathbf{I}(\mathbf{X}'\mathbf{X})^{-1}
\end{aligned}
$$

Here σ^2 is a population parameter, and it is estimated using $s^2 = \frac{1}{N-K-1}\sum_{i=1}^{N}(y_i - \mathbf{x}_i\widehat{\beta}_{\text{OLS}})^2$, where N is the sample size, K is the total number of covariates. For the proof of s^2 as a consistent estimator of σ^2, please see Wackerly et al. (2002, pp. 548-549).

1.4.3 Major Assumptions of OLS Regression

Based on early works by Laplace, Gauss and then Yule (Hald, 1998), modern econometricians, statisticians, and other quantitative researchers usually characterize the classical linear regression model with multiple (five to eight, depending on how the assumptions are defined and grouped) assumptions, including zero conditional mean, linearity, spherical disturbance (homescedasticity and absence of serial correlation), identifiability, nonstochastic covariates, and normality (Johnson and Wichern, 1998; Davidson, 2000; Greene, 2000; Cameron and Trivedi, 2005; Fox, 2008; Wooldridge, 2010). All these assumptions involve the behavior of the disturbance (error) terms to varying degrees. Albeit similar in content, various texts provide somewhat different renditions of the same set of assumptions. Below, our discussion begins with the most important ones and then descend to relatively inconsequential ones.

1.4.3.1   Zero Conditional Mean and Linearity

The most critical assumption of all for the OLS estimator is the zero conditional mean assumption, and it is sometimes grouped with the linearity assumption since linearity is a trivial result of the former. The zero conditional mean assumption states that in a structural model, $y = \mathbf{x}\beta + \varepsilon$, that linearly associates a continuous response variable, y, with a set of covariates $\mathbf{x}$ and a disturbance term ε, we have $E(\varepsilon|\mathbf{x}) = 0$; that is, the expected value of disturbance term ε conditional on $\mathbf{x}$ is zero. Applying the law of iterated expectation, we can get $E(\varepsilon) = E_{\mathbf{x}}(E(\varepsilon|\mathbf{x})) = 0$, one of the three assumptions for the Gauss-Markov theorem to be tenable. The conditional mean assumption is usually viewed as the fundamental assumption of all considered for the OLS regression model in that it ensures that our linear estimator of the parameters of interest is unbiased ($E(\widehat{\beta}) = \beta$).

An important result of the zero conditional mean assumption is that if we have the structural equation as specified just previously, then $E(y|\mathbf{x}) = E(\mathbf{x}\beta + \varepsilon|\mathbf{x}) = \mathbf{x}\beta$, thereby the linearity assumption; the linearity assumption can be stated that the conditional expectation of the continuous response variable, y, is a linear function of covariates, $\mathbf{x}$, consisting of first- and higher-order terms. The concept of linearity (or the superposition principle) comes from linear systems in mathematics (e.g., map in algebra), physics, and engineering (e.g., waves). Linearity usually consists of two properties, namely, additivity and homogeneity; that is, $f(ax + by) = af(x) + bf(y)$, wherein a and b are scale factors and x and y are variables. The additivity property states that the function of a sum of variables is equivalent to the sum of the function applied to the variables separately, and the homogeneity property shows that the function of a variable multiplied by a scale factor is equivalent of the scale factor multiplied by the function of that variable (Zeidler, 1996; Axler, 2015).

It is important to note that the linearity assumption only pertains to the parameters and the error terms. Thus the right-hand equation is usually considered to be linear in β and ε, but not necessarily in $\mathbf{x}$; that is, a higher-order polynomial or other nonlinear transformation of some of the elements in $\mathbf{x}$ would still make this function linear. For example, $y = \beta_0 + \beta_1 x_1 + \beta_2 x_1^2$ can be viewed as linear in β, and thus the linearity assumption is still maintained. But for an equation such as $y = \frac{\exp(\mathbf{x}\beta)}{1+\exp(\mathbf{x}\beta)}$, the linearity assumption is violated, since y is not linear with respect to β.

1.4.3.2   Spherical Disturbance

A second important assumption is the two-fold spherical disturbance assumption. The first part of the assumption is homoscedasticity; that is, the variance of the conditional disturbance term is constant, and notationally, $V(\varepsilon|\mathbf{x}) = \sigma^2$. The second component is independence (absence of serial correlation); that is, the conditional disturbance terms are independent of each other and not correlated across observations, or, $V(\varepsilon_i, \varepsilon_j|\mathbf{x}) = 0 \;\forall\; i \neq j$. These two components, homoscedasticity and independence combined, constitute the spherical disturbance assumption. This assumption ensures that the estimation of the standard errors of the parameter estimates are consistent and the hypothesis testing of these estimates are reliable and robust. The violation of this assumption gives rise to many extensions to the classical ordinary least squares, including but not limited to general linear regression, time series regression, and spatial regression.

1.4.3.3   Identifiability

The identifiability assumption presupposes that none of the covariates are linearly dependent on (i.e., a linear combination of) other covariates in the function. In matrix language, this assumption implies that the covariate matrix $\mathbf{X}$, including the unit column, is full rank; that is, if $\mathbf{X}_{N \times K}$ has full column rank, then it has rank K. If, instead, one of the covariates,

x_k, is a linear combination of one or several other covariates, that would imply that the full set of statistical information about x_k can be extracted from other covariates, thus leading to estimation failure, since statistical software applications cannot distinguish between x_k and the other covariate or covariates that are linearly dependent on x_k. Having a covariate that is a linear combination of another or several other covariates is usually an extreme case, resulted from negligence in data management; the most commonly occurring issue related to this identifiability assumption is multicollinearity wherein one covariate is highly correlated with another or a linear combination of several other covariates. When the latter happens, usually one will get large standard errors with non-significant (large) p-values for covariates in question. Albeit rare in empirical analysis, the identifiability assumption also requires that covariates vary their values for identifiability purposes. In other words, if a covariate is uniform or has little variability in its values, whereas the response variable changes its values in a reasonable range, then that would imply this covariate is, most likely, not associated the response variable and should be dropped from the model.

Based on three of these assumptions, including zero marginal mean (can be easily obtained from zero conditional mean), spherical disturbance, and identifiability conditions, Gauss proved the Gauss theorem that the OLS estimator is the best linear unbiased estimator (BLUE), best in the sense that it has the minimum variance of all candidate linear estimators (Plackett, 1949; Davidson, 2000; Wooldridge, 2010).

1.4.3.4 Nonstochastic Covariates

The nonstochastic covariates assumption presupposes that covariates are fixed with known constants, as they would be in a usual experimental design, so that the response variable, y, is a response to the collective or controlled stimulus of $\mathbf{x}$. This assumption, albeit useful for focusing on the relationship between covariates and the response variable, can be relaxed without much compromise in the estimation.

1.4.3.5 Normality

The normality assumption states that the disturbance term, conditioned on $\mathbf{x}$, follows a normal distribution; or, $\varepsilon|\mathbf{x} \sim N(0, \sigma^2 \mathbf{I})$. This assumption, to some extent, is based on the premise that the conditional disturbance term absorbs many trivial elements that should cancel out each other and sum up to zero; in addition, normal distribution is a natural candidate for distributional assumption since it is the most widely used distribution and many other distributions virtually collapse to normal in asymptotic situations. Based on the previous discussions, we have

$$
\begin{aligned}
\widehat{\beta}_{\mathrm{OLS}} &= (\mathbf{X}'\mathbf{X})^{-1}\mathbf{X}'y \\
&= (\mathbf{X}'\mathbf{X})^{-1}\mathbf{X}'(\mathbf{X}\beta + \varepsilon) \\
&= \beta + (\mathbf{X}'\mathbf{X})^{-1}\mathbf{X}'\varepsilon
\end{aligned}
$$

Note that the last equality from Eq.1.4.3.5 only has ε as a random variable, and both β and $\mathbf{X}$ are considered as fixed. Thus, if $\varepsilon|\mathbf{x}$ is normal, then $\widehat{\beta}_{\mathrm{OLS}}$ should follow a normal distribution, for it is a simple linear transformation of ε. This assumption also makes it possible that one can conduct hypothesis testing of parameter estimates.

1.4.4 Estimation and Interpretation

The frequentist approach to estimating the OLS regression can be implemented quite easily with the `lm` function from the **stats** package that comes with the R installation by default. In the following, we first read in a Stata data file using the `read.dta` function from the `foreign` package, select cases using the `subset` function, and then use the `na.omit` function to drop missing cases using listwise deletion.

```
require(foreign)
# read in recoded cumulative GSS data
readin <- read.dta("data/gssCum7212Teach.dta", convert.factor = F)
# select variables and cases
mydta <- subset(readin, age >= 30 & age <= 60, select = c(physhlth,
    age, educ, impinc, female))
# drop missing cases
usedta <- na.omit(mydta)
```

After reading in the data and selecting cases, we can examine the descriptive statistics of these variables using R's base loop functions such as `lapply` and `sapply`.

```
# descriptive statistics
# -c(2,4) age impinc not tabulated
lapply(mydta[,-c(2,4)], table, useNA="ifany")

## $physhlth
##
##    0    1    2    3    4    5    6    7    8   10   11   14   15   20   21   24
##  202   34   25   20    5    7    5    6    2    8    1    3    4    3    1    1
##   30 <NA>
##   13 2860
##
## $educ
##
##    0    1    2    3    4    5    6    7    8    9   10   11   12   13   14   15
##    7    2   11    8   14   17   30   28  107   83  141  171  956  277  357  131
##   16   17   18   19   20 <NA>
##  442   96  176   62   77    7
##
## $female
##
##    0    1
## 1412 1788

lapply(mydta, mean, na.rm=T)

## $physhlth
## [1] 2.867647
##
## $age
## [1] 43.50438
##
## $educ
## [1] 13.18039
##
## $impinc
## [1] 10.17536
```

Histogram of usedta$physhlth

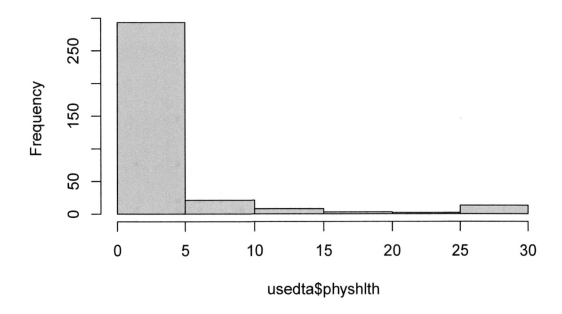

Figure 1.6
Histogram of Physical Unhealthy Days

```
##
## $female
## [1] 0.55875

sapply(mydta, sd, na.rm=T)

## physhlth      age      educ    impinc    female
## 6.5852799 8.8032919 3.1617869 0.8873471 0.4966140
```

In Fig. 1.6 we draw a histogram of the response variable, `physhlth`, using the `hist` function.

```
# doesn't look like a normally distributed variable
hist(usedta$physhlth)
```

Also we can use the `ggplot` function from the `ggplot2` package (Wickham, 2016) to draw a bivariate scatter plot with a fitted regression line,

```
require(ggplot2)
ggplot(usedta,aes(x = educ, y = physhlth)) +
            geom_point() +
            geom_smooth(method = "lm")
```

Fig. 1.7 shows that there is a weak negative bivariate association between years of education (`educ`) and physically unhealthy days (`physhlth`). We can further explore the

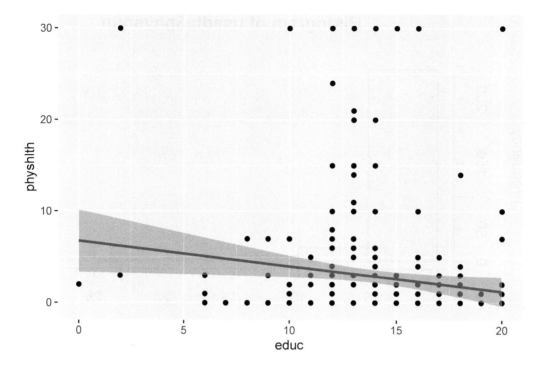

Figure 1.7
Bivariate Regression of Physical Unhealthy Days on Education

association between physical health and its covariates by running a linear regression model
using the `lm` function from the `MASS` package (Venables and Ripley, 2002).,

```
# run ols regression
myols = lm(physhlth ~ age + educ + impinc + female,
                    data = usedta)
summary(myols)

##
## Call:
## lm(formula = physhlth ~ age + educ + impinc + female, data = usedta)
##
## Residuals:
##     Min      1Q  Median      3Q     Max
## -8.3722 -3.0363 -1.6731 -0.0955 28.2018
##
## Coefficients:
##             Estimate Std. Error t value Pr(>|t|)
## (Intercept) 11.27147    4.12849   2.730  0.00666 **
## age          0.07505    0.04148   1.809  0.07129 .
## educ        -0.16031    0.13152  -1.219  0.22377
## impinc      -0.99626    0.42609  -2.338  0.01997 *
## female       1.37151    0.71187   1.927  0.05487 .
## ---
## Signif. codes:  0 '***' 0.001 '**' 0.01 '*' 0.05 '.' 0.1 ' ' 1
```

```
##
## Residual standard error: 6.447 on 335 degrees of freedom
## Multiple R-squared:  0.05291,Adjusted R-squared:  0.0416
## F-statistic: 4.679 on 4 and 335 DF,  p-value: 0.001095
```

To interpret the results, for example, one can say for each year increase in education, we would expect the number of physically unhealthy days decreases by 0.16 day, holding all other variables constant; being female vs. male, ceteris paribus, increases the number of physically unhealthy days by 1.372 days.

To check the model assumption and homoscedasticity, we can have various disgnostics plots using the `plot` function,

```
par(mfrow=c(2,2))
plot(myols)
```

Four diagnostic figures are graphed in Fig. 1.8. The upper left figure plots residuals against fitted values to check the linearity assumption. A horizontal line without a particular pattern implies an adherence to the linearity assumption. The upper right figure graphs a quantile-quantile (Q-Q) plot to check the normality assumption about the residuals. If the points fall on a single straight line (i.e., the dashed line), then there is strong evidence that the normality assumption holds. The lower left figure graphs a scale-location plot to check the homoscedasticity (equal variance) assumption of residuals. Again, if the red line is roughly horizontal and the residual points are randomly situated around the red line, then the homoscedasticity assumption is met. The right figure graphs the residuals vs. leverage plot to find cases with large influence on the results (e.g., the inclusion and exclusion of these cases can change the results dramatically) using Cook's D(istance) (Cook, 1977).

One can also predict the number of physically unhealthy days for all cases in the estimation sample using the `predict` function,

```
usedta$yhat <- predict(myols, type = "response")
summary(usedta$yhat)
```

```
##    Min. 1st Qu.  Median    Mean 3rd Qu.    Max.
## -1.073   1.795   2.884   2.868   3.787   8.372
```

Next, analysts can make predictions for hypothetical cases, such as a white female with average education and sample median (logged) income. The following R code chunk produces this prediction by first using the `data.frame` function to create a hypothetical `x` vector containing the characteristics previously described, feeding this vector into the `predict` function by setting the `newdata` argument to this `x` vector, and printing the output.

```
# hypothetical predictions
female.vec = data.frame(age = 35,
               female = 1,
               educ = mean(usedta$educ),
               impinc = median(usedta$impinc))

ypred <- predict(myols, newdata=female.vec,
            type = "response", # se.fit=TRUE,
```

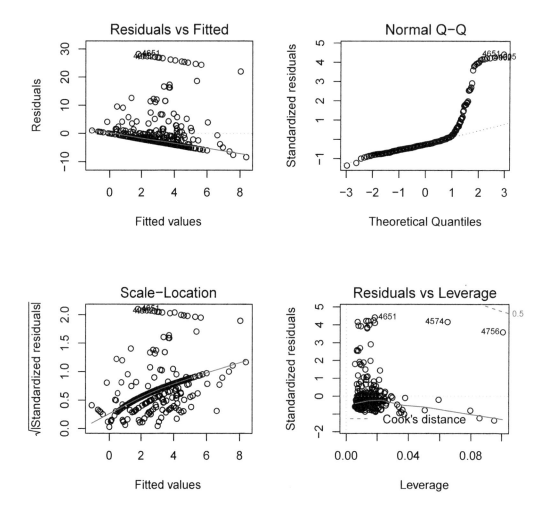

Figure 1.8
Linear Regression Diagnostics Plot

```
                     interval = 'confidence')
ypred

##       fit      lwr      upr
## 1 2.804022 1.601032 4.007011
```

The results show that the predicted number of physically unhealthy days for the hypothetical individual aforementioned is 2.804, and we are 95% confident that such prediction in the population lies somewhere between 1.601 and 4.007. More precisely, if we can re-sample 100 times, then 95 of the 100 confidence intervals so constructed will contain the population prediction. We can also use the `plot` function from the `effects` package (Fox and Weisberg, 2019) to graph an effect-displays plot. Fig. 1.9 plots the predictions of physically unhealthy

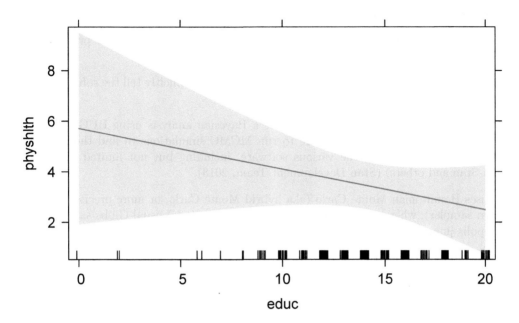

Figure 1.9
Prediction Plot of Physically Unhealthy Days by Education

days with their corresponding confidence bands while varying the values of education and setting age to its sample mean, sex to female, and income to its sample median,

```
library(effects)
plot(effect("educ", myols, xlevels=list(educ=0:20),
            given.values=c(impinc=median(usedta$impinc),
            female=1, age=mean(usedta$age))))
```

1.4.5 A Brief Introduction to Stan and Other BUGS-like Software

User-written packages in R, such as `arm` (Gelman and Su, 2020a), `brms`(Bürkner, 2017), `MCMCpack` (Martin et al., 2011), `nimble` (de Valpine et al., 2017), `rstan` (Stan Development Team, 2020) and `rstanarm` (Goodrich et al., 2018) can estimate a large variety of statistical models using Bayesian methods. For flexible parameterization, however, users often need to use statistical software specially-designed for Bayesian analysis using Markov chain Monte Carlo, or Bayesian inference using Gibbs Sampling (BUGS). Among the most popular ones of various software for Bayesian analysis are BUGS (WinBUGS and OpenBUGS from the BUGS Project), JAGS (just another Gibbs sampler), and Stan (in honor of Stanislaw Ulam). Despite differences in technical details, they all somewhat follow the BUGS syntactic style, which aligns well with the logic of the process for Bayesian analyses:

- First, users need to declare (empirical) data information, such as variable names, the number of covariates/predictors, and the number of levels of the response variable.

- Second, users provide the likelihood function comprising empirical data (variable names and other necessary information from empirical data and needed for the specification of the likelihood function) and the parameters of interest.

- Third, users specify priors for the parameters, unless default options are provided and users feel comfortable about them.

- Next, users link the first two steps by issuing functions to explicitly tell the software which is which between data and parameters

These are the necessary steps/information for a Bayesian analysis using BUGS software, and then users can request the software to run MCMC simulations to find the posterior. Differences also exist among the various software, including but not limited to (mainly between Stan and others) (Stan Development Team, 2018),

- Stan uses Hamiltonian Monte Carlo (aka hybrid Monte Carlo, or more precisely the No-U-Turn sampler), whereas BUGS uses (with a few exceptions) usual Gibbs sampling and Metropolis jumping.

- Stan is compiled through C++ and BUGS (including JAGS) are interpreted, and Stan sometimes adopts C syntax style such as the use of semicolon to end each statement.

- Stan usually requires explicit declaration of data information (i.e., the data code block), and BUGS and JAGS process empirical data more implicitly.

- Stan and BUGS differ in how probability distributions are named.

- Stan usually allows more flexible vector and matrix operations.

There are both limitations and advantages choosing one over the others. This book focuses on Stan for multiple reasons, including large user base and research team, regular maintenance, and clear syntax structure.

1.4.6 Bayesian Approach to Linear Regression

As discussed previously in Section 1.2.2, to obtain the posterior, one needs to provide the likelihood function of the empirical data and the priors. Using conjugate priors or some combinations of the likelihood and priors to analytically obtain the posterior with close-form solutions is mathematically intractable for most users and even many quantitative scholars. The MCMC methods thus provide a convenient and effective alternative. Both the likelihood and priors have to be supplied for using the MCMC methods. One can estimate Bayesian linear regression models using, for example, the custom-made R function, `bayesglm`, from the **arm** package (Gelman and Su, 2020a). Unlike functions from the **rstan** or **rstanarm** package, `bayesglm` uses empirical Bayes (e.g, priors are estimated from the data) methods to approximate the posterior.

```
require(arm)
# run bayes lm using bayesglm
bayeslm <- bayesglm(physhlth ~ age + educ + impinc + female,
# bayeslm$coefficients
display(bayeslm)

## bayesglm(formula = physhlth ~ age + educ + impinc + female, data = usedta,
##      prior.mean = 0, prior.scale = Inf, prior.df = Inf)
```

```
##               coef.est coef.se
## (Intercept) 11.27      4.13
## age           0.08      0.04
## educ         -0.16      0.13
## impinc       -1.00      0.43
## female        1.37      0.71
## ---
## n = 340, k = 5
## residual deviance = 13923.2, null deviance = 14701.0 (difference = 777.9)
## overdispersion parameter = 41.6
## residual sd is sqrt(overdispersion) = 6.45
```

We can also concatenate Stan codes in R to run the model with functions from the **rstan** package (Stan Development Team, 2020). First, one needs to create a model text file, delimited within two quotation marks; this file specifies the Bayesian model to be estimated in the **stan** function of the **rstan** package. The contents in this file begin after the opening quotation mark, and all the strings after are assigned to **modelString**.

```
modelString = "
```

Within the Stan model text file, the first code block is usually the **data** block, which is enclosed by two curly bracket signs and used to declare empirical data information, including the total number of cases, number of covariates, covariate matrix, the response variable vector, and their dimensions. **int<lower=0>** means that the number (N) defined after is an integer and its lowest threshold is 0; **matrix[N,K]** defines a matrix with N rows and K columns. These are the data that will be supplied by the data argument in the **stan** function during estimation. Note that in Stan, each command line usually ends with a semicolon, and one can use two forward slash signs (//) to begin a comment line.

```
data {
// number of cases/observations
int<lower=0> N;
// number of predictors
int<lower=0> K;
// covariate matrix
matrix[N, K] x;
// response variable vector
vector[N] y;
 }
```

The second is the **parameters** code block. In this block, one usually declares parameters to be estimated. For example, **real b0** defines the intercept, β_0; **vector[K] b** declares a parameters vector with K elements, corresponding to the slopes vector, β_{-0} (β excluding β_0), for our covariates; **real<lower=0> sigma** declares a parameter with its lowest threshold set at zero; in other words, this parameter is a positive real number, and we use it for estimating σ.

```
parameters {
// intercept
real b0;
```

```
// slope parameter vector
vector[K] b;
// standard error
real<lower=0> sigma;
}
```

Next is the `model` code block where one specifies priors and the likelihood function comprising data (from the `data` block) and parameters (from the `parameters` block). Usually, at the beginning of this block, priors are supplied by users. If no priors are provided, then uniform priors are used by default.

Below, we specify a normal distribution for the response variable, y, conditional on $\mathbf{x}\beta$.

```
model {
// the likelihood function
y ~ normal(x * b + b0, sigma);
}
```

After the Bayesian model is specified, one can close the model file using the closing quotation mark, and divert it to a text file, `model.txt`, using the `writeLines` function.

```
" # close quote for modelstring
writeLines(modelString,con="model.txt")
```

After completing the Stan model code block, one can start preparing the data. The following few lines extract estimation sample and related model information to be sent to the `stan` function for model estimation. These include the covariate matrix, x, stripped from the `dataMat`, a data matrix converted from the data frame, `usedta`. The first column of `dataMat` is excluded (`dataMat[, -1]`), since that column contains the response variable. `predictorNames` strips all covariates' names from the `dataMat` matrix. `nPredictors` is the number of covariates. `y` comes from the first column of `dataMat`. `predictedName` gets the response variable's name from the first column name of `dataMat`.

```
dataMat = as.matrix(usedta)
nData = NROW(dataMat)
x = dataMat[,-1]
predictorNames = colnames(dataMat)[-1]
nPredictors = NCOL(x)
y = as.matrix(dataMat[,1])
predictedName = colnames(dataMat)[1]
tbl = table(y)
nYlevels = dim(tbl) #nYlevels = max(y)
```

After the data are extracted from the data frame `usedta` and turned into either scalars or matrices, one can create a data objects list. Within the `dataList` list, the symbols on the left are the names of data used previously in the Stan model code block; those on the right of the equal signs are the data objects just obtained in the R code chunk above. This data objects list is prepared to be used for the `data` argument of the `stan` function later.

```
dataList = list(
      x = x ,
      y = as.vector( y ) ,
```

```
       K = nPredictors ,
       N = nData
)
```

Next step is to set up values to be fed into several other important arguments in the stan function, including the number of burn-in/warmup iterations (`burnInSteps`), number of chains (`nChains`), and parameter names (`parameters`). The burn-in steps refer to the trial-and-error process to find the general area of the multivariate posterior distribution more quickly than it would otherwise and discard the draws during this process since they come out of noise for the most part.

```
parameters = c("b0", "b", "sigma")  # The parameter(s) to be monitored.
adaptSteps = 500        # num of steps to "tune" the samplers.
burnInSteps = 500       # num of steps to "burn-in" the samplers.
nChains = 3             # num of chains
numSavedSteps=10000     # num of steps in chains to save.
thinSteps=1             # num of steps to "thin" (1=keep every step).
nPerChain = ceiling( ( numSavedSteps * thinSteps ) / nChains ) # Steps per chain.
```

Below we set up a list of initial values of the parameters to be those from our previous OLS regression. If the initial values are not supplied, then they will be set to their default values, which are random numbers generated from −2 to 2.

```
bInit = myols$coefficients
initsList =  list(
        b0 = bInit[1],
        b = bInit[2:(nPredictors+1)]
)

initsChains <- list()
for (i in 1:nChains) {
        initsChains[[i]] <- initsList
}
```

Now it comes down to estimation finally. In the stan function, the `model_code` argument takes in the Stan model syntax (text) file; the data argument reads in the empirical data objects from the list `dataList` previously defined; the `pars` argument declares parameters defined in the `parameters` list, and these names (`parameters = c("b0", "b", "sigma")`) should be the same as those declared in the `parameters` code block of the Stan syntax file. To make the results reproducible, the `seed` argument is set to be 47306.

```
mcmcSamples <- stan(model_code = modelString, data = dataList,
    seed = 47306, pars = parameters, chains = nChains, iter = nPerChain,
    warmup = burnInSteps, init = initsChains)
```

In the following, a results summary table is requested to show basic descriptive statistics of the posterior, including means, standard errors of the means (standard deviation divided by the square root of corresponding effective sample size), the 2.5%, 50%, and 97.5% percentile of the posterior, the rough estimate of effective sample size (`n_eff`), and the potential scale reduction scale factor (`Rhat`). A value of (close to) one for `Rhat` indicates that different chains converge on the same posterior, and the obtained results are robust (Gelman et al., 2014, pp. 284-285).

```
print(mcmcSamples, pars=c("b0", "b[1]", 'b[2]', "b[3]", "b[4]", "sigma"),
probs=c(.025,.5,.975))
        mean se_mean   sd  2.5%    50% 97.5% n_eff Rhat
b0     14.31    0.08 5.17  4.10  14.41 24.36  4067    1
b[1]    0.00    0.00 0.05 -0.09   0.00  0.10  7655    1
b[2]   -0.10    0.00 0.16 -0.43  -0.10  0.22  6217    1
b[3]   -0.92    0.01 0.50 -1.90  -0.93  0.06  3996    1
b[4]   -0.36    0.01 0.84 -1.97  -0.36  1.30  6740    1
sigma   7.68    0.00 0.31  7.12   7.67  8.31  7823    1
```

As shown in the table above, there is a .95 probability that the regression slope parameter for age (b[1]) is somewhere between −0.09 and 0.10; alternatively, one can say the 95% credible interval for this parameter is in that range. In Bayesian statistics, a credible interval corresponds to a range within which a parameter falls with a certain probability. Thus, one can also state that there is a .95 probability that the regression slope parameter for education (b[2]) is somewhere between −0.43 and 0.22, and so on and so forth.

```
library("bayesplot")
library("ggplot2")
mcmc_areas(mcmcChain, pars = c("b[1]", "b[2]"), prob = 0.95) +
    ggtitle("Plotting Posterior and 95% Credible Interval")
```

We can also use functions such as mcmc_area from the bayesplot package (Gabry and Mahr, 2021) to plot the posterior. Fig. 1.10 graphs the posterior of age (b[1]) and educ.(b[2])along with their respective central 95% credible bands.

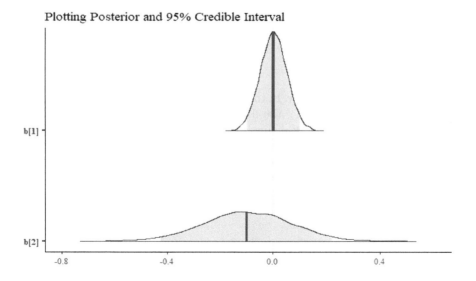

Figure 1.10
Posterior Distribution Plot

2

Binary Regression

> Thus in nothingness emerges existence, and out of existence arises nothing.
>
> ---
>
> in Chapter 2 of *Dao De Jing* by Laozi

2.1 Introduction

Binary regression models are models that take binary indicators as their response variables, and these binary response variables are usually coded as zero and one. Because of the prevalence and ease of the binary coding, there are numerous applications of this type of models. In health research, for example, analysts and scholars often use a binary response variable to measure self-reported health with one denoting poor health and zero for good health, and such coding schema—which explicitly predicts the presence of a negative health outcome—has been applied to a wide array of health conditions; in politics, we could use a binary response variable to measure voter turnout or choice in a two-candidate/party system; in education, researchers are often interested in some level of education attainment and above as opposed to below that level. There are numerous treatments of this regression model, including McCullagh and Nelder (1989), Greene (2007), and Agresti (2013). Binary regression is not only the most frequently used regression model for categorical data but also the foundation for other more complicated models such as ordered and multinomial regression models, item response models, propensity score matching, and some simple forms of survival regression models (e.g., discrete time survival analysis).

2.1.1 A Brief History of Binary Regression

Arguably the earliest application of binary regression analysis, or more precisely the probit version of it, can date back to Bliss (1934) in analyzing the dosage effect of nicotine on the percentage of asphids killed. In that example, the dosage was measured by milligram of nicotine per 100 milliliters of spay. It can be shown from Fig. 2.1 that was originally published in Bliss (1934) and replicated here, that the relationship between the dosage and percent killed follows an S-shaped curve; so a simple straight line would miss the pattern of diminishing returns on the high end of dosage.

Note that all the data points in this graph are approximated using digitized graph data from the original publication, and several cumulative normal densities with different means and standard deviations are experimented with to fit the data. Unlike the usual data for binary regression models where the probabilities are unobserved, this application uses the inverse of the cumulative density function of a standard normal (Φ^{-1}) and linearizes proportion data, thus turning them into the values of an observed continuous response variable.

DOI: 10.1201/9780429056468-2

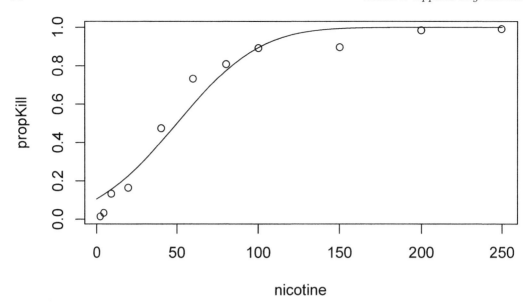

Figure 2.1
Approximated Bliss (1934) Fig. 1

The discovery of the logistic function dates back to the 19th century when scientists, both in natural and social sciences, were studying population growth, and it was usually credited to Pierre-Francois Verhulst, a Belgian mathematician, who found this function while solving a differential equation and named it logistic function (Verhulst, 1838). This function was rediscovered by Pearl and Reed in 1920 in their study of the population growth in the United States (Pearl and Reed, 1920). It is widely accepted that Joseph Berkson formally proposed the use of logistic function as an alternative of standard normal in the statistical analysis of bioassay, and coined the term logit as an analogue of probit (Agresti, 2013; Cramer, 2003).

2.1.2   Linear Probability Regression

Linear probability regression (LPR) models had not phased out until the mid to late 1990s when the binary regression became the dominant tool for analyzing binary response variables. Note that the linear probability model is simply a term for applying the classical OLS regression to binary response data. There are good reasons as to why LPR was so popular even when the estimation of binary regression already became computationally inexpensive and readily accessible in most software packages. The linear probability model treats the one-zero coding of binary response variables as if it were continuous measures, and presupposes that all the assumptions in the classical OLS regression still hold. The conditional expectation for the binary response variable is,

$$
\begin{aligned}
E\left(y|x\right) &= \sum y \times p(y|\mathbf{x}) \\
&= 1 \times p(y = 1|\mathbf{x}) + 0 \times p(y = 0|\mathbf{x}) \\
&= p(y = 1|\mathbf{x})
\end{aligned}
\tag{2.1}
$$

Using this result, one can set up the structural model to be $p(y = 1|\mathbf{x}) = \mathbf{x}\beta$, linking $\mathbf{x}$ and β to the conditional expectation, $E\left(y|\mathbf{x}\right) = p(y = 1|\mathbf{x})$. Below, data from the General Social Survey (GSS) cumulative file 1972–2012 are used to illustrate how a LPR model is estimated in R. The response variable is an ordinal measure of self-reported health, and it is recoded into a binary response variable, `hlthc2`, with one and zero denoting having poor (fair or poor) and good health (excellent or good) respectively; the predictors include age in years (`age`), education in years (`educ`), logged family income (imputed; `impinc`), and gender (female = 1; `female`). One can use the `lm` function from the `MASS` package (Venables and Ripley, 2002) for estimating a linear probability model by regressing `hlthc2` on the four predictors.

```
# load libraries
require(foreign)
require(MASS)
# read in recoded cumulative GSS data
readin <- read.dta("data/gssCum7212Teach.dta", convert.factor=F)
# create list of variable names used for variable selection below
usevar <- c("hlthc2", "age", "educ", "impinc", "female")
# subset the data (select variables)
mydta <- subset(readin[complete.cases(readin[usevar]),],
        select=c(hlthc2, age, educ, impinc, female))
lpm <- lm(hlthc2 ~ age + educ + impinc + female, data = mydta)
summary(lpm)$coefficients

##                   Estimate   Std. Error     t value      Pr(>|t|)
## (Intercept)    1.1636937349 0.070057548  16.61054051 4.463772e-60
## age            0.0051825990 0.000357181  14.50972928 1.352919e-46
## educ          -0.0212695231 0.002104573 -10.10633615 9.601347e-24
## impinc        -0.0889768525 0.007163069 -12.42160964 7.999648e-35
## female         0.0007864096 0.012352068   0.06366623 9.492390e-01

plot(mydta$educ, jitter(mydta$hlthc2, 0.3),
        xlab="Education", ylab="Health")
abline(lm(mydta$hlthc2 ~ mydta$educ))
```

The interpretation of the results from this multiple regression is quite straightforward. It can be shown, for example, that for each year increase in education, the expected/predicted probability for $y = 1$ (poor health) decreases by 0.021, holding other variables constant; being female vs. male, ceteris paribus, increases the predicted probability by 0.001. There are however several major concerns that would make LPR a much less attractive model than a binary regression model (BRM). First, the predictions can be out of a reasonable range; as shown in Fig. 2.2, the prediction line can go above and below the zero-one range. Second, the normality assumption is usually violated. This is clear from the graph that the conditional distributions of y, the observed binary indicator of self-reported health, is not normally distributed vertically around the prediction points, especially for the values of education that are further away from the center. Third and relatedly, since the conditional y follows a Bernoulli distribution, the variance of conditional y, $V(y|\mathbf{x}) = p(y|\mathbf{x})(1 - p(y|\mathbf{x})) = \mathbf{x}\beta(1 - \mathbf{x}\beta)$, is heteroscedastic. Last is about the linearity assumption that, in most cases, the relationship between the response variable and its covariates does not follow a simple linear pattern. It is quite common that as the value of some covariate increases or decreases, the association tends to become weak or tenuous, thus demonstrating a pattern of diminishing return. Given these limitations about the LPR model, analysts usually prefer BRM that largely relies on maximum likelihood estimation (MLE) while analyzing binary response data.

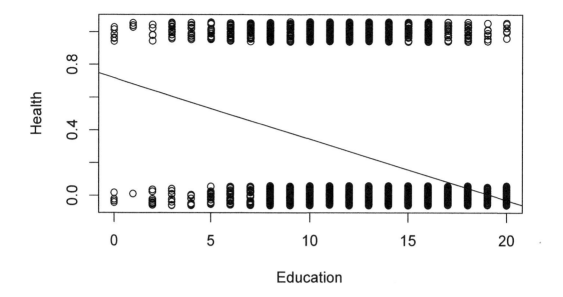

Figure 2.2
Linear Probability Regression of Health on Education with Jittering

2.2   Maximum Likelihood Estimation

Maximum likelihood estimation (MLE) is a widely-used method for estimating regression models with categorical and limited response data. Building on the scholarships by prominent mathematicians and statisticians such as Bennett, Chauvenet, Edgeworth, and Gauss, Ronald A. Fisher spearheaded the introduction, formalization, and popularization of MLE in statistics and other related fields. Fisher (1912) critiqued the least squares method as well as the method of moments for arbitrariness, and he proposed using an "absolute criteria" to find the solution to the problem of "fitting a theoretical curve" (p. 155). Along with his innovative work on MLE, Fisher (1922) introduced several other influential statistical concepts as well as techniques that are very important for modern statistics, including consistency, efficiency, likelihood, and sufficiency. However, it was not until the breakthroughs in computational power and numerical methods in the 1960s and 1970s that MLE became a widely used estimation tool (Eisenpress, 1962; Eisenpress and Greenstadt, 1966; Rao, 1973; Berndt et al., 1974). Essentially, MLE is an estimation method to find the set of parameters that provide the "maximal" probability of observing the data, based on parameterization, statistical models, and sufficient assumptions.

The MLE method requires the construction of a likelihood function. Many confuse likelihood with probability, and this is true even in Fisher's early work on MLE. According to Fisher (1922), "The likelihood that any parameter (or set of parameters) should have any assigned value (or set of values) is proportional to the probability that if this were so, the totality of observations should be that observed" (p. 310). In non-technical terms, probability refers to the chance of observing an event or series of events given the population parameters, whereas likelihood is proportional to the probability of observing the population parameters given some sample data. If one uses D to denote sample data and θ to denote the parameter vector, the probability and likelihood can be expressed as $p(D|\theta)$ and $L(\theta|D)$, respectively. There is almost trivial difference in the expanded mathematical notations between these two, except that they correspond to conceptually related yet distinct ideas; that is, the knowns and unknowns in these two functions are assumed to be different. Because MLE typically involves maximizing a function based on population parameters, the parts of the likelihood function that do not contain these parameters (e.g., constants) are usually dropped. It is in this sense that the likelihood is proportional to the probability.

2.2.1 Simple MLE Examples

In order to explain the MLE procedure in simple terms, one can start with the classic example of tossing a coin with the outcome being either head or tail, assuming that the probability of observing head is π and tail is $1 - \pi$ in the population. Let a discrete random variable Y to be the number of times head turning up out of n flips. For Y to follow a binomial distribution, the probability of observing s heads out of n trials is,

$$P(Y = s) = \frac{n!}{s!(n-s)!}\pi^s(1-\pi)^{n-s} \tag{2.2}$$

If one sets $\pi = 0.5$, $n = 10$, and $s = 8$, then $P(Y = 8|n = 10, \pi = 0.5) = \frac{10!}{8!(10-8)!}0.5^8(1-0.5)^{(10-8)} \approx 0.044$. In this example, the population parameter, π, is set to be 0.5, and one can easily compute the probability with the supplied sample data information, $P(D|\theta)$, as discussed in the previous section. In most empirical studies, however, the population parameters are unknown, and one usually begins with sample data, and use them to draw inferences about the population parameters. Then what would the observation of 8 heads out of 10 flips imply about the probability of flipping a coin and getting heads in the population (e.g., among an infinite series of flips)? One can tackle this problem using a few alternatives, but the MLE method suggests that the best estimate of this population parameter, π, is the probability that maximizes the likelihood of observing the data (i.e., $n = 10$, $s = 8$). To solve this problem, one can construct a likelihood function for π given the data, $L(\pi|n = 10, s = 8) = \frac{10!}{8!(10-8)!}\pi^8(1-\pi)^{(10-8)}$. In this likelihood, π becomes a variable and L is a function of π. One can graph this likelihood function using the `function` function from the `base` package and the `curve` function from the `graphics` package, both of which come with the R base installation (R Core Team, 2020). `function` defines the likelihood function and x is a name holder, corresponding to the parameter variable π in this case. Then this `function` object (i.e., `binom`) is fed as the first argument in the `curve` function, followed by the `from` and `to` arguments to set the range over which this function is plotted.

```
binom <- function(x) {
    factorial(10)/(factorial(8) * factorial(2)) * (x^8) * (1 - x)^2
}
curve(binom, from = 0, to = 1, xlab = expression(pi), ylab = "Likelihood")
```

By examining Fig. 2.3, one can visually estimate that the likelihood function probably reaches its maximum of about 0.3 when π is about 0.8. To analytically find the value of π that maximizes this likelihood function, one needs to take the first derivative of this likelihood function with respect to π as follows,

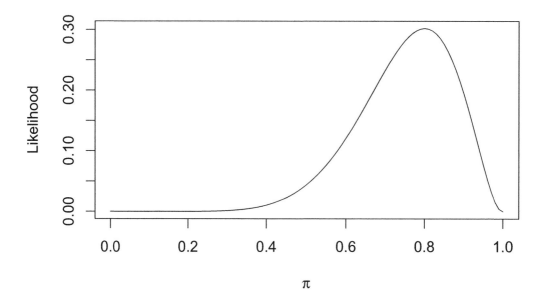

Figure 2.3
Maximum Likelihood Estimation of Mean for Binomial Distribution

$$dL\left(\pi|n=10, s=8\right)/d\pi$$

$$= d\left(\frac{10!}{8!(10-8)!}\pi^8(1-\pi)^2\right)/d\pi \qquad (2.3)$$

$$= \frac{10!}{8!(10-8)!}\left(8\pi^7(1-\pi)^2 - 2\pi^8(1-\pi)\right)$$

Then after setting the last line of the equation to zero and solving for $\widehat{\pi}_{ML}$, the ML estimate of π, one gets $5\widehat{\pi}^2 - 9\widehat{\pi} + 4 = 0 \Rightarrow \widehat{\pi} = 0.8$.

If this first warm-up example appears primitive and has little to do with empirical data analysis, the second example provides some flavor of empiricism. This particular example uses intelligence quotient (IQ) with only two observed values of 80 and 120. Without loss of generality and for illustrative purposes, one can first assume that IQ follows a normal distribution. The question now becomes how to estimate μ, the population mean for IQ, using MLE. Note that the probability density function for a continuous random variable that follows a normal distribution is $f(x|\mu,\sigma) = \frac{1}{\sqrt{2\pi}\sigma}\exp\left(-\frac{(x-\mu)^2}{2\sigma^2}\right)$. With the usual assumption that the cases are independently and identically distributed (i.i.d.), the likelihood of observing these two cases then is

$$L = \prod_{i=1}^{2}$$

$$= \frac{1}{\sqrt{2\pi}\sigma}\exp\left(-\frac{(80-\mu)^2}{2\sigma^2}\right)\frac{1}{\sqrt{2\pi}\sigma}\exp\left(-\frac{(120-\mu)^2}{2\sigma^2}\right) \qquad (2.4)$$

$$= \frac{1}{2\pi\sigma^2}\exp\left(-\frac{(80-\mu)^2 + (120-\mu)^2}{2\sigma^2}\right)$$

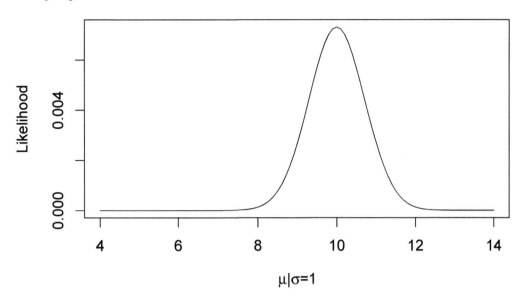

Figure 2.4
Maximum Likelihood Estimation of Mean for Normal Distribution

One can graph this likelihood function using the same `function` and `curve` functions with the following R code chunk,

```
# sigma = 1
std = 1
mvnorm <- function(m){1/(sqrt(2*pi)*std)*exp(-((8-m)^2+(12-m)^2)/2*std^2)}
curve(mvnorm, 4, 14,
      xlab=expression(paste(mu, "|", sigma, "=1")),ylab="Likelihood")
```

Fig. 2.4 shows the likelihood function graphed against μ re-scaled to be a multiple of 10 points while σ is fixed to be 1. Here only μ is treated as a random variable. It can be easily shown that the distribution centers around 100, and the density curve tapers off very quickly when it hits 8 in the left and 12 in the right tail.

Fig. 2.5 uses the `wireframe` function from the `lattice` package (Sarkar, 2008) to draw a three-dimensional graph of the likelihood function, μ, and σ, with the latter two being random variables. The `expand.grid` creates a data frame from all combinations of the values of vectors or factors listed within the function. The `drape = F` argument requests not to be draped in color. The `aspect` argument determines the relative size of each dimension. The `scales` argument sets scales and decides the appearances of axis labels and tick marks. the `screen` argument adjusts the angle and side at which we look at the cube. It can be visually estimated that the parameter maximizing the value of the likelihood is when μ is around 100.

```
require(lattice)
s <- seq(1, 1.6, by = 0.01)
m <- seq(4, 14, by = 0.25)
z <- expand.grid(s = s, m = m)
z$lf <- (1/z$s) * exp(-((8 - z$m)^2 + (12 - z$m)^2)/2 *
    z$s^2))
```

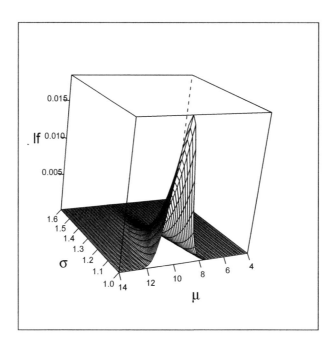

Figure 2.5
Maximum Likelihood Estimation in Normal Distribution

```
wireframe(lf ~ s * m, z, ylim = range(4, 14), xlab = expression(sigma),
    ylab = expression(mu), drape = FALSE, aspect = c(1, 1),
    scales = list(arrows = FALSE, cex = 0.6, col = "black",
        font = 10, tck = 1), screen = list(z = 100, x = -70,
        y = 20))
```

Analytically maximizing this function with respective to μ in this particular case is relatively straightforward since the two exponential terms can be combined. In more realistic cases, however, the likelihood function, which is usually composed of a long series of multiplicative terms when there are a large number of independently distributed random variables, is often intractable for taking derivatives. It is helpful to simplify the target likelihood function by taking a natural log of it, which turns multiplicative terms into additive ones. Then the log likelihood (LL) function can be maximized with respect to the population parameters. Because the LL function is a monotonic increasing function, the solution that maximizes the LL function should also maximize the likelihood function. Taking the partial derivative of the LL function with respect to μ in the IQ example, one should have

$$
\begin{aligned}
\frac{\partial LL}{\partial \mu} &= \frac{\partial}{\partial \mu} \ln \left(\frac{1}{2\pi\sigma^2} \exp \left(-\frac{(80-\mu)^2 + (120-\mu)^2}{2\sigma^2} \right) \right) \\
&= \frac{\partial}{\partial \mu} \left(-\frac{(80-\mu)^2}{2\sigma^2} \right) + \frac{\partial}{\partial \mu} \left(-\frac{(120-\mu)^2}{2\sigma^2} \right) \\
&= \frac{200 - 2\mu}{\sigma^2}
\end{aligned}
\tag{2.5}
$$

and setting up the first-order condition for the above equation to zero results in $200 - 2\widehat{\mu}_{\mathrm{ML}} = 0 \Rightarrow \widehat{\mu}_{\mathrm{ML}} = 100$.

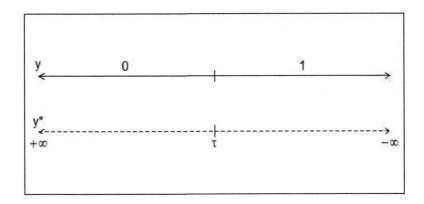

Figure 2.6
Latent Variable Approach

2.2.2 MLE for Binary Regression

Binary regression models (BRM) can be derived using multiple approaches that are largely based on the ideas about probability and likelihood discussed in the previous sections. The most popular approach, the latent variable approach, introduces a latent continuous variable linking a systematic (structural) component or what is often called linear predictor, $\mathbf{x}\beta$, to an observed binary response variable, y, and a probability distribution. Suppose there is an observed binary response variable, $y = 0, 1$. This variable can be a binary indicator of health status, for example, with one denoting having poor health and zero good health. Conceptually, for every binary response variable, one could construct a latent unobserved continuous variable, y^*, to denote the degree to which the status of zero and one is distanced from the threshold/cut-off point usually denoted by τ, around which the status changes from zero to one or vice versa. Fig. 2.6 provides a graphical presentation of this latent variable approach.

For example, among those claiming to have good health, there is still some variation in the degree of "goodness." Some have perfect health without any diagnosed or self-identified disease or ailment, whereas others have experienced or are experiencing health issues, but still self-perceive to be healthy. Without loss of generality, it is usually assumed that $y^* = \mathbf{x}\beta + \varepsilon$, and if $y^* > \tau$, then $y = 1$; otherwise $y = 0$. One can then link probability with these two assumptions,

$$
\begin{aligned}
P(y = 1) &= P(y^* > 0) \\
&= P(\mathbf{x}\beta + \varepsilon > 0) \\
&= P(\varepsilon < \mathbf{x}\beta) \\
&= F(\mathbf{x}\beta)
\end{aligned}
\tag{2.6}
$$

where $F(\mathbf{x}\beta)$ is some generic cumulative density function. Depending on the distributional assumption one makes about ε, for example, standard normal or standard logistic, one can have either probit or logit models. For binary logit models, $F(\mathbf{x}\beta) = \frac{\exp(\mathbf{x}\beta)}{1+\exp(\mathbf{x}\beta)} = \Lambda(\mathbf{x}\beta)$, and for probit models, $F(\mathbf{x}\beta) = \int_{-\infty}^{x\beta} \frac{1}{\sqrt{2\pi}} \exp\left(-\frac{t^2}{2}\right) dt = \Phi(\mathbf{x}\beta)$. Thus, using this unobserved continuous variable y^* and the error probability distribution as a bridge, one can link $\mathbf{x}\beta$ with the observed binary response variable y. The challenge next is how to capitalize on this relationship with some estimation method to find β. As discussed previously, one can use MLE to estimate the parameter vector, β in the case of BRM, by setting up a likelihood as a function of observed data and parameters. Since for each observed y, one has $P(y = 1) = F(\mathbf{x}\beta)$ for $y = 1$ and $P(y = 0) = 1 - F(\mathbf{x}\beta)$, and if the i.i.d. assumption holds, then the likelihood of observing all these y's is a product of the probabilities of observing y, conditional on $\mathbf{x}$ and β,

$$L(\beta|\mathbf{x}, y) = \prod_{i=1}^{N} [F(\mathbf{x}\beta)]^y [1 - F(\mathbf{x}\beta)]^{1-y} \tag{2.7}$$

Because the likelihood function is a product of a series of terms and computationally intractable, one usually takes the log of this likelihood function, and turns it into a log likelihood function,

$$LL(\beta|\mathbf{x}, y) = \sum_{y=1} \ln F(\mathbf{x}\beta) + \sum_{y=0} \ln [1 - F(\mathbf{x}\beta)] \tag{2.8}$$

What is left then is to numerically find $\hat{\beta}_{ML}$ that maximizes the log likelihood function. To give a jump start, the following R code chunk illustrates how to get estimation results using the logit link,

```
mylogit <- glm(hlthc2 ~ age + female + educ + impinc,
        data=mydta, family=binomial(link="logit"))
summary(mylogit)$coefficients
```

```
##                  Estimate  Std. Error    z value     Pr(>|z|)
## (Intercept)    4.01982014 0.437204248   9.194376 3.771782e-20
## age            0.02758393 0.002183067  12.635401 1.347061e-36
## female         0.01641373 0.079525021   0.206397 8.364808e-01
## educ          -0.12523004 0.013424118  -9.328735 1.071418e-20
## impinc        -0.50327168 0.044616319 -11.279991 1.647628e-29
```

The option, `family=binomial(link="logit")`, requests the `glm` function from the R base package `stats` to set the binary response variable to follow a binomial distribution with the logit link (R Core Team, 2020). In the results section, based on the signs of the coefficients and disregarding the significance level, one can see that `age` and `female` have positive associations with health status reversely coded; that is, age and being female (as opposed to male) increase the likelihood of having poor health ($y = 1$), whereas education and income are negatively associated with poor health status. One can follow the same routine as it is in linear regression models to determine if the predictors are associated with the binary response variable at the desired significance level, usually set to .05.

2.2.3   Numerical Methods for MLE

This section conceptually discusses how MLE is executed numerically. ML estimators belong to the general M estimation framework, under which estimators are to minimize or maximize a given function, and thereby are also called as "extremum estimators" more generally (Amemiya 1985). Among all sets of candidate parameter vectors θ's, let us define θ_0 as the population parameter vector, $U(\theta)$ as the target function for minimization or maximization,

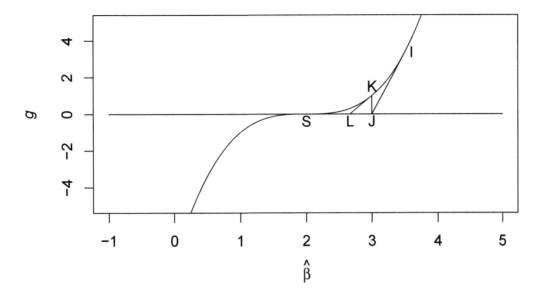

Figure 2.7
Newton-Ralphson Method

and $g(\theta) = \nabla_\theta(U)$ as the gradient[1] containing the first partial derivatives of $U(\theta)$ with respect to elements in θ. In order to find the minima or maxima of the target function, one usually sets the first partial derivatives to zero,

$$g\left(\widehat{\theta}|\mathbf{x}, y\right) = \mathbf{0} \tag{2.9}$$

where $\widehat{\theta}$ provides the extrema (e.g., minimum or maximum) for $U(\theta)$. We can also define $H = \frac{\partial^2 U(\theta)}{\partial\theta\partial\theta'}$, the Hessian matrix, if $U(\theta)$ is twice differentiable. To obtain ML estimates, one can rely on closed-form solutions if the gradient and the roots for the first-order condition (FOC:$g\left(\widehat{\theta}|\mathbf{x}, y\right) = \mathbf{0}$) are easy to compute. If the FOC however contains polynomials or nonlinear terms with respect to θ, then the mathematics involved can be intractable. As a result, statistical packages typically rely on numerical iterative root-finding methods, of which the most commonly known is the Newton's or Newton-Raphson method. Newton's method goes through successive approximations of the roots of $\widehat{\theta}$ under the FOC and stops root-searching when two consecutive approximations are very close to each other according to some set convergence criterion.

Fig. 2.7 gives a graphical illustration of this iterative process for a single-parameter model. The horizontal and vertical axes correspond to the variable $\widehat{\beta}$ and the first derivative of the log likelihood function g (i.e., the gradient) concerning $\widehat{\beta}$ respectively. Suppose point I in Fig. 2.7 supplies a reasonable initial value for $\widehat{\beta}$ and its corresponding g, one can draw a tangent line of the curve at point I that crosses the horizontal axis at point J, whose value for the horizontal coordinate then becomes an updated guess of $\widehat{\beta}$. Through J, one can draw a line perpendicular to the horizontal axis and touches the curve at K, where one can create another tangent line that meets the horizontal axis at L. The process can be iteratively repeated until the difference between the two consecutive intersection points on the horizontal coordinate (e.g., J and L) is reasonably small, viz., smaller than

---

[1]Gradient usually refers to a vector containing first partials of a vector-valued function of several variables.

some set tolerance level, and one can claim to find the root for the equation of the first-order condition. With a Taylor series expansion, it can be proven that the Newton-Raphon method can comfortably reach convergence under certain regularity conditions.

To generalize the Newton-Raphson method to multivariate case more formally, let θ^i be the approximation in the ith iteration, and θ^{i+1} be the approximation in the next iteration. Then one can apply the mean value theorem around $g\left(\theta^{i+1}|\mathbf{x}, y\right)$ by writing,

$$g\left(\theta^{i+1}|\mathbf{x}, y\right) = g\left(\theta^i|\mathbf{x}, y\right) + H\left(\theta^i|\mathbf{x}, y\right)\left(\theta^{i+1} - \theta^i\right) + R^i \tag{2.10}$$

Assuming $\theta^{i+1} = \widehat{\theta}_{\text{ML}}$, then under FOC (i.e., $g\left(\widehat{\theta}|x, y\right) = \mathbf{0}$) and ignoring the remainder term R^i, we can have $\theta^{i+1} \simeq \theta^i + \left(-H\left(\theta^i|\mathbf{x}, y\right)\right)^{-1}\left(g\left(\theta^i|\mathbf{x}, y\right)\right)$. In each iteration, we simply examine the magnitude of the second term on the right-hand side, $\left(-H\left(\theta^i|\mathbf{x}, y\right)\right)^{-1}\left(g\left(\theta^i|\mathbf{x}, y\right)\right)$. If one approaches the solution, then $g\left(\theta^i|\mathbf{x}, y\right)$ should be close to zero, and the difference between θ^{i+1} and θ^i will be trivial. The derivation of the Hessian matrix, however, can be computationally expensive. Therefore, researchers often rely on quasi-Newton methods, including the Berndt-Hall-Hall-Hausman (BHHH) algorithm (Berndt et al., 1974), the Davidon-Fletcher-Powell (DFP) algorithm (Davidon, 1959; Fletcher and Powell, 1963; Fletcher, 1970), and the Broyden-Fletcher-Goldfarb-Shanno (BFGS) algorithm (Broyden, 1967; Fletcher, 1970; Goldfarb, 1970; Shannon, 1970). Note that all three algorithms use a function of either the gradient or a combination of the gradient and the parameter vector from the current and next iterations to replace the Hessian matrix in the current iteration. For example, the BHHH algorithm uses the sample analogue of outer product (the variance-covariance matrix) of the gradient to replace the negative Hessian matrix.

2.2.4   Normality, Consistency, and Efficiency

Along with a few others, main results in calculus and statistics such as the mean value theorem, the central limit theory, and the fact that $V\left(g\left(\theta|\mathbf{x}, y\right)\right) = -E\left[H\left(\theta|\mathbf{x}, y\right)\right]$ in MLE lead to the asymptotic normality, $\sqrt{n}\left(\widehat{\theta} - \theta_0\right) \xrightarrow{d} \mathcal{N}\left(0, -E\left[\frac{1}{n}H\left(\theta_0|\mathbf{x}, y\right)\right]^{-1}\right)$, where $\widehat{\theta}$ is our ML estimate vector, and θ_0 is our population parameter vector. With simple algebraic rearrangements, we can have

$$\widehat{\theta} \xrightarrow{d} \mathcal{N}\left(\theta_0, -E\left[H\left(\theta_0|\mathbf{x}, y\right)\right]^{-1}\right) \tag{2.11}$$

Of Fisher's (1922) three criteria for good estimators (consistency, efficiency, and sufficiency), consistency is the most important to establish in empirical analyses because it is closely related to the accuracy of the results. Fisher (1922) initially relied on what he called "the common-sense criterion" that "when applied to the whole population the derived statistic should be equal to the parameter" (p. 316). The proof of the consistency property of ML estimators is quite technical in nature (see Amemiya 1985:115-120, for example) and is beyond the scope of this book. Here we simply present the result:

$$\lim P\left(\widehat{\theta} = \theta_0\right) = 1 \tag{2.12}$$

In other words, as the number of data points approaches infinity, the ML estimator converges in probability to the true parameter, θ_0, in the population. Alternatively but equivalently, we have $\lim P\left(|\widehat{\theta} - \theta_0| < \varepsilon\right) = 1$ for any arbitrarily small constant vector ε. The efficiency of statistical estimation refers to the property that "in large samples, when the distribution of the statistics tend to normality, that statistic is to be chosen which has

the least probable error" (Fisher 1922, p. 316). In other words, efficiency (or asymptotic efficiency, to be precise) requires that the consistent estimator produces the smallest standard errors. Although efficiency is not viewed as important as consistency, it is a desirable property for statistical estimators. Under certain regularity conditions for the applicability of the Lindeberg-Levy version of the central limit theorem (Theil, 1971), the variance of the unbiased estimator will be at least as large as the inverse of the Fisher information,
$V(\theta) \geq \left(-E\left(\frac{\partial^2 LL(\theta)}{\partial\theta\partial\theta'}\right)\right)^{-1}$.

2.2.5 Nonlinear Probability

When the domain of the predictors does not match the range of the response variable, a common practice in statistics is to apply some function to transform either side to match the other. Bliss(1934) is probably among the first few to capitalize on this idea in analyzing a response variable bounded between zero and one. So a second approach that people sometimes use to construct a binary regression is to transform the linear predictor, $\mathbf{x}\beta$, which is usually assumed to take on any value on the real line, to match the range of $P(y = 1)$, which is bounded between zero and one. For example, taking an exponential function of $\mathbf{x}\beta$ leads to $\exp(\mathbf{x}\beta)$, thus expanding the domain of $\mathbf{x}\beta$ to include all positive real numbers. If one applies a second function to $\exp(\mathbf{x}\beta)$, for example, $F(\mathbf{x}\beta) = \frac{\exp(\mathbf{x}\beta)}{1+\exp(\mathbf{x}\beta)} = \frac{1}{1+\exp(-\mathbf{x}\beta)}$, then the domain of the resultant function can be constrained to be between zero and one, thereby perfectly matching the theoretical range of $P(y = 1)$. This result can be reversely obtained by transforming $P(y = 1)$ using the inverse functions of $\exp(\mathbf{x}\beta)$ and $\frac{\exp(\mathbf{x}\beta)}{1+\exp(\mathbf{x}\beta)}$, including the natural log $\left(\ln\left(\frac{P(y=1)}{1-P(y=1)}\right)\right)$ and the odds function $\left(\frac{P(y=1)}{1-P(y=1)}\right)$ respectively. Using this setup, one can get a logit model. Similarly, if the cumulative density function of the standard normal is applied to $\mathbf{x}\beta \to \Phi(\mathbf{x}\beta)$, then one can turn $\mathbf{x}\beta$ into a function bounded between zero and one. In this case, one has a probit model.

2.3 Hypothesis Testing and Model Comparisons

Hypothesis testing, or null hypothesis significance testing (NHST), and relatedly, the use of p-value and confidence intervals, are a dominant suite of packaged tools of statistical inference in the field for the past century or so. The invention and practice of NHST was ironically established on the feud between Ronald A. Fisher, who systematized the idea of p-value first introduced by Pearson (1900), and Neyman and Pearson (1933) and Neyman (1937), who expanded on and transformed the concept of p-value to include other important concepts in NHST, such as alternative hypothesis, type I and type II errors, and the power as well as the significance level of a test.

Ronald A. Fisher is often credited for formalizing the idea of p-value in the 1920s. Like most of Fisher's innovations in statistics, p-value hardly appears very complicated, and it even seems quite straightforward conceptually in hindsight. The general idea is, after one gets some sample statistic or parameter estimate, such as a sample mean or regression coefficient, this statistic can be turned into a test statistic based on some hypothetical distributional properties about this statistic/estimate, assuming the null hypothesis is true. The null hypothesis is usually set up as a target for refutation with empirical data; for example, there is no difference in certain attribute between two groups (e.g., $H_0 : \mu_1 - \mu_2 = 0$) or no association between two variables in the population (e.g., $H_0 : \beta = 0$). One

Distribution of Test Statistic

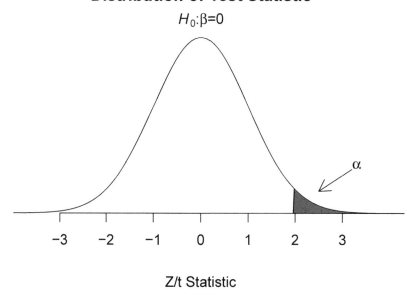

$H_0:\beta=0$

Z/t Statistic

Figure 2.8
Hypothesis Testing

can then calculate how likely we are to get this test statistic and equivalently the sample statistic/estimate that large or larger while assuming the null is true, and this is called the p-value. If the p-value is small, we then reject the null hypothesis. We usually want to measure the calculated p-value against a hypothetical p-value that is presupposed to be the threshold, under which we can claim that the p-value is too small for us to dismiss the test statistic as resulted from a simple chance. This hypothetical p-value is also called significance level, often denoted as α. Fig. 2.8 provides a graphical presentation of the general ideas of significance testing.

The cut-off point for p-value, however, is arbitrary. Probably in the initial proposal and first application of this kind in his seminal work, *Statistical Methods for Research Workers*, Fisher (1934, p. 82) states that,

"In preparing this table we have borne in mind that in practice we do not want to know the exact value of P for any observed χ^2, but, in the first place, whether or not the observed value is open to suspicion. If P is between .1 and .9 there is certainly no reason to suspect the hypothesis tested. If it is below .02 it is strongly indicated that the hypothesis fails to account for the whole of the facts. We shall not often be astray if we draw a conventional line at .05, and consider that higher values of χ^2 indicate a real discrepancy," wherein P refers to the p-value.

Fisher (1934, p. 45) also added that choosing this number is convenient since it roughly corresponds to the probability of exceeding two standard deviations from the mean under a normal distribution. This convenience was especially true when the transformation from probabilities to statistics still relied on primitive computational methods without computers.

Usually, Fisher's version of NHST is called significance test, with a falsifiable null hypothesis only. Based on Fisher's test of significance, the approach that Neyman and Pearson propose incorporates two hypotheses, including the null and the alternative hypotheses, and formulates the testing process as making a decision in choosing between the null and the alternative. Associated with these two decisions are two types of errors, type I and type II errors. Type I errors (or false positive) are errors made when we reject a null hypothesis when it is true, and type II errors (false negative) are those when we accept a false null hypothesis. p-value is usually equivalent of the probability of committing a type I error, or α in the terminology of hypothesis testing. In NHST, β usually corresponds to the probability of committing a type II error, and $(1 - \beta)$ is defined the power of a test, or the probability of avoiding a type II error.

2.3.1 Wald, Likelihood Ratio, and Score Tests

The Wald, likelihood ratio (LR), and score tests are three most commonly used hypothesis testing procedures for models predicting categorical and limited response variables. All three tests are based on results from ML estimation, and are asymptotically equivalent. The Wald test was designed to assess linear constraints on statistical parameters in large samples (Wald, 1943). When there is a single parameter in the usual pair of hypotheses, for example, $H_0 : \beta = \beta_0$ and $H_a : \beta \neq \beta_0$, the test statistic can be computed as the z statistic, $z = \frac{\hat{\beta} - \beta_0}{SE(\hat{\beta})}$. For a linear combination of a composite set of parameters, we have

$$W = \left(\mathrm{T}\hat{\beta}_F - \mathrm{c}\right)' \left(\mathrm{T}\hat{V}\left(\hat{\beta}_F\right)\mathrm{T}'\right)^{-1} \left(\mathrm{T}\hat{\beta}_F - \mathrm{c}\right) \sim \chi_Q^2 \qquad (2.13)$$

for the null hypothesis $H_0 : \mathrm{T}\beta = \mathrm{c}$, where T is a linear transformation matrix, $\hat{\beta}_F$ contains the ML estimates from the full model, $\hat{V}\left(\hat{\beta}_F\right)$ is the estimated variance-covariance matrix for $\hat{\beta}_F$, and c is a $Q \times 1$ constraints vector. To conduct a Wald test for a single parameter, one could calculate the results using some built-in functions in R, or more conveniently the `linearHypothesis` function from the `car` package (Fox and Weisberg, 2019). The following few R code chunks illustrate different ways to conduct the Wald test using the age variable as an example, and the hypothesis under investigation is a simple single-parameter hypothesis, $H_0 : \beta_{\mathrm{age}} = 0$

```
summary(mylogit)$coefficients
```

```
##                 Estimate  Std. Error    z value      Pr(>|z|)
## (Intercept)   4.01982014 0.437204248    9.194376  3.771782e-20
## age           0.02758393 0.002183067   12.635401  1.347061e-36
## female        0.01641373 0.079525021    0.206397  8.364808e-01
## educ         -0.12523004 0.013424118   -9.328735  1.071418e-20
## impinc       -0.50327168 0.044616319  -11.279991  1.647628e-29
```

In the second row of the results section, we can see that the z-value (statistic) is 12.635, and its associated p-value is very small, close to be zero. So we can confidently infer that the effect of age on self-rated health is statistically significant ($p < .001$). The next few lines of R codes illustrate how the p-value and z-value are calculated using the age coefficient estimate and its standard error.

```
zval <- (coef(mylogit)[2] - 0 )/ sqrt(vcov(mylogit)[2,2])
pval <- 2*pnorm(abs(zval), lower.tail=FALSE)
c(zval=zval, pval=pval)
```

```
##      zval.age       pval.age
## 1.263540e+01  1.347061e-36
```

We can also use the custom-made function, `linearHypothesis`, from the `car` package to conduct the test. Here if we square the z statistic (12.635) calculated using either method shown immediately above, we will get the χ^2 produced by `linearHypothesis` since the chi-squared distribution with one degree of freedom is the standard normal squared.

```
library(car)
linearHypothesis(mylogit, c("age=0"))

## Linear hypothesis test
##
## Hypothesis:
## age = 0
##
## Model 1: restricted model
## Model 2: hlthc2 ~ age + female + educ + impinc
##
##   Res.Df Df  Chisq Pr(>Chisq)
## 1   4240
## 2   4239  1 159.65  < 2.2e-16 ***
## ---
## Signif. codes:  0 '***' 0.001 '**' 0.01 '*' 0.05 '.' 0.1 ' ' 1
```

One can also conduct a Wald test for every coefficient using the `Anova` function from the `car` package,

```
Anova(mylogit, test='Wald')

## Analysis of Deviance Table (Type II tests)
##
## Response: hlthc2
##          Df    Chisq Pr(>Chisq)
## age       1 159.6534    <2e-16 ***
## female    1   0.0426    0.8365
## educ      1  87.0253    <2e-16 ***
## impinc    1 127.2382    <2e-16 ***
## ---
## Signif. codes:  0 '***' 0.001 '**' 0.01 '*' 0.05 '.' 0.1 ' ' 1
```

For a linear combination of multiple coefficients, for example, $H_0 : \beta_{\text{age}} = 0 \ \& \ \beta_{\text{educ}} = 0$, we can again use the `linearHypothesis` function, which was written to test both single-parameter hypotheses separately and multiple hypotheses simultaneously.

```
linearHypothesis(mylogit, c("age=0", "educ=0"))

## Linear hypothesis test
##
## Hypothesis:
## age = 0
## educ = 0
##
## Model 1: restricted model
## Model 2: hlthc2 ~ age + female + educ + impinc
```

```
##
##    Res.Df Df  Chisq Pr(>Chisq)
## 1    4241
## 2    4239  2 297.49  < 2.2e-16 ***
## ---
## Signif. codes:  0 '***' 0.001 '**' 0.01 '*' 0.05 '.' 0.1 ' ' 1
```

The results show that for two degrees of freedom (two constraints), we get a $\chi^2 = 297.49$. Thus we can safely reject the null hypothesis, and conclude that at least one of the coefficients is statistically different from zero.

Compared with the Wald test that uses information from the full model only, the likelihood ratio (LR) test absorbs information from both the full and the restricted models and adjudicates between the two. The restricted model is restricted inasmuch as the predictors in the restricted model compose a subset of the covariates in the full model. Similar to the Wald test, the LR test also computes a chi-squared statistic, which is based on the likelihoods from both models,

$$LR = 2 \left(\ln(\widehat{L})_F - \ln(\widehat{L})_R \right) \sim \chi^2_Q \tag{2.14}$$

where LL_F is the log-likelihood for the full model, and LL_R is the likelihood from the restricted model. The formula above shows that the twice of the difference between the two log-likelihoods follows a chi-squared distribution with Q degrees of freedom, which usually corresponds to the number of constrains.

To conduct an LR test of single coefficients, one can simply use the **anova** function from the **stats** package installed with R by default. An LR test needs to run both the full and restricted models. For presentational convenience, here let us assume that the full model is **mylogit**, a logit estimation object already created with **age**, **female**, **educ**, and **impinc** as predictors and **hlthc2** as the binary response variable. Then we only need to run the restricted model. Instead of typing out the function name, model type, and variable names, we could simply use the **update** function, also from the **stats** package, to quickly estimate the restricted model since the predictors in the restricted model are a subset of the covariates from the full model. For example, if we want to test the effect of age on self-rated health, we can exclude **age** in the restricted model. In the **update** function, we first specify the object name for the full model, **mylogit**, and then we type ".~.", corresponding to the original full model. The minus sign in front of the age variable is used to denote that the updated model excludes age. To calculate the LR test statistic, one uses the **anova** function, with the object name for the restricted model specified as the first argument, and the full model second, followed by the test option set to **Chisq**.

```
noAge <- update(mylogit, .~.-age)
summary(noAge)$coefficients
```

```
##                 Estimate Std. Error    z value      Pr(>|z|)
## (Intercept)  5.64202718 0.41193537  13.6963893 1.066998e-42
## female       0.05890093 0.07747682   0.7602393 4.471115e-01
## educ        -0.16126273 0.01310936 -12.3013465 8.907666e-35
## impinc      -0.49275728 0.04345912 -11.3384079 8.467083e-30
```

```
anova(noAge, mylogit, test="Chisq")
```

```
## Analysis of Deviance Table
##
```

```
## Model 1: hlthc2 ~ female + educ + impinc
## Model 2: hlthc2 ~ age + female + educ + impinc
##   Resid. Df Resid. Dev Df Deviance  Pr(>Chi)
## 1      4240     4254.5
## 2      4239     4089.8  1   164.68 < 2.2e-16 ***
## ---
## Signif. codes:  0 '***' 0.001 '**' 0.01 '*' 0.05 '.' 0.1 ' ' 1
```

The results show that with one constraint, we get a $\chi^2 = 164.7$, which is quite close to the one produced by the Wald test (159.653). Similarly, because of the large χ^2 relative to one degree of freedom, we reject the null. For testing multiple coefficients or some linear combination of multiple coefficients, such as $H_0 : \beta_{age} = 0$ & $\beta_{educ} = 0$, we can also use the anova function to conduct an LR test. To do that, one simply "subtracts" these multiple coefficients in the original model statement using the update function like we just did for the single-coefficient LR test, and then applies the anova function. Please also be aware of the difference between the lower case anova function from the stats package and the upper case Anova function from the car package; they serve different purposes.

```
noAgeEduc <- update(mylogit, .~. - age - educ)
summary(noAgeEduc)$coefficients
```

```
##                  Estimate Std. Error     z value      Pr(>|z|)
## (Intercept)    5.62723172 0.41265892  13.6365203  2.428760e-42
## female        -0.01375585 0.07553539  -0.1821113  8.554954e-01
## impinc        -0.68625249 0.04147758 -16.5451430  1.735520e-61
```

```
anova(noAgeEduc, mylogit, test="Chisq")
```

```
## Analysis of Deviance Table
##
## Model 1: hlthc2 ~ female + impinc
## Model 2: hlthc2 ~ age + female + educ + impinc
##   Resid. Df Resid. Dev Df Deviance  Pr(>Chi)
## 1      4241     4416.7
## 2      4239     4089.8  2   326.87 < 2.2e-16 ***
## ---
## Signif. codes:  0 '***' 0.001 '**' 0.01 '*' 0.05 '.' 0.1 ' ' 1
```

Unlike the Wald test that only replies on estimates and statistics from the full model or the LR test that requires running both full and restricted models, the score test runs the restricted model, albeit only in theory. There are multiple names used in different fields for this test, for example, the Rao, the score, or the Lagrange multiplier (LM) test. The score test statistic is usually computed as

$$S = g\left(\widehat{\beta}_R\right)' \widehat{I}\left(\widehat{\beta}_R\right)^{-1} g\left(\widehat{\beta}_R\right) \sim \chi_Q^2 \tag{2.15}$$

with Q restrictions, where $\widehat{\beta}_R$ contains estimates from the restricted model when the null hypothesis is true, $g\left(\widehat{\beta}_R\right)$ is the score function or what is often called the gradient of the likelihood function with respect to $\widehat{\beta}_R$. In the middle is the variance-covariance matrix of the gradient, and is derived based on the (Fisher) information matrix. To conduct a score test in R, we can use the same anova function, the original full model object, and the

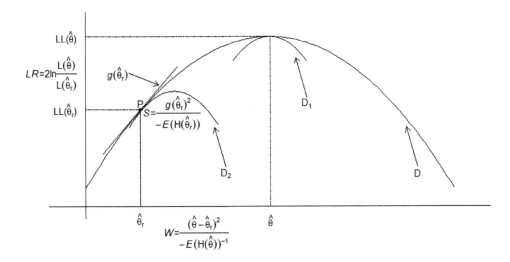

Figure 2.9
Graphical Comparion of Wald, LR, and Score Tests

estimation object obtained previously, a binary logit regression of self-rated health without age or education as its predictors; the only difference between this `anova` and the one for the LR test is to replace the test type to "Rao".

```
anova(noAgeEduc,mylogit,test="Rao")

## Analysis of Deviance Table
##
## Model 1: hlthc2 ~ female + impinc
## Model 2: hlthc2 ~ age + female + educ + impinc
##   Resid. Df Resid. Dev Df Deviance    Rao  Pr(>Chi)
## 1      4241     4416.7
## 2      4239     4089.8  2   326.87 324.72 < 2.2e-16 ***
## ---
## Signif. codes:  0 '***' 0.001 '**' 0.01 '*' 0.05 '.' 0.1 ' ' 1
```

The results suggest that with two degrees of freedom and a $\chi^2 = 324.72$, we can again safely reject the null. Based on the results from the Wald, LR, and score tests, we can see that the χ^2's calculated from the three tests for the same models are quite close to each other, albeit non-identical, and that provides some empirical evidence for the asymptotic equivalence of the three tests.

2.3.1.1 Graphical Comparison of Wald, LR, and Score Tests

Sometimes it can be challenging to distinguish among the Wald, LR, and score tests, to understand the logic behind their corresponding statistical computations, and to beware both their advantages and limitations. Based on Buse (1982) and Engle (1984), we reproduce a graphical rendition of the three tests in one graph (reproduced from Fullerton and Xu 2016:132).

It can be shown in Fig. 2.9 that the LR test (LR) compares the two log likelihoods; the one with a greater log likelihood, measured by the vertical distance on the y-axis scaled by the log function, is considered to be a better-fitting model. Note that in this graph symbols with the subscript r denote parameter estimates from the restricted model, and the ones without it correspond to those based on a full model.

The Wald test (W), on the other hand, gauges the difference between the estimated and the hypothesized value, measured by the horizontal distance on the x-axis and normalized by the negative expected inverse of the Hessian matrix ($-E\left[H\left(\theta\right)\right]^{-1}$) from the full model. The Hessian matrix contains the second partial of the log likelihood function with respect to parameters of interest (θ), and it roughly tells the curvature (i.e., how fast the curve changes) of the function. When the curvature of the function is large, that would imply a large Hessian and accordingly a small inverse of it, roughly speaking; under that circumstance, the Wald test statistic is more likely to be rejected, corresponding to a statistically significant result. The score test (S) takes a somewhat different approach. Theoretically, it only estimates the restricted model once and then computes the gradient (first partials) of the log likelihood function with respect to the parameters (θ), normalized by the negative expected Hessian from the restricted model. Converse to the scenario discussed for the Wald test, when the Hessian matrix is small, denoting a small curvature (little curve change in the vicinity of the restricted parameter estimate), one is more likely to have a large test statistic and accordingly significant result.

Based on Engle (1984), the three tests are asymptotically equivalent. But it is also important to note that the three tests test virtually different hypotheses, and there exist differences in their testing properties. The LR test needs to estimate two models with one nested within the other. Since it uses the maximum amount of information, the LR test is usually deemed most robust of all three and is invariant to re-parameterization. The Wald test, on the other hand, only estimates the full model. However, this efficiency does not come without its cost that the test is sensitive to re-parameterization and is least reliable of the three, especially when the sample size is small to moderate (Agresti, 2013, p. 12). Although the score test is both reliable and efficient, the theoretical ease does not lend itself to computational efficiency, and it has to involve estimating the full model with constraints and obtaining both the gradient and Hessian matrices.

2.3.2 Scalar Measures

Scalar measures are usually single-number model fit statistics used for model comparisons, for example, the Akaike information criterion (AIC) (Akaike, 1974), the Bayesian Information Criteria (Schwarz, 1978; Raftery, 1995, 1999), and various analogues of the R-squared measure in the linear regression model. Both the AIC and the BIC statistics are founded on information theory. Hirotugu Akaike proposed the AIC in his 1974 IEEE paper, in response to issues associated with hypothesis testing in time series analysis and ML estimation in general (Akaike, 1974). This paper turns out to be one of the most cited papers in the history of science ranked by Nature in 2014 (Van Noorden et al., 2014), and probably remains so as of today. The AIC statistic is usually calculated as,

$$\text{AIC} = -2\ln(\widehat{L}) + 2K \tag{2.16}$$

where $\widehat{L}$ stands for the estimated likelihood and K is the total number of parameter estimates. So among models estimated using the ML method, the one with a higher likelihood produces a smaller but positive $-2\ln(\widehat{L})$ and accordingly a smaller AIC. The second term $2K$ in the formula penalizes model complexity. Thus, the model with the smallest AIC is the best-fitting model.

The BIC is a model fit measure closely related to the AIC, and it was first proposed by Gideon E. Schwarz, hence also called the Schwarz information criterion (Schwarz, 1978). The BIC is usually formulated as follows,

$$\text{BIC} = -2\ln(\widehat{L}) + K\ln(N) \tag{2.17}$$

where $\widehat{L}$ stands for the estimated likelihood, K for the total number of parameter estimates and N for the sample size. The BIC weighs both the explanatory power $\left(-2\ln\left(\widehat{L}\right)\right)$ and the model complexity $(K\ln(N))$ of a model and provides an overall assessment measure. Unlike the AIC, the BIC multiplies K by a factor of $\ln(N)$, a much greater number than the factor of 2 in most cases, thus imposing a greater penalty for model complexity. With R, one can invoke the `AIC` and `BIC` functions from the `stats` package and have the estimation object names supplied within the parentheses, for example,

```
AIC(mylogit)
```

```
## [1] 4099.81
```

```
BIC(mylogit)
```

```
## [1] 4131.576
```

When one has several models for comparison, multiple estimation objects can be listed,

```
AIC(mylogit, noAge, noAgeEduc)
```

```
##             df      AIC
## mylogit      5 4099.810
## noAge        4 4262.487
## noAgeEduc    3 4422.676
```

```
BIC(mylogit, noAge, noAgeEduc)
```

```
##             df      BIC
## mylogit      5 4131.576
## noAge        4 4287.900
## noAgeEduc    3 4441.736
```

It can be shown from the results that the full model with `age` and `educ` is preferred by both AIC and BIC since the full model has the smallest value for both measures. In this case, both AIC and BIC converge on the same model.

In addition to AIC and BIC, there is an array of pseudo R-squareds (R^2) added to the long list of scalar measures of model fit statistics, especially for binary regression models, including but not limited to the Cox & Snell's, Efron's, McFadden's, Nagelkerke Cragg & Uhler's, McKelvey & Zavoina's, and Tjur's R^2s that are usually named after the creators, and the count R^2 (Menard, 2000; Allison, 2013), of which the McFadden's and the recently proposed Tjur's R^2s are probably preferred (Allison, 2013). The McFadden's R^2 is highly recommended by many statistical practitioners and is provided by most statistical software packages by default. The McFadden's R^2 is computed as

$$R^2_{\text{McFadden}} = 1 - \frac{\ln\left(\widehat{L}_F\right)}{\ln\left(\widehat{L}_C\right)} \tag{2.18}$$

where $\widehat{L}_F$ is the estimated likelihood from the model of interest, and $\widehat{L}_C$ represents the estimated likelihood for the constant only model with the same data and model. Below, we use the base `function` function and the `logLik` function from the `stats` package to write a short R function, `mfR2`, to compute the McFadden's R^2 for commonly used binary regression models. In the R code chunk, `modobj` is a placeholder for a binary regression estimation object, the `logLik` function extracts the log likelihood from a model object, and the `return` function returns the results. Once we have the function written, we can run through all three models and compare. The results show that the full model, `mylogit`, has the largest R^2, and thus is the best-fitting model.

```
# pseudo R2 (McFadden)
mfR2 <- function(modobj) {
        full <- logLik(modobj)
        constant <- logLik(update(modobj, .~ 1))
        pseudor2 <- 1 - full/constant
        return (pseudor2)
}
mfR2(mylogit)

## 'log Lik.' 0.1351001 (df=5)

mfR2(noAge)

## 'log Lik.' 0.1002747 (df=4)

mfR2(noAgeEduc)

## 'log Lik.' 0.06597557 (df=3)
```

We can also program and calculate Tjur's R^2, which is defined as,

$$D = \bar{\hat{\pi}}_1 - \bar{\hat{\pi}}_0 \tag{2.19}$$

where $\bar{\hat{\pi}}_1$ and $\bar{\hat{\pi}}_0$ denote model based averages of fitted values (i.e., predicted probabilities for $y = 1$) for the success ($y = 1$) and failure ($y = 0$) group respectively, based on cases from the estimation sample (Tjur, 2009). Both the idea and computation are surprisingly simple and yet compelling that the best fitting model should maximally discriminate between the success and failure group. So if a model is effective in predicting the binary response, then on average the probability for $y = 1$ in the success group, for example, should be clearly higher than that in the failure group. Since the Tjur's R^2 is quite straightforward, analysts can program by themselves; or more conveniently, one can use the `PseudoR2` function from the `DescTools` package to calculate a long list of pseudo-R^2, including the Cox & Snell, Efron, McFadden, McKelvey & Zavoina, Nagelkerke, and Tjur's R^2 (Signorell et al., 2020).

```
library(DescTools)
PseudoR2(mylogit, c("McFadden", "Nagel", "Tjur"))

##   McFadden Nagelkerke       Tjur
##  0.1351001  0.2080114  0.1551199
```

It appears that results from all scalar measures of model fit converge to the same conclusion that the full model fits the data best. Thus, we feel quite confident about the statistical inference we make in this case.

2.3.3 ROC Curve

As a nomenclatural residual from signal detection research, the receiver operating characteristic (ROC) curve was first used in WWII to improve predictive accuracy of incoming aircraft through radar signals. The term receiver operating characteristic was initially used for measuring radar operators' ability to make such distinction. Albeit its various extensions and advanced applications, the ROC curve is usually applied to binary classifiers, wherein the true positive rate (sensitivity) is plotted against the false positive rate (1 - specificity). Take a classic binary classification problem as an example. After running a binary regression model, one can use estimated parameters, combined with observed values of covariates, to predict the probabilities of $\widehat{P}(y = 1|\mathbf{x})$ for all cases. A typical threshold such as 0.5 can be used to classify cases with the predicted class variable $\widetilde{y} = 1|\mathbf{x}$ if $\widehat{P} \geq 0.5$ and $\widetilde{y} = 0|\mathbf{x}$ if $\widehat{P} < 0.5$. While cross-examining y and $\widetilde{y}$, one can calculate the true positive rate as a ratio of the number of cases that are correctly re-classified as positive ($y = 1|\mathbf{x}$ & $\widetilde{y} = 1|\mathbf{x}$) over the total number of cases with observed positive results ($y = 1|\mathbf{x}$). Similarly, one can calculate the false positive rate by taking a ratio of the number of cases that are wrongly re-classified as positive ($y = 0|\mathbf{x}$ & $\widetilde{y} = 1|\mathbf{x}$) over the total number of cases with observed (true) negative results ($y = 0|\mathbf{x}$). The true positive rate is also called sensitivity, a measure to gauge how accurately our model predicts the "positive" outcome ($y = 1$). Specificity, in statistics, usually refers to how well the same model predicts the negative outcome ($y = 0$); that is, the extent to which cases with true negative outcomes are predicted to have negative outcomes.

Note that the choice of 0.5 appears fair and reasonable, but it sometimes only provides a cross-sectional view of the full picture. A thorough analysis would choose multiple thresholds and plot their corresponding true positive rates against the false positive rates. This would lead to an ROC curve. The area under ROC curve (AUC) is a discriminant measure of a model to distinguish between (true) positive and negative outcomes. In our example, it is to correctly distinguish between cases reporting to have poor vs. good health. The greater the AUC number (area) is, the more accurately the model predicts; this curve has also religiously demonstrated the trade-off between sensitivity and specificity that as one increases, the other usually decreases.

One can use multiple packages to plot ROC curves. Below, we use the `ROCR` package (Sing et al., 2005). The first function `predict` (`predict.glm`) is from the `MASS` package, and it is used to calculate predicted probabilities for all cases in the estimation sample. The `prediction` function from the `ROCR` package is to use the predicted probabilities and true class variable (observed binary response variable) to produce statistics for a comprehensive ROC performance analysis later; the `performance` function serves to produce performance objects and statistics.

```
mylogit.prob = predict(mylogit, type="response")

library(ROCR)
mylogit.pred <- prediction(mylogit.prob, mydta$hlthc2)
mylogit.perf <- performance(mylogit.pred,"tpr","fpr")
mylogit.auc <- performance(mylogit.pred, measure = "auc")
mylogit.auc@y.values[[1]]

## [1] 0.7516058
```

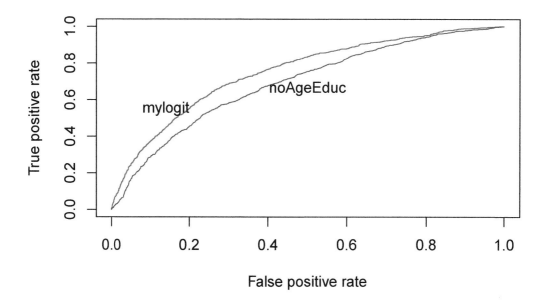

Figure 2.10
Model Comparisons with ROC Curves

It can be shown from the results that the area under the ROC curve is 0.752. Hosmer et al. (2013, p. 177) suggests several cut-off points (0.5, 0.7, 0.8, 0.9) for predictive performance evaluation, the intervals between which are indicative of poor, acceptable, and excellent discrimination of the model under study. A value of 0.5 suggests that the model has little discriminant power, and it fares as well as random guesses without a priori information. A value above 0.9 is considered to be unusually powerful in prediction.

```
noAgeEduc.prob = predict(noAgeEduc, type="response")
noAgeEduc.pred <- prediction(noAgeEduc.prob, mydta$hlthc2)
noAgeEduc.perf <- performance(noAgeEduc.pred,"tpr","fpr")
noAgeEduc.auc <- performance(noAgeEduc.pred , measure = "auc")
noAgeEduc.auc@y.values[[1]]

## [1] 0.6906249
```

Then, we can proceed to calculate the AUC for the `noAgeEduc` model, and it is equal to 0.691. Again, the results show that the full model, `mylogit`, fares better than the restricted model since `mylogit` has a greater AUC. An easier and probably aesthetically pleasing way to present and compare the results is to plot the two ROC curves for their corresponding models,

```
plot(mylogit.perf, col="red")
plot(noAgeEduc.perf, add= TRUE, col="blue")
text(x = 0.14, y = 0.55, labels = c("mylogit"))
text(x = 0.50, y = 0.66, labels = c("noAgeEduc"))
```

2.3.4   Goodness of Fit Measures: The Hosmer-Lemeshow Test

The previously discussed, either the NHST statistics or scalar measures, largely rely on or can realize their utility via model comparisons. Goodness of fit measures, however, focus on how the model predictions fit observations. Hosmer and Lemeshow (1980) and Hosmer, Lemeshow, and Sturdivant (2013), proposed various similar forms of the Hosmer-Lemeshow tests. The most discussed and used one follows the following procedure: First, we rank all observations in deciles based on the values of predicted probabilities calculated from a target model; second, we calculate the test statistic with the following formula,

$$\widehat{\text{HL}}_{\text{bin}} = \sum_{g=1}^{G} \left[\frac{(o_{1g} - \widehat{e}_{1g})^2}{\widehat{e}_{1g}} + \frac{(o_{0g} - \widehat{e}_{0g})^2}{\widehat{e}_{0g}} \right] \tag{2.20}$$

where o_{1g} and o_{0g} denote the number of cases for $y = 1$ and $y = 0$ in the gth group, respectively; and $\widehat{e}_{1g}$ and $\widehat{e}_{0g}$ are the sum of predicted probabilities calculated from the model, corresponding to o_{1g} and o_{0g} across all cases in the same gth group. Hosmer and Lemeshow (1980) show that this test statistic approximately follows a χ^2 distribution with $G-2$ degrees of freedom. Note that for a goodness of fit measure, usually the better the model fit, the smaller the difference between the observed data and the model based predicted data, thus leading to a small test statistic and accordingly a high p-value. The null hypothesis, H_0, is usually set up as there is no or trivial difference between the observed and predicted data.[2] To implement, one can use the `logitgof` function from the `generalhoslem` package (Jay, 2019) to calculate the HL test statistic as follows,

```
library(generalhoslem)
(HLtest = logitgof(mydta$hlthc2, fitted(mylogit), g=10))


##
##  Hosmer and Lemeshow test (binary model)
##
## data:  mydta$hlthc2, fitted(mylogit)
## X-squared = 17.428, df = 8, p-value = 0.02595
```

The first argument of `logitgof` is to specify the binary response variable used in the estimation, and the second is to specify the estimation object, and the g argument is to set the number of groups to be divided for calculating the HL χ^2 test statistic. The results show that for a χ^2 statistic of 17.428 with 8 degrees of freedom, we need to reject the null at the conventional .05 level., implying that our initial model fails to fit the data that well.

2.3.5   Limitations of NHST

Since its initial design and ensuing developments for several decades, null hypothesis significance testing (NHST) has sustained almost a century's scientific inquiries and made paramount influence in the scientific community. Despite probably thousands of citations and millions of applications, the NHST framework has not gone unchallenged (Meehl, 1967, 1997; Nuzzo, 2014; Blume et al., 2019), and scholars have proposed several alternatives, including the highest density interval (HDI) and region of practical equivalence (ROPE)

---

[2]Note that the formula (same) and its interpretation (different) for the chi-squared test statistic discussed here is equivalent of, albeit slightly different from, the one used in Hosmer and Lemeshow (2013, pp. 154–169). The interpretation by Hosmer and Lemeshow is better suited for a model with a majority of categorical covariates, and the m_j therein refers to the number of cases having the same covariate pattern in the same decile group.

under the Bayesian inference framework (Kruschke and Liddell, 2018; Xu et al., 2019) and the second-generation p-value (Blume et al., 2019), such as using interval as opposed to point null hypothesis.

There are two major types of criticisms of the NHST framework, with one focusing on its misuse and abuse and the other on the test design itself. Based on numerous studies and opinion pieces, Nuzzo (2014), for example, summarizes the problems with p-value in many scientific fields, and bemoans the file-drawer effect that statistically significant results are much more preferred than null effects and p-hacking has become a tacit trick in publication and research in general. Such misuse and abuse of p-value has led to the compromise of scientific integrity and innovation; that is, well-established but imperfectly contended theories/hypotheses get reinforced, and it now requires much higher cost to advance, question, challenge, and "usurp" traditional wisdom with fresher and frequently more accurate/comprehensive ideas, which is at the core of scientific discovery and innovation.

There are also legitimate concerns about how the classical NHST is designed and executed. For example, most social science research would test if some effect/coefficient is exactly zero or equal to each other in the null hypothesis. The chances are such null hypotheses are unlikely to hold a priori. Germanely, the calculation of p-value is conditional on the null being true ($P(D|H_0)$), and in many cases social scientists actually want to know if some hypothesis holds given the empirical data ($P(H_0|D)$. Thus, the NHST framework has become controversial, and the tide has risen to potentially shake or even crumble its foundational work. Statistical practitioners might need to beware of its limitations and alternatives, such as the Bayesian inference and the second-generation p-value.

2.4   Interpretation of Results

A great challenge in applying categorical and limited response variable models lies with interpretation. Unlike in linear models where the types of statistics are few and often ready to be used for interpretation, the parameter estimates in nonlinear models usually are not that intuitively comprehensible and cannot be directly used for interpretation. For example, the regression coefficients from binary logit and probit models can only show the direction of the association between predictors and the likelihood of the response variable, and they are a few steps away from providing meaningful interpretation about the effects of predictors. Several texts on categorical and limited response variable models, especially those in social sciences, have provided cogent discussion about miscellaneous ways to approach interpretations in such models (Liao, 1994; Long, 1997; Greene, 2000; Liao, 2002; Cameron and Trivedi, 2005; Powers and Xie, 2008; Wooldridge, 2010). Following Xu and Fullerton (2013), below we lay out a foundational framework for interpreting results from binary regression models, which, we believe, can also provide guidelines for other types of categorical and limited response variable models in this text. In short, we argue that there are usually multiple methods to interpret and present results from such nonlinear models, and most of these methods rely on some transformation of raw estimation results; usually, it is a good idea to choose several statistics to triangulate results and enrich statistical communication.

Based on Xu and Fullerton (2013), there are in general three major pairs of statistics that can be used for post-estimation analysis and interpretation, including effect/prediction, local/global, and ratio/difference (relative/absolute). For describing the associations between covariates and the response variable, we can use measures of effect, prediction, or both to be most effective. For measures of effect, they are statistics that specifically characterize how the response variable reacts to (or covary with when the causation is unclear) the stimuli from the predictors (assuming there is a temporal order), or to put it concretely, the amount of change in the response variable associated with some change in a predictor. For the models discussed in this text, examples may include discrete changes (or the first difference used in King et al. (2000)), odds ratio coefficients, marginal effects, and raw coefficients from binary regression models on the scale of logit or probit. These statistics are to describe some dynamic relationship between predictors and the response variable. Measures of prediction are statistics used to predict the value of the response variable, represented in various forms, such as odds and predicted probabilities.

For the second pair, local vs. global measures, the focus is on the amount of the data that a measure is based on and its generalizability. For local measures, we refer to statistics that are calculated based on the values of predictors at a "local" area, for example, around the middle range of a distribution or typical values of some predictors. Sometimes these values are from sample observations and other times they are hypothetical. Global measures are usually statistics calculated based on a significant proportion or all of sample observations, for example, average marginal effect or average predicted probability.

The last pair, ratio vs. difference, are two ways to measure the difference in two statistics. The first, ratio, is to take a ratio of two statistics, and the second is to take a difference. Taking a ratio of two statistics is used usually when the two statistics are small in absolute terms, for example, the probabilities of rare events; the probability of a diagnosis of pancreatic cancer is roughly 0.00017 (56770/327200000) in 2019, and thus it would be hair-splitting to compare absolute differences. If somehow this incidence rate or probability increases to 0.017, then there is a one hundred fold increase; but in absolute terms, this is only a 0.01683 increase in probability. This is exactly the rationale for using odds ratios in health research. There are a few examples for ratio measures, including odds ratios and risk ratios. There are also cases in which knowing absolute differences (i.e., taking a difference in two statistics) is substantively more meaningful than taking a ratio. In the previously discussed case, for example, one might want to know the total number of additional individuals affected after the probability or rate change, for a given base population.

Note that there could be any combination of the three pairs; for example, one can have a difference in local prediction measure, such as the difference in predicted probabilities of two hypothetical cases (i.e., a 35-year old female with college education and sample average income and an otherwise similar male), a ratio of global effect measure, such as the ratio of the average marginal effects of education between males and females. The following few sections illustrate how to calculate these measures using concrete examples, with a focus on the dyad of prediction vs. effect and group comparisons. Also note that the discussion and guidelines provided below pertain to almost all models covered in this text.

2.4.1 Precision Estimates

As discussed previously, the precision estimates (e.g., standard errors and/or confidence intervals) of raw coefficients from categorical and limited response variable models covered in this text can be obtained using the ML or M-estimation method (Wooldridge, 2010). Deriving the precision estimates of those statistics based on raw parameter estimates for post-estimation analyses and interpretations in these models, however, can be quite convoluted. In general, there are three ways to calculate these precision estimates, including the end-point transformation, the delta method, and simulation methods (Xu and Long, 2005a). For simulation methods, one could use either case-bootstrapping, which re-samples a data set with replacement to calculate statistics, or parametric bootstrapping that re-samples from a theoretical distribution of raw parameter estimates produced by the ML estimation of the model and empirical data, and then construct an empirical distribution of the quantities of interest.

2.4.1.1   End-Point Transformation

For some of the statistics for post-estimation analysis and interpretation, one can directly use the end-point transformation method; that is, if some statistic is a function of another statistic with known bounds, and if this function is a monotonic increasing function, then one can simply apply the function to get the upper and lower bounds for the first statistic (Xu and Long, 2005a). For example, the precision estimates for predicted probabilities in binary logit or probit models can be constructed by applying either the cumulative density function of the standard logistic, $\frac{\exp(\mathbf{x}\beta)}{1+\exp(\mathbf{x}\beta)}$, or standard normal distribution, $\Phi(\mathbf{x}\beta)$, both of which are monotonic increasing functions of the linear predict, $\mathbf{x}\beta$, whose interval estimates can be obtained through the ML estimation. Another group of statistics that commonly use this method to compute their corresponding precision estimates are those that take an exponential function of raw parameter estimates since the exponential function is a monotonic increasing function, for example, odds ratio coefficients, $\exp(\beta)$, in binary, ordered, and multinomial regression models, and the expected rate, $\exp(\mathbf{x}\beta)$, in count regression models.

2.4.1.2   Delta Method

Since most statistics of interest are not simple monotonic functions of other statistics, one has to turn to other techniques than the end-point transformation method. The delta method builds on the well-known properties of the Taylor series, and approximates a non-linear function of a random variable with usually up to the second order term in the Taylor series expansion (Xu and Long, 2005b). Usually, the delta method uses a first-order Taylor series expansion to estimate the variance of a nonlinear function of ML estimates with their known property of asymptotic normality. Without loss of generality and under standard regularity conditions, if $\widehat{\theta}$ is a vector of ML estimates, then

$$\sqrt{n}\left(\widehat{\theta}-\theta_0\right) \xrightarrow{d} \mathcal{N}\left(0, nV\left(\widehat{\theta}\right)\right) \tag{2.21}$$

where $V\left(\widehat{\theta}\right)$ is the asymptotic variance-covariance matrix for $\widehat{\theta}$. Let $G(\widehat{\theta})$ be some function of $\widehat{\theta}$, such as a predicted probability from a binary regression model. After the first-order Taylor series expansion of $G(\widehat{\theta})$ and some simple algebraic rearrangements, we can get

$$G\left(\widehat{\theta}\right) \xrightarrow{d} \mathcal{N}\left(G\left(\theta_0\right), g\left(\theta_0\right)' V\left(\widehat{\theta}\right) g\left(\theta_0\right)\right) \tag{2.22}$$

where $g\left(\theta_0\right) = \nabla G\left(\theta_0\right)$ is the gradient of $G\left(\theta_0\right)$ with respect to θ_0, and θ_0 can be evaluated at $\widehat{\theta}$. Note that $G\left(\widehat{\theta}\right)$ can be any function of $\widehat{\theta}$, such as predicted probabilities and differences in or ratios of predicted probabilities. For model specific derivations of the variance of most models covered in this text, please refer to (Xu and Long, 2005b).

2.4.1.3   Re-sampling Methods

Because the delta method is used mostly for approximation, it may have computational bias and even out-of-range predictions. When that happens, simulation methods can be effective alternatives. With simulation, one is barely consumed with complex closed-form analytic solutions. Generally, two simulation methods have been proposed to derive precision estimates, and both fall under the bootstraping framework. The first is case bootstrap, with which one re-samples from the original data with replacement, runs the same statistical model with a different sample each time, and then calculates quantities of interest. After repeating this process for R times, one can construct empirical distributions of quantities of interest and provides interval estimates based on these distributions. A second simulation method is parametric bootstrapping, which relies on the asymptotic normality assumption of the ML estimates by re-sampling parameter estimates from the estimated multivariate normal distribution, calculate quantities of interest, and then construct their empirical distributions. In both cases, after the empirical distributions of quantities of interest are constructed, one can use multiple methods to calculate the standard errors and construct confidence intervals, including the normal, percentile, and bias-corrected methods, the choice of which depends on the assumptions made about the empirically constructed distributions (Xu and Long, 2005a). Below, all three methods for computing precision estimations are illustrated in various places, wherever they are deemed appropriate.

2.4.2 Interpretation Based on Predictions

For binary regression models, predictions may include the latent response variable, y^*, and predicted probabilities (equivalently odds). Since the predicted values of y^* are of little substantive interest, the following discussion focuses on predicted probabilities. There are several R packages that can be used to output predicted probabilities, including **car** (Fox and Weisberg, 2019), **VGAM** (Yee, 2015), and **zelig** (Imai et al., 2008). But let us begin with the **predict** function from the **stats** package that comes with R installation by default. Below is an example to illustrate how to calculate the predicted probability of having poor health for a hypothetical 35-year old woman with college education and the sample average income, using the sample model estimates from the binary logit model estimation object, **mylogit**, which was already produced and used for testing purposes in previous sections. To calculate this prediction using **predict**, one needs to first set up a new data frame vector using the **data.frame** function as follows

```
x.vector = data.frame(age=35,
                      female=1,
                      educ = 16,
                      impinc = mean(mydta$impinc))
x.vector

##   age female educ   impinc
## 1  35      1   16 9.954116
```

Then one can invoke the **predict** function to carry out the calculation. Below, the **newdata** argument feeds some hypothetical data created by users, **x.vector** in this case, to the model estimation object **mylogit**, and then uses parameter estimates from **mylogit** to calculate the prediction. The **type** argument is set to **response** such that the calculated prediction is predicted probability in binary logit.

```
prob = predict(mylogit, newdata=x.vector, type="response")
prob

##         1
## 0.1179868
```

Next step is to calculate the confidence interval for this predicted probability. As discussed in previous sections, there are generally three methods one can choose from. The following example uses end-point transformation for illustration. To use the end-point transformation method, one needs to first calculate the confidence interval of the linear predictor, $\mathbf{x}\hat{\beta}$, and then use the cumulative density function of the standard logistic distribution, or the inverse logit function, $\frac{\exp(\mathbf{x}\beta)}{1+\exp(\mathbf{x}\beta)}$, to transform the lower and upper bounds of $\mathbf{x}\hat{\beta}$ to the lower and upper bounds of the predicted probability. To calculate the linear predictor and its standard error, one can use the `predict` function and set the `type` argument to `link` and `se.fit` to `TRUE` so that both the linear predictor and its standard error can be calculated. $\hat{\beta}_{\mathrm{ML}}$ is assumed to be normally distributed, so does the linear predictor, $\mathbf{x}\hat{\beta}$, since it is a simple linear transformation of $\hat{\beta}$. Then one can easily constructed the confidence interval for the linear predictor like what is usually done for a normally distributed random variable.

```
link = predict(mylogit, newdata=x.vector, type="link", se.fit=TRUE)
zval = qnorm(0.975)
link.lb = link$fit - zval*link$se.fit
link.ub = link$fit + zval*link$se.fit
```

After one obtains the lower and upper bounds of the linear predictor, these two numbers can be plugged into the inverse logit function to produce the lower and upper bounds of the predicted probability. Below, we first define an inverse logit function, `invLogit`, to transform linear predictor to probability. `input` is a placeholder for input data, which can be one number or a data array.

```
invLogit <- function(input){
        invLogit <- 1/(1 + exp(-1*input))
        return(invLogit)
}
```

Then one can apply this function to the two bounds

```
prob.lb = invLogit(link.lb)
prob.ub = invLogit(link.ub)
print(c(prob.lb, prob.ub), digits=3)
```

```
##     1     1
## 0.103 0.134
```

The last step is to print all x values that are used in computing the linear predictor, the predicted probability, and its the lower and upper bounds,

```
print(cbind(x.vector, prob.lb, prob, prob.ub))
```

```
##   age female educ  impinc    prob.lb      prob    prob.ub
## 1  35      1    1      16 9.954116 0.1033779 0.1179868 0.1343509
```

Thus, the results show that for a 35-year-old female with roughly college education (or 16 years of education) and sample average income, we would expect her predicted probability of having poor health to be 0.118, and we are 95% confident that this prediction is somewhere between 0.103 and 0.134 in the population.

One could also use the delta method with the `deltaMethod` function from the `car` package to calculate the same prediction with its associated precision estimates, as illustrated in the following R code chunk.

```
deltaMethod(mylogit,
  "1/(1+exp(-1*(Intercept + 35*age + 1*female + 16*educ + 9.954116*impinc)))")
```

Re-sampling (bootstrap) methods can also be used to compute quantities of interest. The R codes below use case resampling bootstrap to calculate the predicted probabilities for the same hypothetical female and an otherwise similar male, and then take a difference in these two predicted probabilities. The workhorse of bootstrapping in R is the `boot` function from the `boot` package (Canty and Ripley, 2019; Davison and Hinkley, 1997). To make it function properly, one needs to first define a function so that the `boot` function to be used later would know what to be bootstrapped. The three arguments, `formula`, `data`, and `selectCase`, in the parenthesis right after `function` are placeholders for model formula, data, and a Boolean vector of logic operator to be specified later in the `boot` function. Other than that, one can proceed as usual to calculate quantities of interest, including the predicted probability for the previously specified hypothetical female, the predicted probability for an otherwise similar male, and their difference and ratio. The `return` function returns statistics for the `boot` suite of functions to display.

```
require(boot)
bs <- function(formula, data, selectCase) {
  # allow boot to sample
  d <- data[selectCase,]
  nowfit <- glm(formula, family= "binomial", data=d)
  dataF <- data.frame(age = 35, female = 1, educ = 16, impinc=mean(mydta$impinc))
  dataM <- data.frame(age = 35, female = 0, educ = 16, impinc=mean(mydta$impinc))
  predF = predict(nowfit, dataF, type="response")
  predM = predict(nowfit, dataM, type="response")
  predD = predF - predM
  predR = predF/predM
  retnlist = c(predD, predF, predM, predR)
  return(retnlist) }
```

After the bootstrap function `bs` is defined, one can invoke the `boot` function and set all arguments to preferred conditions. In this case, the replication number is 500.

```
# bootstrapping with 500 replications
set.seed(47306)
results <- boot(data=mydta, statistic=bs, R=500,
                formula = hlthc2 ~ age + female + educ + impinc)
```

The last step is to display the results,

```
# print the predicted prob for the hypothetical female
quantile(results$t[, 2], c(0.025, 0.975))

##      2.5%      97.5%
## 0.1010057 0.1324679

# print the difference between P(female|x) and P(male|x)
quantile(results$t[, 1], c(0.025, 0.975))

##        2.5%       97.5%
## -0.01406464  0.01717895
```

The results show that the 95% percentile confidence interval of the predicted probability for the hypothetical female from the case-resampling bootstrap is quite close to that from the end-point transformation and the delta method, In addition, the difference in the predicted probability is not significantly different from zero at .05 level.

King and his colleagues have written several R packages, including `Zelig` and `ZeligChoice`, to provide a variety of statistics for post-estimation analyses, including first and second difference as well as relative risk ratio (King et al., 2000; Imai et al., 2008). Below, the few lines of R codes calculate the same statistics as those in previous sections, but with `Zelig`'s functions.

```
library(Zelig)
z.out <- zelig(hlthc2 ~ age + female + educ + impinc, data = mydta, model = "logit")
x.out <- setx(z.out, age = 35, female = 1, educ = 16)
s.out <- sim(z.out, x=x.out) #plot(s.out)
summary(s.out)
```

The line that begins with `z.out` re-estimates the same binary logit model. As the name of the function signifies, `setx` sets the values for a specific x vector. When a predictor's name is not specified, then that variable is set to its sample mean; so `impinc` (imputed family income) is set to its sample mean. Unlike the case-reampling bootstrapping that re-samples from the estimation sample, `Zelig` uses parametric bootstrapping to (1) directly re-sample from the estimated multivariate normal distribution of parameter estimates, yielded often from an ML estimation, (2) calculate the linear predictor (in generalized linear models or almost equivalently, categorical and limited response variable models) or some linear combination of the structural component, (3) use the result to re-sample again from the stochastic component of the same model, and (4) last to compute the quantity of interest and its precision estimates from the resultant empirical distribution. In calculating such post-estimation statistics, the `Zelig` package also differentiates between predicted and expected values. For predicted values, the second re-sampling step (i.e., re-sampling from the stochastic component) only re-samples once, whereas in computing expected values this step re-samples multiple times and then averages across (please see King et al. 2000). The following few lines of R codes calculate the difference in the same pair of predicted probabilities between a male and a female as discussed in previous sections,

```
x.start <- setx(z.out, age = 35, female = 1, educ = 16)
x.end <- setx(z.out, age = 35, female = 0, educ = 16)
s.out <- sim(z.out, x = x.start, x1 = x.end)
summary(s.out)
plot(s.out)
```

2.4.3   Interpretations Based on Effects

As discussed previously, there are several types of quantities to characterize the effects predictors have on response variables. One can use, for example, odds ratio coefficients, discrete changes/first difference in predictions, marginal effects, and some variants or extensions to these three types to measure effects on different scales. Effects statistics are measures of how the response variables, possibly measured on different scales (e.g., the latent variable scale, odds, or predicted probabilities in binary logit models), respond to changes in predictors.

2.4.3.1   Odds Ratios

For models parameterized with odds comparisons, such as binary logit, ordered logit, multinomial logit models, using odds ratio coefficients for interpretation is a natural choice. In binary logit models, it is well known that, $\ln \frac{P(y=1|\mathbf{x})}{1-P(y=1|\mathbf{x})} = \mathbf{x}\beta$. Exponentiating both sides, we get $\Omega = \frac{P(y=1|\mathbf{x})}{1-P(y=1|\mathbf{x})} = \exp(\mathbf{x}\beta)$. If one increases the value of a generic predictor, x_k, by one unit from any possible value of x_k to $x_k + 1$, then we would have $\frac{\Omega_{\mathbf{x},x_k+1}}{\Omega_{\mathbf{x},x_k}} = \frac{\exp(\beta_0+\beta_1 x_1+\beta_2 x_2+...+\beta_k x_k+...+\beta_K x_K)}{\exp(\beta_0+\beta_1 x_1+\beta_2 x_2+...+\beta_k (x_k+1)+...+\beta_K x_K)} = \exp(\beta_k)$. For interpretation, one can say for each unit increase in x_k, we would expect the odds (of $y = 1$ in binary logit models) to vary by a factor of $\exp(\beta_k)$, and $\exp(\beta_k)$ is commonly known as an odds ratio (or sometimes factor change) coefficient. The following two lines of R codes illustrate how to get odds ratio coefficients and their associated confidence intervals after running a binary logit model. As discussed previously, one can use the end-point transformation to calculate the precision estimates since the exponential function is a monotonic increasing function. This is actually how it is done by the `confint` function (from the `stats` package) indeed.

```
# CIs using profiled log-likelihood
confint(mylogit)
```

```
##                  2.5 %      97.5 %
## (Intercept)  3.16589679  4.88048938
## age          0.02332059  0.03188007
## female      -0.13934309  0.17246669
## educ        -0.15171252 -0.09907100
## impinc      -0.59121104 -0.41623385
```

```
exp(mylogit$coefficients)
```

```
## (Intercept)        age     female       educ     impinc
## 55.6910884  1.0279679  1.0165492  0.8822939  0.6045495
```

```
exp(confint(mylogit))
```

```
##                  2.5 %       97.5 %
## (Intercept) 23.7099974 131.6950965
## age          1.0235946   1.0323937
## female       0.8699295   1.1882322
## educ         0.8592353   0.9056784
## impinc       0.5536564   0.6595260
```

With these results, one can interpret the effect of, for example, age, on the binary health measure as: For each year increase in age, the odds of having poor health will increase by a factor of 1.028, holding all other covariates constant, and one can be 95% confident that this odds ratio coefficient lies somewhere in between 1.024 and 1.032 in the population.

2.4.3.2 Discrete Rates of Change in Prediction

In this text, discrete rates of change (DRC) are to describe the ratio of change in the response variables over some discrete change in a generic predictor. There are several different names floating around, including discrete change and first difference (King et al., 2000). We use discrete rate of change as an analogue of instantaneous rate of change, or marginal effect to be discussed in the next section. The DRC, for the models considered in this text, is defined as

$$\text{DRC} = \frac{\triangle F(\mathbf{x}\beta)}{\triangle x_k}$$
$$= \frac{F(\mathbf{x}\beta, x_k + \triangle_{x_k}) - F(\mathbf{x}\beta, x_k)}{\triangle_{x_k}} \qquad (2.23)$$

where Δ_{x_k} is a discrete change in the generic predictor of interest, x_k, and $F(\mathbf{x}\beta)$ is usually a cumulative density function of the linear predictor $\mathbf{x}\beta$, such as the inverse logit function. For example, if one is interested in the effect of gender on reporting poor health status for a hypothetical case, the following R codes can be used to calculate such effect. The dc function from the `glm.predict` package can calculate predictions and discrete rates of changes for a wide array of categorical and limited response variable models, and it uses simulation (parametric bootstrapping) to calculate precision estimates (i.e., confidence intervals) (King et al., 2000; Zelner, 2009; Schlegel, 2019). Note that the two `values` (i.e., `values1` and `values2`) arguments are to set the values of two hypothetical cases (vectors) for comparison.

```
library(glm.predict)
# values have to be specified for predictors, including the unit column
dc(mylogit, values1 = c(1, 35, 1, 16, mean(mydta$impinc, na.rm=T)),
         values2 = c(1, 35, 0, 16, mean(mydta$impinc, na.rm=T)),
         set.seed=47306)

##                    Mean          2.5%        97.5%
## Case 1      0.118158828   0.10455806   0.13342010
## Case 2      0.116341371   0.10014606   0.13320065
## Difference  0.001817457  -0.01377777   0.01830225
```

Results show again that the difference in the predicted probabilities between a 35-year old female with college education and average income, and an otherwise similar male is not statistically significant since the 95% confidence interval (the 2.5% and 97.5% columns) contains zero. Although DRC can precisely measure the association between predictors and response variables at two specific locations, it falls short of providing a summary of the general association. Thus, a summary measure of discrete rates of changes, such as average discrete rates of changes (ADRC), may better serve the needs. Without loss of generality, the ADRC can be defined as,

$$\text{ADRC} = \frac{1}{N} \sum_{i=1}^{N} \frac{\triangle F(\mathbf{x}\beta)}{\triangle x_{ik}} \tag{2.24}$$

Although it appears that there is only one formula for ADRC, the way how $\triangle x_{ik}$ is defined can change the result for ADRC. For example, one could define $\triangle x_{ik}$ to be the difference between the current value of x_k and one unit above that value for case i. Or rarely used, but the start value can be set to a fixed number such as the minimum of x_k and the end value to be the maximum of it. In brief, $\triangle x_{ik}$ can be defined differently, depending on the needs. The `margins` package provides a variety of R functions to conduct post-estimation analyses of models covered in the first five chapters of this book, including the `dydx` function as the low-utility workhorse function and its wrapper, the `margins` function (Leeper, 2018). What `dydx` does is to rely on the limit definition of the marginal effect (partial effect or derivative), $M = \lim_{h \to 0} \frac{f(x+h)-f(x-h)}{2h}$, to calculate marginal effects by tweaking h to be small enough such that marginal effects can be reasonably approximated. When the `change` argument is used, the `dyxy` function actually calculates the difference in the predictions of two x vectors. In the following case, education is set to half unit below and half unit above its sample mean, and all other variables are set to their observed values. The `dydx` function carries out the same calculation for all observations in the estimation sample. The `sum` and `length` functions are used to sum across cases and take an average.

```
library(margins)
ardcEduc = sum(dydx(mydta, mylogit, "educ", change = c(mean(mydta$educ) -
    0.5, mean(mydta$educ) + 0.5)))/length(mydta$age)
ardcEduc
```

```
## [1] -0.02027683
```

The results show that the predicted ADRC for education, so specified, is roughly -0.02. To get its precision estimates, one can use either the delta method or simulation, the latter of which appears to be easier computationally. The following R code chunk illustrates how this can be done. Results (not shown here) indicate that the ADRC for education for the two locations compared (i.e., half unit below and above the sample mean of education) is roughly between -0.023 and -0.0138.

```
adrc <- function(formula, data, selectCase) {
    # allow boot to sample
    d <- data[selectCase,]
    nowfit <- glm(formula, family= "binomial", data=d)
    avg <- sum(dydx(mydta, nowfit, "educ",
            change = c(mean(mydta$educ)-0.5,
                mean(mydta$educ)+0.5) ))/length(mydta$age)
    # retnlist = c(avg)
    return(avg)
}
set.seed(47306)
res <- boot(data=mydta, statistic=adrc, R=500,
            formula = hlthc2 ~ age + female + educ + impinc)
boot.ci(boot.out = res, type = c("norm", "basic", "perc"))
```

2.4.3.3 Marginal Effects

Marginal (partial) effects can be viewed as special cases of discrete rates of changes when the changes in x become infinitesimally small, and they are usually defined as the instantaneous rates of change, or the instantaneous effects of predictors on the response variable in a regression-type model. In theory, marginal effects exist only for continuous variables with well-defined functions. In binary regression models, one can have marginal effects of predictors on binary response variables represented on the scale of a latent continuous variable, whose unit is unknown. Thus, it is of little substantive meaning to focus on such marginal effects, which are actually equivalent of $\hat{\beta}_{\mathrm{ML}}$, the raw regression coefficients yielded from an ML estimation. When marginal effects are used in most models covered in this text, they usually refer to the marginal effects of predicted probabilities with respect to x_k. Since predicted probabilities are usually calculated using some cumulative density function, $P = F(\mathbf{x}\beta)$, the marginal effect of a generic variable x_k on the predicted probability can be defined as,

$$\mathrm{M} = \frac{\partial F(\mathbf{x}\beta)}{\partial x_k} \tag{2.25}$$
$$= f(\mathbf{x}\beta)\beta_k$$

where $\mathbf{x}\beta$ is assumed to be a linear function of x_k, without nonlinearity terms such as polynomials or interactions, and $f(\mathbf{x}\beta)$ is the corresponding probability density function, or the first partial derivative of the CDF, $F(\mathbf{x}\beta)$. This formula can be applied to all marginal effects of x_k on response probabilities in the probability models discussed in this text. If one uses a binary logit, for example, then the marginal/partial effect of x_k on the predicted probability, $P = \frac{\exp(\mathbf{x}\beta)}{1+\exp(\mathbf{x}\beta)}$, is $\left(\frac{\exp(\mathbf{x}\beta)}{(1+\exp(\mathbf{x}\beta))^2}\right)\beta_k$. Using the following R codes, we can manually calculate the marginal effect of age, given the x vector defined previously, and then use the `dydx` function to check the answer,

```
linePred = predict(mylogit, newdata=x.vector, type="link")
exp(linePred)/((1+exp(linePred))^2)*mylogit$coefficients["age"]
```

```
##            1
## 0.002870547
```

```
# use the dydx function from the margins package
dydx(x.vector, mylogit, "age")
```

```
##      dydx_age
## 1 0.002870547
```

We can also use the custom-made function `logitmfx` from the `mfx` package to calculate marginal effects at means (MEM), the marginal effects by setting all predictors at their sample means (Fernihough, 2019),

```
library(mfx)
logitmfx(hlthc2 ~ age + female + educ + impinc, data = mydta, atmean=T)
```

```
## Call:
## logitmfx(formula = hlthc2 ~ age + female + educ + impinc, data = mydta,
##      atmean = T)
##
## Marginal Effects:
##                dF/dx    Std. Err.          z  P>|z|
## age      0.00461734  0.00036336   12.7075 <2e-16 ***
## female   0.00274615  0.01329784    0.2065 0.8364
## educ    -0.02096258  0.00223009   -9.3999 <2e-16 ***
## impinc  -0.08424394  0.00740245  -11.3806 <2e-16 ***
## ---
## Signif. codes:  0 '***' 0.001 '**' 0.01 '*' 0.05 '.' 0.1 ' ' 1
##
## dF/dx is for discrete change for the following variables:
##
## [1] "female"
```

Although the marginal effects at a certain data point/vector in a multidimensional space, be it hypothetical or observed, can be interesting sometimes, one might have to turn to some summary measure to describe the effects of predictors since summary measures are assumed to average out the variation in marginal effects caused by variations in x. This often leads to the average marginal effect (AME), which is usually defined as follows in most models covered in this text,

$$\text{AME}_{x_k} = \beta_k \times \sum_{i=1}^{N} \frac{f(\mathbf{x}\beta)}{N} \tag{2.26}$$

where $f(\mathbf{x}\beta)$ is the probability density function of the linear predictor, $\mathbf{x}\beta$, β_k is the regression coefficient for the predictor, x_k, and AME_{x_k} is the average marginal effect for x_k. Note that for the `margins` function, because none of the categorical covariates below are declared to be factors, all covariates are treated as continuous variables in computing their average marginal effects.

```
summary(margins(mylogit))
```

```
## factor     AME     SE         z        p     lower     upper
##    age  0.0043  0.0003  13.4787  0.0000   0.0037   0.0049
##   educ -0.0195  0.0020  -9.6326  0.0000  -0.0235  -0.0155
## female  0.0026  0.0124   0.2064  0.8365  -0.0217   0.0268
## impinc -0.0784  0.0066 -11.8423  0.0000  -0.0914  -0.0655
```

One can also set some of the values to hypothetical ones, for example, age at 30, and then calculate AME_{x_k},

```
summary(margins(mylogit, at = list(age = 30)))
```

```
## factor     age     AME     SE        z        p     lower     upper
##    age 30.0000  0.0036  0.0002  16.5862  0.0000   0.0032   0.0040
##   educ 30.0000 -0.0163  0.0018  -8.8665  0.0000  -0.0199  -0.0127
## female 30.0000  0.0021  0.0104   0.2064  0.8365  -0.0182   0.0225
## impinc 30.0000 -0.0656  0.0059 -11.1611  0.0000  -0.0772  -0.0541
```

The precision estimates for AMEs can also be derived, as discussed previously, using either the delta or bootstrap method. The `margins` function uses the delta method.

2.4.4 Group Comparisons

There is a burgeoning line of research focusing on group comparisons, especially in recent years. Early and recent important studies include, Chow's test for comparing regression coefficients from different data sets (Chow, 1960), Clogg, Petkova, and Haritou's work (Clogg et al., 1995) for testing the equality of regression coefficients for nested models, Liao's comprehensive research on group comparisons in a wide array of statistical models (Liao, 2002), Allison (1999) and Williams (2009) studies about group comparisons with heterogeneous choice models, and Xu and Fullerton's (2013) proposed framework for group comparisons in generalized linear models and nonlinear models. The techniques are of a variety and also abound with assumption pitfalls. This text in general suggests that statistical analysts be clear about their purposes, make their goals transparent to the targeted audience, and use multiple, not single, measures to make group comparisons. To explore group differences, one could refer to and apply simple algebraic reformulation of the measures discussed in the previous subsections and construct measures for group comparisons accordingly. For example, one could simply compare average marginal effects between two groups by taking a difference/ratio of the two marginal effects,

$$\text{AME}_{g1} - \text{AME}_{g0} = \beta_k \times \sum_{i=1}^{N_{g1}} \frac{f(\mathbf{x}\beta, x_{ik})}{N_{g1}} - \beta_k \times \sum_{i=1}^{N_{g2}} \frac{f(\mathbf{x}\beta, x_{ik})}{N_{g0}} \qquad (2.27)$$

where AME_{g1} and AME_{g0} denote the average marginal effects for group 1 and group 0, and N_{g1} and N_{g0} correspond to the two group sizes, respectively. Alternatively, one can calculate $\text{AME}_{g1}/\text{AME}_{g0}$. Compared below are the two marginal effects of education between males and females in the binary logit model used throughout this chapter. As shown by the R code chunk (results not shown), case bootstrapping can be used to calculate precision estimates. So one can first re-sample the data, run the logit, compute the marginal effect of education, then average across within males and females respectively, and lastly, take a difference in the two average marginal effects. One can repeat this process for R times and then construct the empirical distribution of the difference in average marginal effects.

```
bs <- function(formula, data, index) {
        # allow boot to sample
        d <- data[index,]
        nowfit <- glm(formula, family= "binomial", data=d)
        fmle = margins(nowfit, data=mydta[mydta["female"] == 1,])
        male = margins(nowfit, data=mydta[mydta["female"] == 0,])
        mdif = mean(fmle$dydx_educ) - mean(male$dydx_educ)
        mrat = mean(fmle$dydx_educ) / mean(male$dydx_educ)
        retnlist = c(mdif, mrat)
        return(retnlist)
}
# bootstrapping with 1000 replications
set.seed(47306)
results <- boot(data=mydta, statistic=bs, R=1000,
        formula = hlthc2 ~ age + female + educ + impinc)
boot.ci(boot.out = results, type = c("norm", "basic", "perc"))
```

This is one way (case bootstrapping) to make group comparisons and one angle (average marginal effect) to examine group differences. Given the non-linear nature of the models considered in this text, it is suggested that multiple comparisons be made to triangulate results and substantiate conclusions. Limitations also abound with the practice of group comparisons described previously. For example, a common quantity of interest is either group difference in the prediction or effect at means (i.e., calculate the quantity for both groups while holding all other covariates at their sample means and then take a difference or ratio) or other hypothetical locations. Despite their popularity, the main problem with such quantities is that the likelihood of having such cases with values set at the means (or other hypothetical locations) of covariates is rather low. For example, the average years of education usually corresponds to a below average income. It is only when covariates follow multivariate normal, then such quantities/statistics can roughly approximate real cases from empirical data. If one uses averaged quantities, for example, average marginal effects, then group differences in some attributes might confound with effects in such comparisons, and it becomes hard to disentangle one from the other. For example, it is well known that women receive lower pay than men do, and if income is one of the covariates, it has to be involved in the computation of average marginal effects in binary regression models. Thus, the calculated group difference in average marginal effects of any covariate could be misleading since income is not controlled for in a strict sense. Such problems warrant further research, and one may turn to some newer methods in the last chapter of this text.

2.5 Bayesian Binary Regression

As discussed in the first chapter, the Bayes' theorem states that $P(\theta|D) = \frac{P(D|\theta)P(\theta)}{P(D)}$. Since $P(D)$ does not involve θ and can be viewed as a constant or normalizing factor, we then have $P(\theta|D) \propto P(D|\theta)P(\theta)$ by dropping the $P(D)$ term, or in other words, $P(\theta|D)$ is proportional to the product of $P(D|\theta)$ and $P(\theta)$. Since $P(D)$ is computationally expensive to calculate and it only scales up or down the posterior by a constant, the process of deriving the posterior $P(\theta|D)$ usually drops the $P(D)$ term. To get the posterior $P(\theta|D)$, one needs to provide priors for parameters, which is the primary source of grievances and paradoxically the main advantage for using Bayesian methods.

2.5.1 Priors for Binary Regression

To get the posterior, one has to provide priors. There are several conceptually distinct pairs of priors as discussed in the first chapter, including informative vs. non-informative (or weakly informative) priors and conjugate vs. non-conjugate priors, as well as other types, such as flat priors or hyper-priors. As we recall, a prior is called a conjugate prior for a likelihood function, if the posterior, or the product of the prior and the likelihood function, is from the same probability distribution as the prior. Conversely, if the prior and the posterior are not from the same probability distribution, then the prior is non-conjugate for the likelihood function. Conjugate priors provide closed-form solutions of the posteriors and largely save computational cost that would otherwise be required of using non-conjugate priors. For example, the beta distribution is a conjugate for binomial, Bernoulli and geometric likelihoods, and the normal is its own conjugate. Conjugate priors have been used less frequently, or are not prerequisite, in recent years due in large part to the exponential growth in computing power that has made MCMC sampling procedures less time consuming.

Informative priors, as the name suggests, are priors that provide clear and specific information about the probability distributions of parameters. For example, for the parameter of education in the binary logit regression of SRH discussed previously, we can provide a normal prior with its mean centered around -0.2 and standard deviation/scale set to be 0.05, thus constraining the range of this parameter to be between -0.3 to -0.1 for roughly 95% of its distribution. This prior is informative insomuch as one can identify the direction and even magnitude of the effect within a small range. In contrast, one could choose a uniform or normal prior with its mean centered around zero and standard deviation/scale set to be $1e + 6$, thus providing a non-informative prior, a near-flat prior from which any value is almost equally likely to be sampled. Usually, using an informative prior is indicative of allocating more confidence to the information provided in the prior relative to what additional information is obtained in the data at hand, compared with the almost total reliance on the data by using a non-informative prior. In most cases, the results from using non-informative priors are very similar to those from using their frequentist counterparts of estimation methods. For example, results from the Bayesian estimation of a binary logit model with non-informative priors are quite close to those from a binary logit model using the ML estimation.

In empirical research, scholars have suggested using weakly informative priors, priors that include less information than one actually obtains (Gelman et al., 2008, 2014; Gelman, 2019). As argued in Gelman et al. (2008), some carefully chosen, weakly informative priors can produce "stable and regularized estimates" (p. 1361) while still retaining their generality. For the binary logit model, for example, Gelman and his colleagues suggest using independent Student-t priors for the coefficients, and by default using the Cauchy prior distributions with zero means and a scale parameter of 2.5 (Gelman et al., 2008). The standard Cauchy distribution is a special case of the Student-t distribution with one degree of freedom. Such priors can remedy commonly encountered issues in the estimation of binary logit models with the classical frequentist approach, such as separation, which occurs when a binary response variable separates a covariate (almost) completely such that there is an unusually high or even perfect bivariate correlation between the response variable and the covariate in question; in simple terms, the problem of separation means that knowing the value of the covariate in question can perfectly predict the binary response variable. Separation is a common problem in binary or multi-category regression models, when there are usually multiple sets of binary/dummy variables or data sparsity (Gelman et al., 2008).

There is a long list of R packages that can provide various capabilities for Bayesian estimation and inference, most notably `arm` (Gelman and Su, 2020b), `bayesm` (Rossi, 2019), `brms` (Bürkner, 2017), `MCMCpack` (Martin et al., 2011), and `rstanarm` (Stan Development Team, 2020). Below we first re-estimate the same binary logit model as discussed previously using the `bayesglm` function from the `arm` package by Gelman and Su (2020b). The syntactic structure of `bayesglm` is almost identical to that of the `glm` function, except that `bayesglm` has a few additional options for MCMC simulations. In general terms, this function uses an approximate EM algorithm and adapts the classical estimation methods for generalized linear models with the prior distribution using augmented regression to obtain results (Gelman et al., 2008). The default prior is a Cauchy distribution with mean, scale, and degrees of freedom set to 0, 2.5, and 1 respectively. In the R code chunk below, we specify these prior options explicitly. The `coefplot` function graphically displays parameter estimates and their corresponding 95% (default) interval bars, and the results are presented in Fig. 2.11. One can also set `prior.df` to be `Inf` to have a normal prior.

```
library(arm)
bayeslogit.cauchy <- bayesglm(hlthc2 ~ age + female + educ + impinc,
        data=mydta, family=binomial(link="logit"),
        prior.mean=0, prior.scale=2.5, prior.df=1)
display(bayeslogit.cauchy, digits=3, detail=T)

## bayesglm(formula = hlthc2 ~ age + female + educ + impinc,
## family = binomial(link = "logit"),
##      data = mydta, prior.mean = 0, prior.scale = 2.5, prior.df = 1)
##             coef.est coef.se z value Pr(>|z|)
## (Intercept)   4.013    0.437   9.189   0.000
## age           0.028    0.002  12.631   0.000
## female        0.017    0.079   0.208   0.835
## educ         -0.125    0.013  -9.333   0.000
## impinc       -0.503    0.045 -11.279   0.000
## ---
## n = 4244, k = 5
## residual deviance = 4089.8, null deviance = 4728.7 (difference = 638.8)

coefplot(bayeslogit.cauchy)
```

Next, we estimate two binary logit regression models with the same health data that were used previously, except that a separation issue is introduced to illustrate the utility of having weakly informative priors. In the slightly edited new data, only females with poor health status and males are kept such that being female can perfectly predict the binary response variable.

```
library(arm)
sep <- mydta[ which((mydta$female==1 & mydta$hlthc2==1) | mydta$female==0), ]
```

With such ill-structured data, a classical binary logit model is estimated using the ML method. One can find that all coefficients fare normally, except the one for female that has a very large standard error.

Regression Estimates

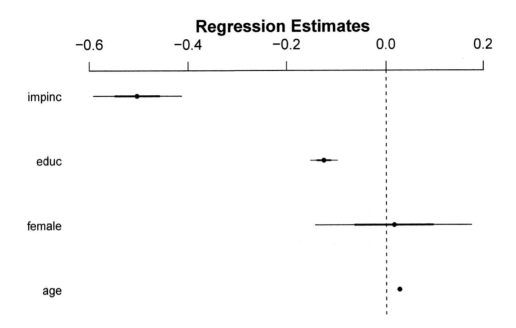

Figure 2.11
Plotting the Bayesian Estimation of Binary Regression Models Using the Cauchy Priors

```
mysep <- glm(hlthc2 ~ age + female + educ + impinc,
        data=sep, family=binomial(link="logit"))
display(mysep)

## glm(formula = hlthc2 ~ age + female + educ + impinc,
## family = binomial(link = "logit"),
##     data = sep)
##              coef.est coef.se
## (Intercept)   3.55     0.67
## age           0.03     0.00
## female       20.45   407.67
## educ         -0.08     0.02
## impinc       -0.54     0.07
## ---
##   n = 2498, k = 5
##   residual deviance = 1780.3, null deviance = 3393.4 (difference = 1613.1)
```

Below, the same separation data are used to estimate a Bayesian binary logit model.

```
bayeslogit.sep <- bayesglm(hlthc2 ~ age + female + educ + impinc,
            data=sep, family=binomial(link="logit"),
            prior.mean=2.5, prior.scale=2.5, prior.df=1)
display(bayeslogit.sep)

## bayesglm(formula = hlthc2 ~ age + female + educ + impinc,
```

```
## family = binomial(link = "logit"),
##      data = sep, prior.mean = 2.5, prior.scale = 2.5, prior.df = 1)
##              coef.est coef.se
## (Intercept)  3.53     0.67
## age          0.03     0.00
## female       8.77     1.82
## educ        -0.08     0.02
## impinc      -0.54     0.07
## ---
## n = 2498, k = 5
## residual deviance = 1780.8, null deviance = 3393.4 (difference = 1612.6)
```

The results show much more reasonable estimates for the gender variable, including its standard error.

2.5.2   Bayesian Estimation of Binary Regression

Customized R functions for Bayesian estimation, albeit convenient, can be problematic when analysts want to experiment with different model specifications, including priors, or to conduct post-estimation analysis. The R code chunk below illustrates how to code the model using Stan (Stan Development Team, 2018) along with functions from the `rstan` package (Goodrich et al., 2018). `rstan` is an R package that interfaces with Stan. It sends data and the model specification information to Stan, which then processes the data, estimates the model, and transmits results back to R via `rstan`. As illustrated in the first chapter, there are usually three steps to run Bayesian regression models using the `rstan` package. The first step is to clean and prepare the data to be sent to Stan, including dropping missing cases. The second step is to specify the model, including the `data`, `parameters`, `model`, and `generated quantities` blocks. Below, we begin with the second step since the first step has been executed at the very beginning of this chapter, and we will continue using the subset of the cumulative GSS data object in R, `mydta`, with `hlthc2` as the binary response variable and `age`, `female`, `educ`, and `impinc` as the predictors. To illustrate the codes more effectively, we cut R code blocks into even smaller chunks. The first line of the second step usually begins with a string block with a starting double quotation mark, declaring that what follows is viewed as strings/characters to be processed by Stan.

```
modelString = "
```

The following few sections specify the Bayesian binary logit model, beginning with the `data` block. Note that the symbols appearing in the block all refer to the elements in existing data, not parameters, and their names have to be linked to the names or information used in the data. In this (data) block, Stan usually recognizes real-valued arrays, scalars, vectors, or matrices. `N`, `D`, `y`, and `x` are real scalar or variable names that will be linked to their counterparts in the data, `mydta`, in later steps.

```
data {
  # pound sign comments out annotations
  # N for total number of cases/row number
int<lower=0> N;
// two slashes can be used for comments too
// D for total number of predictors, except the unit column
int<lower=1> D;
// y for binary response variable and its values set to be 0 and 1
```

```
int<lower=0, upper=1> y[N];
// the predictor matrix with N rows and D columns
row_vector[D] x[N];
}
```

The second program block is to declare parameters to be estimated. b0 is the intercept, and b is the column vector of slope parameters with its dimension set to D.

```
parameters {
real b0;
vector[D] b;
}
```

The next is the model block to specify the priors and likelihood function. eta is the linear predictor and prob is the cumulative density function for a standard logistic distribution. There are two parts in this block. The first part declares eta and prob, the two intermediate quantities, to be real-numbered scalars, and then supplies priors for all parameters. In this case, the prior for all regression coefficients, including the intercept, follows a normal distribution with its mean and scale parameters set to be zero and 10 respectively. This prior is usually viewed as a weakly informative prior. The second part, a for loop, defines the likelihood function; y follows a Bernoulli distribution with its parameter set to be prob. In this block, Stan recognizes N, D, y, and x as real-valued scalars or data that are already defined in the data block.

```
model {
real eta;
real prob;
b0 ~ normal(0, 10);
for (d in 1:D) {
b[d] ~ normal(0, 10);
}

for (n in 1:N) {
eta <- b0 + x[n]*b;
prob <- 1/(1+exp(-1*eta));
y[n] ~ bernoulli(prob);
// y[n] ~ bernoulli_logit(alpha + beta * x[n]);
}
}
```

For a simple Bayesian estimation, we could stop here. But if one intends to conduct post-estimation analyses, such as calculating predicted probabilities or model fit statistics, then a generated quantities block is needed. Below, we include the calculation of four quantities of interest, including the predicted probability for a 35-year old female with college education and average income, the predicted probability for an otherwise similar male, the difference in these two predicted probabilities, and vector-wise log likelihoods for measuring model predictive accuracy.

```
generated quantities {
real PredF;
real PredM;
```

```
real PredD;
real etaPred;
vector[N] log_lik;
etaPred <- b0 + 35*b[1] + 1*b[2] + 16*b[3] + 9.954116*b[4];
PredF <- inv_logit(etaPred);
PredM <- inv_logit(b0 + 35*b[1] + 0*b[2] + 16*b[3] + 9.954116*b[4]);
PredD = PredF - PredM;
for (n in 1:N) {
log_lik[n] = bernoulli_logit_lpmf(y[n] | b0 + x[n] * b);    }
}
```

Below, the closing quotation mark ends all the program blocks, matching the beginning quotation mark. Once all the program blocks are created, one can use the `writeLines` function to save all the syntax to a text file (`model.txt`). This step is not necessary for Stan, but can be essential for other BUGS-like software, such as JAGS.

```
" # close quote for modelstring
writeLines(modelString,con="model.txt")
```

Next step is to link existing data to those names and symbols specified in the Stan program blocks previously annotated. We first turn the data frame into a matrix, and then extract all necessary information out of this matrix.

```
# dataMat: data matrix from data frame
dataMat = as.matrix(mydta)
# nData: sample size
nData = NROW(dataMat)
# x: the predictor matrix, excluding the unit column
x = dataMat[,-1]
# predictorNames: predictor names
predictorNames = colnames(dataMat)[-1]
# nPredictors: the number of predictors, excluding y
nPredictors = NCOL(x)
# y: the response variable column
y = as.matrix(dataMat[,1])
# predictedName: the response variable name
predictedName = colnames(dataMat)[1]
# tabl: tabulate y
tbl = table(y)
# K: dimension of table y
K = dim(tbl)
#nYlevels: the nunmber of response levels for y
nYlevels = max(y)
```

Then we need to package the necessary vectors, matrices, and scalars into a list to be fed into Stan. Note that the symbols to the left of the equal sign are those declared in the data block in the Stan codes discussed previously, and those on the right are what we just extract in the previous R code chunk from the existing data object, `mydta`. By setting up this list and specifying it in the `stan` function from the `rstan` package, we can link empirical data with those mere symbols in the Stan codes, thus turning the latter into data for Bayesian estimation in Stan.

```
dataList = list(
        x = x,
        y = as.vector( y ),
        D = nPredictors,
        N = nData )
```

Until this step, we could directly request to run the `stan` function from the `rstan` package that communicates with the Stan program. However, it would be "tidy" in coding to set the few arguments in the `stan` function outside the function first. There are a few technical terms commonly used in the MCMC process. Adaptation or tuning refers to the process of finding a good proposal distribution.[3] The burn-in number specifies the iterations that the whole MCMC process needs to "warm up" in the testing stage, and all these iterations will be dropped in making Bayesian inferences. Since auto-correlation is a common phenomenon for a series of iterations next to each other, thinning is sometimes required for complex models. The thinning number refers to the *n*th iteration/step to be retained in the MCMC process for making Bayesian inferences later. Having multiple MCMC chains with different starting values is usually required so that one can see if different chains converge to the same posterior.

```
# parameters to be presented after running the stan function
parameters = c("b0", "b", "PredF", "PredM", "PredD", "log_lik")
# the number of adaptation/tuning steps
adaptSteps = 500
# the number of burnin steps
burnInSteps = 500
# the number of chains
nChains = 3
numSavedSteps=10000
thinSteps=1
nPerChain = ceiling( ( numSavedSteps * thinSteps ) / nChains )
```

Once all the sampling parameters are set, we can invoke the `stan` function and run the model. Note that for reproducibility, the seed is set to be a fixed number of 47306.

```
mcmcSamples <- stan(model_code = modelString, data = dataList,
    seed = 47306, pars = parameters, chains = nChains, iter = nPerChain,
    warmup = burnInSteps)  # init=initsChains
```

After prolonged elapse of time, usually longer than what one would otherwise spend using the ML estimation, we can summarize and print out the results.

```
# summary(mcmcSamples)
print(mcmcSamples, digits=3)
```

Below is a summary of the MCMC sampling statistics. The first three columns are self-explanatory, corresponding to the means, their standard errors, standard deviations, 2.5, 25,

---

[3]Depending on the software used for the MCMC sampling, the adapt stage can be separate like that in JAGS, or incorporated in the burn-in stage, like in Stan, OpenBUGS, and WinBUGS. During the adapt or tuning stage, the program usually attempts to find the appropriate proposal distribution for sampling, monitors acceptance rate, and adjustes step size. Once the sampling process stabilizes, the sampling process moves into the warm-up/burn-in stage, in which the samples will be thrown away for quality control purposes.

50, 75, and 97.5 percentile of the posterior distribution of our parameters of interest. `n_eff` refers to the effective sample size (ESS; effective independent draws from the posterior) for each parameter. In general, an ESS number greater than 100 is considered as sufficient, but a couple hundred would be safer. The `Rhat` statistic is the potential scale reduction statistic, or the split R as it is called sometimes. It is a "ratio of the average variance of draws within each chain to the variance of the pooled draws across chains" (Stan Development Team, 2018). When `Rhat` is equal/close (<1.10 for single parameters or <1.01 for all) to one, one can safely claim that these chains reach equilibrium or convergence.

	mean	se_mean	sd	2.5%	25%	50%	75%	97.5%	n_eff	Rhat
b0	2.963	0.007	0.423	2.122	2.684	2.960	3.256	3.785	3548	1.000
b[1]	0.024	0.000	0.002	0.020	0.023	0.024	0.025	0.028	8301	1.000
b[2]	0.039	0.001	0.079	-0.119	-0.014	0.039	0.090	0.194	4601	1.001
b[3]	-0.112	0.000	0.013	-0.138	-0.121	-0.112	-0.104	-0.087	5042	1.000
b[4]	-0.399	0.001	0.043	-0.484	-0.428	-0.399	-0.371	-0.312	3480	1.000

If one compares the results here with those produced previously by the ML estimation, it can be shown that they are very similar to each other.

2.5.3 Bayesian Post-estimation Analysis

As illustrated in previous sections, post-estimation analysis comes in many forms. In this section, we focus on predictions. Recall that `PredF` is the predicted probability of having poor health for a 35-year old female with college education and average income, `PredM` is the probability for an otherwise similar male, and `PredD` is the difference in the two predicted probabilities. From the results below, it can be shown that the distributions for the two probabilities are similar, suggesting that the two hypothetical individuals aforementioned are equally likely to have poor health. Results from the third row indicate that the difference in the two probabilities is probably not different from zero since the 95% (2.5% and 97.5%) centered credible interval contains zero.

	mean	se_mean	sd	2.5%	25%	50%	75%	97.5%	n_eff	Rhat
PredF	0.128	0.000	0.008	0.112	0.122	0.128	0.133	0.145	6563	1.000
PredM	0.124	0.000	0.009	0.107	0.118	0.124	0.129	0.142	5927	1.000
PredD	0.004	0.000	0.009	-0.013	-0.001	0.004	0.010	0.021	4605	1.001

We can also use some `rstan` package functions to conduct post-estimation analysis. For example, we can use the `stan_plot` function to plot the posterior distributions and `stan_trace` to investigate issues such as chains convergence graphically.

```
stan_trace(mcmcSamples)
```

It appears that in Fig. 2.12 for all eight quantities of interest, the three chains overlap for the most part, which provides additional confidence that results from different chains converge.

Like how AIC and BIC are used for the frequentist ML estimation, one can turn to the Watanabe-Akaike information criterion (WAIC; or sometimes the widely applicable Bayesian information criterion) (Watanabe, 2010; Gelman et al., 2014) and the leave-one-out cross-validation statistic (LOO-CV) (Gelman et al., 2014; Vehtari et al., 2017) for model comparison and selection. WAIC is defined as

$$-2\left(\widehat{\text{elppd}}_{\text{WAIC}}\right) = -2\left(\text{lppd} - p_{\text{WAIC}}\right) \tag{2.28}$$

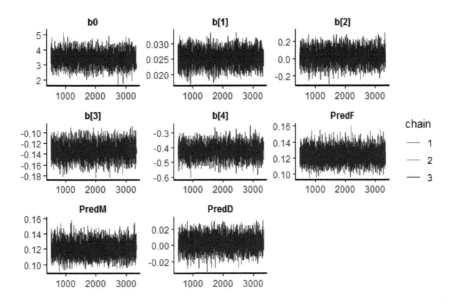

Figure 2.12
MCMC Trace Plot

where elppd is the expected log pointwise predictive density, lppd is the log pointwise predictive density, and p_{WAIC} is the effective number of parameters (Gelman et al., 2014, 166-81). Because WAIC is fully Bayesian, averaging over the posterior distribution, and is invariant to reparameterization, scholars usually prefer it over AIC or DIC (deviance information criterion) (Xu et al., 2019).

```
log_lik <- extract_log_lik(mcmcSamples)
waic(log_lik)

Computed from 8502 by 4244 log-likelihood matrix

          Estimate    SE
elpd_waic  -2049.9  35.1
p_waic         5.0   0.1
waic        4099.8  70.3
```

The Bayesian LOO-CV is defined as

$$-2\left(\widehat{\text{elpd}}_{\text{LOO}}\right) = -2\sum_{i=1}^{N}\log p\left(y_i|y_{-i}\right) \tag{2.29}$$

where epld is the expected log predicted density, and $p\left(y_i|y_{-i}\right)$ is the leave-one-out density given the data without the ith data point (Gelman et al., 2014; Vehtari et al., 2017). Note that by multiplying the expected log (with or without pointwise) predictive density term by -2, both WAIC and LOO-CV are brought to the same deviance scale as that of AIC and DIC. Therefore, they are similar to AIC and DIC in interpretation that smaller values are indicative of better model fit (especially predictive accuracy).

```
loo(log_lik)

Computed from 8502 by 4244 log-likelihood matrix
          Estimate   SE
elpd_loo  -2049.9  35.1
p_loo         5.1   0.1
looic      4099.8  70.3
------
Monte Carlo SE of elpd_loo is 0.0.

All Pareto k estimates are good (k < 0.5).
See help('pareto-k-diagnostic') for details.
```

To compare models using WAIC and LOO, one can use functions, such as `loo_compare`, or simply `loo` and `waic`, from the `loo` package for the few models under consideration (Vehtari et al., 2017).

2.5.4   Bayesian Assessment of Null Values

There could be multiple ways to investigate null values under the Bayesian framework. Below, the text focuses on the use of highest density interval (HDI) and region of practical equivalence (ROPE) in Bayesian assessment of null values (Kruschke and Liddell, 2018; Kruschke, 2018). HDI is a special type of credible interval, the counterpart of confidence interval in Bayesian statistics. Credible intervals are usually constructed in such a way that the probability below the interval is equivalent to the probability above that interval, or equal-tailed credible interval (ETI). HDI refers to the part of a probability distribution—usually corresponding to the posterior of parameter values—that has higher density than those outside. When one has a symmetric distribution, then ETI and HDI should match exactly. But if the posterior is asymmetric or multi-modal, then ETI and HDI will be different. Similar to confidence intervals under the frequentist framework, HDI is conventionally set at the threshold of 95%.

ROPE sets up a decision rule about the assessment of null values. In general, discrete decisions about point estimates, as practiced in most frequentist statistics, can be misleading or even inaccurate on several levels. For example, a typical null hypothesis (with any meaningful alternative) that a difference or an effect is exactly zero is simply unrealistic and can be rejected a priori in many cases. Thus, the real challenge is how to circumscribe the range of values such that the values within that range are practically equivalent to the null value. Cohen (1988) defines the effect size, $d = \frac{\mu_A - \mu_B}{\sigma}$, to be a unit-free measure of the degree to which the alternative hypothesis deviates from the null hypothesis, and he suggests that a value of 0.2 or smaller can be viewed as small enough to be practically equivalent to trivial. Kruschke (2018, p. 277) proposes that for a regression coefficient, if the effect is within the range of $\pm 0.1 S_y / 2 S_x$, then one can safely infer that the effect/coefficient is practically equivalent to zero. Although there are general guidelines, the choice of radius (half of the usually centered ROPE) for the ROPE's has to be contexualized with the substantive problems of interest, and the guideline and choices accordingly may vary across different problems and are still under development. In general, if the 95% HDI falls within a reasonably chosen ROPE, then one can accept the null. If however the 95% HDI falls outside the ROPE, then one can safely reject the null hypothesis. When there is non-trivial overlap between the HDI and ROPE, then a discrete decision has to be withheld.

Going back to the binary logit example that has been used throughout this chapter, below we use functions from the BEST package to illustrate how to combine the use of HDI

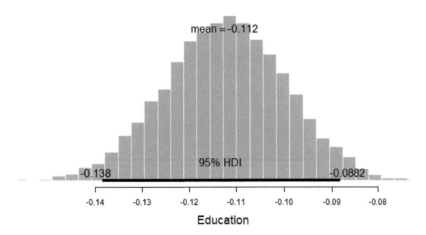

Figure 2.13
HDI and ROPE for Education Coefficient

with ROPE to assess the null value of the effect of education (Kruschke and Meredith, 2020). First, the `rstan`-produced estimation object, `mcmcSamples`, is turned into a matrix. Then the `plotPost` function from the BEST package graphs the HDI + ROPE plot. Following Kruschke's $\pm 0.1 S_y/2S_x$ rule, we calculate the standard deviation of the latent variable y^* (for SRH) to be $S_{y^*} = 2.04$ and the standard deviation of education, $S_{\text{educ}} = 3.22$. So $0.1 S_y/2S_x = 1/32$. This means that if the coefficient estimate is between $-1/32$ to $1/32$, it can be viewed as practically equivalent to zero. Based on how the data are read into Stan, we know that `b[3]` corresponds to the coefficient for education,

```
mcmcChain = as.matrix(mcmcSamples)
library(BEST)
plotPost(mcmcChain[, "b[3]"], ROPE=c(-1/32, 1/32),
        xlab="Education", col="grey")
```

It can be shown from Fig. 2.13 that the 95% HDI of the regression coefficient parameter for education is completely outside the chosen ROPE, which is not shown in the graph because it is far from the center of the parameter distribution. Thus, one can conclude that the null value (zero) can be rejected. One can continue to examine if the predicted probability of having poor health for a 35-year old female with college education and average logged family income is different from of an otherwise similar male. Roughly following Cohen's ideas about effect size, the ROPE's radius for assessing this difference is set as one fifth of the greater standard deviation of the two probabilities' respective distributions.

```
plotPost(mcmcChain[, "PredD"], ROPE=c(-0.0018, 0.0018),
        xlab="Female-Male", col="grey")
```

Fig. 2.14 shows that there is 15% of the 95% HDI falls within the chosen ROPE. Thus, we have to withhold our decision as to whether the difference in the two predicted probabilities are different from zero. It is also without doubt that whether one accepts, rejects, or withholds decisions about the ROPE'd null values appears to rely on the radius. To make

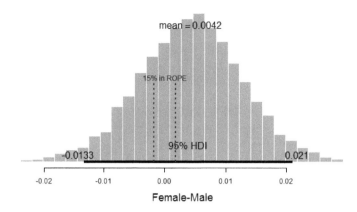

Figure 2.14
HDI and ROPE for Gender Difference

the whole process transparent, Kruschke (2018) proposes graphing the area of HDI in ROPE
as a function of radius. Below the `plotAreaInROPE` function from the BEST package is used
to plot such a graph.

```
plotAreaInROPE(mcmcChain[, "PredD"],
               credMass = 0.95, compVal = 0, maxROPEradius=0.03,
               xlab="Radius of ROPE around 0 for Female-Male")
```

It can be shown in Fig. 2.15 that the curve goes up quickly without much staggering
in the lower end, providing some evidence that, across the board, there is some non-trivial
proportion of the posterior overlaps with the ROPE'd region as the radius of the ROPE
increases. Thus, it is prudent to withhold decision as to whether this difference in predicted
probabilities is significantly different from zero.

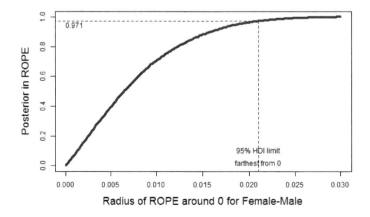

Figure 2.15
Posterior in ROPE as Function of ROPE Radius

3

Polytomous Regression

> Simplicity aids, and multiplicity ails.
>
> ———————————————————
> in Chapter 22 of *Dao De Jing* by Laozi

Binary outcomes are often idealized portraits of our empirical world replete with much richer and even irregular, chaotic data, such as multi-category choice (polytomous), count, time-to-event, or mixture (generated through mixing different distributions) data. For example, albeit artificial and subjective in measurement, binary indicator of self-rated health (SRH) is frequently obtained by dichotomizing multi-category responses, such as excellent, very good, good, fair, and poor, numericalized from one to five. The original SRH variable is viewed as an ordinal response variable, which is usually measured on an ordinal scale and is one of the most recurring type of data in social sciences. Numerous applications involving ordinal response variables span the scientific fields from health research to education, political science, psychology, and sociology. The Likert scale, which usually has response levels of strongly disagree, disagree, neutral, agree, and strongly agree, is a typical ordinal measure, created by Rennis Likert first as a psychometric scale, and now it has become an almost predominant measurement scale, widely used in behavioral, health, and social sciences (Likert, 1932). As shown in the SRH example, ordinal measures are also frequently used in health research to identify the severity level or prognosis stage of diseases, health conditions, or risk factors, such as blood pressure, obesity, pain, physical exercise, and cancer, just to name a few.

In general, the different response levels of an ordinal variable represent the ranking or ordering of such responses, and the distance between these responses usually does not have numerical meaning and thus is immeasurable. For example, the assignment of 4 to fair health and 2 to very good for the SRH variable does not imply that having fair health is twice that of very good in measuring the severity of poor health status on a scale from excellent to poor health; instead, it simply means fair is ranked higher than very good in describing one's health status on that particular ordinal scale. An important property of ordinal data is reversibility (palindromic) invariance. McCullagh (1980, p. 116) argues, for example, that "A more appealing requirement for ordinal data is that the model should in some sense be invariant under a reversal of category order but not under arbitrary permutation" (palindromic invariance).

Albeit simple and ostensibly straightforward, the use and modeling of ordinal measures can involve considerable complication. Body mass index (BMI), for example, can be classified into into underweight, normal, overweight, and obesity (different stages), but it is plausible that the underlying process that would lead to underweight (e.g., genetics, malnutrition, destitution, and morbidity) can be quite different from the ordinal range from normal to obesity (e.g., genetics, diet, and physical activity). It is also quite common to find that for some ordinal measures, a disproportionate number of cases concentrate in the tails (e.g., no drug-use or physical exercise), center (e.g., neutral views; central tendency or social desirability bias) or moderate choices of either side of the ordinal scale/distribution. Sometimes these anomalies exist as a matter of fact, and other times they result from response biases for various reasons.

DOI: 10.1201/9780429056468-3

Multi-category variables can also be measured on nominal scales, wherein each category is sufficiently distinctive and permutation-invariant. For measuring distinctiveness, one popular principle is the independence of irrelevant alternatives (IIA); that is, the introduction of new categories does not affect the relationship (e.g., relative odds) among existing categories. For example, Arrow defines IIA to be "...that the choice between x and y is determined solely by the preferences of the members of the community as between x and y" (Arrow, 1963, p. 28) along with Condition 3 on the preceding page of Arrow (1963). The permutation invariant standard presupposes that any arbitrary reordering of the categories is not going to change the way nominal variables are analyzed and its associated results.

3.1   Ordered Regression

3.1.1   Types of Ordinal Measures and Regression Models

A variety of ordinal/ordered regressions were devised to model ordinal response variables that are apparently similar in measurement, albeit distinct in their data generation processes. O'Connell (2006), Fullerton (2009), and Fullerton and Xu (2016), for example, classify ordinal regression models into three major types based on the nature of ordinal response variables and how comparisons of ordered levels are set up, including cumulative, stage/sequential/continuation ratio, and adjacent models. Fullerton (2009) and Fullerton and Xu (2016) cogently discuss the extensions to these classic ordered regression models by allowing some of the coefficients to vary freely, with proportionality constraints, or restricting them to be the same across equations. Yee (2015) proposes the vector generalized linear and additive models to include main ordered regression models and their variants.

Based on Fullerton and Xu (2016), an ordered response variable with L levels, for example $y = 1, ..., l, ..., L$, can be modeled with three different types of parameterization under the general regression framework, depending on the nature of ordinal response variables. In general, most ordered responses can be modeled using the cumulative model, wherein the probability that the response variable set to a certain response level and below, $P(y \leq l)$, is the initial quantity of interest for deriving level-specific probabilities ($P(y = l)$) to set up the model for ML estimation. This type of ordered response variables usually has no additional mechanisms in generating the response levels other than simple ordering, such as the Likert scale.

In the second type of ordered responses, not only do the response levels have ordering, they also represent steps in a well-defined sequence (stage or a uni-directional process), in which a response level of l corresponds to a stage that has progressed through earlier ones, including $1, ..., l-1$. Typical examples include promotion, educational attainment, offspring size, disease progression, and organism growth. For this type of models, in general, the initial quantity of interest to set up a probability model is the probability of staying at some stage of the whole sequence given having progressed through all the previous steps, beginning from $y = 1$ to the preceding level $l-1$, $P(y = l|y \geq l)$; this type of ordered regression models is usually called the stopping ratio model. If however the probability of the ordered response y continuing past stage l is the quantity of interest, then the probability model needs to be set up reversely (in odds comparison) or complementarily (in probability). For models in the latter case, they are called the continuation ratio model. Note that the stopping and continuation models differ in that the stopping model is interested in $y = l$ and continuation is in $y > l$, for $y \geq l$. The stopping ratio and continuation ratio models are the flip side of one another of the same coin and are therefore equivalent mathematically in essence. The model can also be set up reversely as regards the ordering levels for $y \leq l$, for example, $P(y = l | y \leq l)$ (Yee, 2015, p. 404). As the ways to parameterize this particular type of model are many, the statistical literature differs sometimes even diametrically in naming the same parameterization. This text generally follows the nomenclature used in Yee (2015). Sometimes researchers are more interested in the probability of the ordered response y at some level l given y is either l or its adjacent level $l + 1$, $P(y = l | y = l \vee y = l + 1)$. In such cases, one has an adjacent category ordered regression model.

As stated previously, models can also vary based on different link functions, such as the logit and probit link functions. One needs to be cautious about using the complementary log-log link function; unlike the logit and probit links, the complementary log-log is asymmetric and can lead to mathematically different results for reversely-coded ordinal response variables. Moreover, models can vary based on different types of parameter constraints, leading to parallel (all slopes are equivalent across cut-point equations), partial (some slopes vary across cut-point equations), and proportional constraint models (a subset of slopes differ by the same constant across cut-point equations) (Fullerton and Xu, 2016).

3.1.2 A Brief History of Ordered Regression Models

It is widely acknowledged that Aitchison and Silvey (1957) were the first to construct an ordered regression model. Like many early statistical breakthroughs, this new modeling technique was conceived to solve empirical problems in entomological studies. Aitchison and Silvey (1957) devised the ordered probit model in its prototypical form while attempting to estimate the mean time that a primitive wingless insect (bristletail) spent in each stage (instar) of its growth using random sampling at multiple time points after hatching. Snell (1964) was probably among the first to formulate a primitive form of the ordered regression model by assigning the numerical values for ordinal scales using the logit link function. In the model, Snell's goal was to estimate the mid-point of the boundary values of adjacent ordinal levels, and it turns out that these boundary values, which are estimated as the intermediate parameters for constructing the final scores (i.e., taking an average of two boundary values), happen to be the thresholds or cut-points in the modern form of the classic proportional odds (ordered logit/cumulative logit) model. McKelvey and Zavoina (1975) probably first established the standard modern formulation of the ordered regression model using the latent variable approach along with a probit link function. McCullagh (1980) generalized a variety of regression models for ordinal responses, provided a formal and rigorous treatment of such models, and thereby popularized this general type of models. Recent advances in ordered regression models have been largely built on these early developments; these efforts include the stereotype logit model (Anderson, 1984), the partial proportional odds model (Peterson and Harrell, 1990), the generalized ordered logit model (Clogg and Shihadeh, 1994; Fu, 1998; Williams, 2006), and the vector generalized linear and additive models (Yee, 2015).

3.1.3 Cumulative Regression

3.1.3.1 Model Setup and Estimation

The cumulative logit is probably the most frequently used ordered regression model in the literature and practice. This model can be derived using the latent variable approach. As shown in Fig. 3.1, for an ordinal response variable, $y = 1...l...L$, we can safely assume that there is a corresponding latent continuous variable, y^* such that $\tau_{l-1} \leq y^* \leq \tau_l \iff y = l$, where τ's are called cut-points or thresholds.

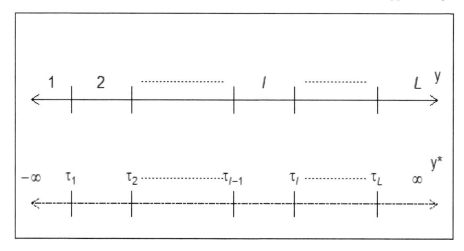

Figure 3.1
Latent Variable Approach to Ordered Regression

As illustrated in Fig. 3.2, if one further assumes one further assumes the structural equation $y^* = \mathbf{x}\beta + \varepsilon$, linking the latent continuous variable y^* and a set of covariates, $\mathbf{x}$, and the error term, ε, then one can have the following,

$$
\begin{aligned}
P(y = l|x, \beta) &= P(\tau_{l-1} < y^* \leq \tau_l | \mathbf{x}, \beta) \\
&= P(\tau_{l-1} < \mathbf{x}\beta + \varepsilon \leq \tau_l) \\
&= P(\tau_{l-1} - \mathbf{x}\beta < \varepsilon \leq \tau_l - \mathbf{x}\beta) \\
&= F(\tau_l - \mathbf{x}\beta) - F(\tau_{l-1} - \mathbf{x}\beta)
\end{aligned}
\tag{3.1}
$$

where $F(\mathbf{x}\beta)$ is a generic cumulative density function, and one can assign $\tau_0 = -\infty$ and $\tau_L = +\infty$, without loss of generality. Depending on the distributional assumption one makes about ε, for example standard normal or standard logistic, either the probit or logit model is established. Other link functions, such as the complementary log-log

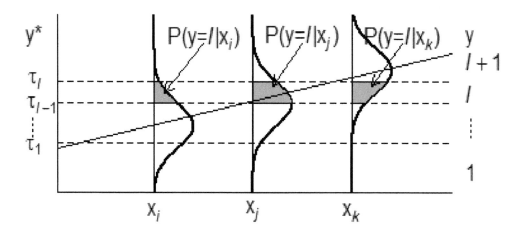

Figure 3.2
Latent and Observed Ordinal Response Variable

(cloglog) transformation, are also possible. As noted previously, the cloglog link function presupposes an asymmetric error distribution $(F(\mathbf{x}\beta) = 1 - \exp(-\exp(\mathbf{x}\beta)))$, as opposed to the logit or probit link functions, wherein the error distributions are symmetrical. For binary logit models, $F(\mathbf{x}\beta) = \frac{\exp(\mathbf{x}\beta)}{1+\exp(\mathbf{x}\beta)} = \Lambda(\mathbf{x}\beta)$, and for probit models, $F(\mathbf{x}\beta) = \int_{-\infty}^{\mathbf{x}\beta} \frac{1}{\sqrt{2\pi}} \exp\left(-\frac{\varepsilon^2}{2}\right) d\varepsilon = \Phi(\mathbf{x}\beta)$.

Based on the observed y and Eq. 3.1 , one can set up the log likelihood function and apply ML estimation to derive the parameter estimates, assuming the data points are independent and identically distributed, conditional on covariates. The estimation procedures for clustered, multilevel, or panel data are usually more convoluted and are not considered in this text.

Below, the data from the cumulative file of the General Social Survey database are used to illustrate the estimation of a cumulative ordered logit model using R. The ordered response variable of interest is self-rated health (SRH), `health`, with one to four denoting excellent, good, fair, and poor health, respectively. It is a common practice in health research that a high value corresponds to a relative poor health condition.

```
require(foreign)
# load the MASS package with the polr function
require(MASS)
# read in recoded cumulative GSS data
readin <- read.dta("data/gssCum7212Teach.dta", convert.factor=F)
# create list of variable names used for variable selection
usevar <- c("health", "age", "female", "white", "educ", "impinc")
# subset the data (select variables)
mydta <- subset(readin[complete.cases(readin[usevar]),],
        select=usevar)
```

After reading in the data, one can label the values of the ordinal response variable—SRH—using the `factor` function, and examine descriptive statistics by, for example, tabulating the new `health` variable.[1] One can also use the `hist` function to graph a histogram of the ordinal response variable, as shown in Fig. 3.3.

```
# create nomial variable hlthFac
mydta$hlthFac <- factor(mydta$health, levels = c(1, 2, 3,
    4), labels = c("1exltHlth", "2goodHlth", "3fairHlth",
    "4poorHlth"))
# create ordinal variable health
mydta$health <- factor(mydta$health, levels = c(1, 2, 3, 4),
    labels = c("1exltHlth", "2goodHlth", "3fairHlth", "4poorHlth"),
    ordered = T)
table(mydta$health, useNA = c("ifany"))

##
## 1exltHlth 2goodHlth 3fairHlth 4poorHlth
##      1288      1915       809       232

hist(as.numeric(mydta$health), xlim = c(1, 4), breaks = 36)
```

---

[1]Here we create two health variables, one ordered and the other nominal, for analysis in ordered and multinomial regression models respectively.

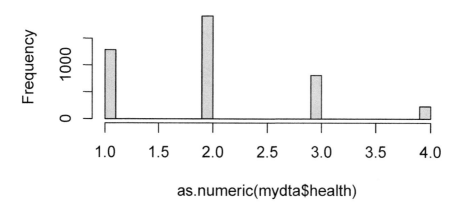

Figure 3.3
Barplot of Health

Statisticians and data analysts have written several packages to estimate various kinds of ordered regression models, including `MASS` (Venables and Ripley, 2002), `mvord` (Hirk et al., 2020), `oglmx` (Carroll, 2018), `ordinal` (Christensen, 2019), and `VGAM` (Yee, 2015), just to name a few. Below, the `polr` function from the `MASS` package is to estimate a cumulative logit model.

```
ordLogit <- polr(health ~ age + female + white + educ + impinc,
       data=mydta, method="logistic")
summary(ordLogit)

## Call:
## polr(formula = health ~ age + female + white + educ + impinc,
##     data = mydta, method = "logistic")
##
## Coefficients:
##            Value Std. Error   t value
## age      0.026262   0.001756  14.95855
## female   0.005214   0.059015   0.08835
## white   -0.342852   0.075999  -4.51129
## educ    -0.109079   0.010245 -10.64739
## impinc  -0.395350   0.035488 -11.14054
##
## Intercepts:
##                       Value   Std. Error t value
## 1exltHlth|2goodHlth   -5.3828   0.3464   -15.5379
## 2goodHlth|3fairHlth   -3.1436   0.3394    -9.2627
## 3fairHlth|4poorHlth   -1.2144   0.3391    -3.5808
##
## Residual Deviance: 9391.051
```

```
## AIC: 9407.051
```

```
# summary(ordLogit)$coefficients
```

The results section lists the ML estimates and their associated standard errors of the parameters in Eq. 3.1. The challenge for interpreting these coefficients is that they are directly linked to the latent continuous variable, y^*, and their connection with the observed ordinal response variable, y, is not so obvious.

3.1.3.2 Hypothesis Testing and Model Comparison

Before engaging in interpretation, one usually needs to examine the statistical significance of parameter estimates and make model comparisons. Since the preceding chapter elaborates on how to conduct the Wald, LR, and score tests and make model comparisons using scalar measures of model fit statistics, this and next few chapters only sample some of these testing and model comparison techniques for illustrative purposes. Caveats and exceptions, if any, will be noted whenever appropriate. To use the LR test to compare the baseline cumulative logit model and a second one without age and gender, for example, one can have the following R code chunk. The **update** function requests to estimate a new constrained model by removing age (**age**) and gender (**female**), and the **anova** function lists the estimation object names for the two models of interest,

```
noAgeFem <- update(ordLogit, .~. - age - female)
summary(noAgeFem)$coefficients
```

```
##                         Value Std. Error    t value
## white              -0.2097166 0.07514205  -2.790936
## educ               -0.1353273 0.01006217 -13.449114
## impinc             -0.3906572 0.03500743 -11.159264
## 1exltHlth|2goodHlth -6.7105020 0.32860133 -20.421408
## 2goodHlth|3fairHlth -4.5607747 0.31890946 -14.301158
## 3fairHlth|4poorHlth -2.7073710 0.31757811  -8.525056
```

```
lrtest = anova(noAgeFem, ordLogit, test="Chisq")
print(lrtest)
```

```
## Likelihood ratio tests of ordinal regression models
##
## Response: health
##                               Model Resid. df Resid. Dev  Test   Df
## 1             white + educ + impinc      4238   9620.796
## 2 age + female + white + educ + impinc      4236   9391.051 1 vs 2    2
##    LR stat. Pr(Chi)
## 1
## 2 229.7453       0
```

The results show that for 2 degrees of freedom with a χ^2 of 229.745, one can safely infer that the two models are statistically different from one another at the conventional significance level ($p < 0.05$) and the baseline/full model outperforms the restricted model without age and gender. One can also use scalar measures, such as pseudo-R^2

```
library(DescTools)
PseudoR2(ordLogit, c("McFadden", "Nagel"))
```

```
##    McFadden Nagelkerke
```

```
## 0.07476732 0.18022561
```

```
PseudoR2(noAgeFem, c("McFadden", "Nagel"))
```

```
##    McFadden Nagelkerke
## 0.05213217 0.12902338
```

or adjusted McFadden's pseudo $R^2_{\text{adj}} = 1 - \frac{\ln(\widehat{L}_F) - K}{\ln(\widehat{L}_C)}$, where K is the effective degrees of freedom of the model

```
# adjusted McFadden's pseudo R2
amfR2 <- function(modobj) {
        fullmod <- logLik(modobj)
        consmod <- logLik(update(modobj, .~ 1))
        k = modobj$edf # effective degrees of freedom
        pseudor2 <- 1 - (fullmod-k)/consmod
        return (pseudor2)  }
amfR2(ordLogit)
```

```
## 'log Lik.' 0.07319095 (df=8)
```

```
amfR2(noAgeFem)
```

```
## 'log Lik.' 0.0509499 (df=6)
```

and information measures such as AIC and BIC,

```
AIC(ordLogit, noAgeFem)
```

```
##            df      AIC
## ordLogit   8 9407.051
## noAgeFem   6 9632.796
```

```
BIC(ordLogit, noAgeFem)
```

```
##            df      BIC
## ordLogit   8 9457.877
## noAgeFem   6 9670.916
```

One can also have an analog of the Hosmer-Lemeshow (HL) test in binary regression models for the ordered regression. Fagerland and Hosmer (2013, 2017) propose the ordinal HL test, in which the observed and expected frequencies in each response category are compared for groups with roughly equal number of cases ordered based on predicted ordinal score (Fagerland and Hosmer, 2017, p. 671),

$$\widehat{\text{HL}}_{\text{ord}} = \sum_{g=1}^{G} \sum_{l=1}^{L} \left(o_{gl} - \widehat{e}_{gl} \right)^2 / \widehat{e}_{gl} \qquad (3.2)$$

where g and l denote a specific group and the response level respectively, and o_{gl} and $\widehat{e}_{gl}$ correspond to observed and expected frequencies in group g and level l across all combinations of groups and levels. Fagerland and Hosmer (2013, 2017) show that $\widehat{\text{HL}}_{\text{ord}}$, so constructed, follows a chi-squared distribution with $(G-2)(L-1) + (L-2)$ degrees of freedom, wherein

G and L denote the total number of groups and response levels respectively, and $G = 10$ is usually recommended. One can again use the `logitgof` function from the `generalhoslem` package to calculate the ordinal HL test statistic as we do for binary regression models (Jay, 2019) ,

```
library(generalhoslem)
logitgof(mydta$health, fitted(ordLogit), g=10, ord=T)

##
##  Hosmer and Lemeshow test (ordinal model)
##
## data:  mydta$health, fitted(ordLogit)
## X-squared = 68.178, df = 26, p-value = 1.217e-05
```

In the `logitgof` function, the `ord` argument is set to `TRUE` to denote that the model, for which the HL test statistic is calculated, is a cumulative logit ordered regression model. The results show that the fitted and observed frequency data are statistically different at a conventional significance level, thus providing clear evidence for lack of fit. This is not that surprising since the model examined here is a naive ordered regression model for illustrative purposes.

3.1.3.3   Interpretation

Based on the signs of the raw coefficients from the estimation object, `ordLogit`, one can easily discover the direction of the associations. So a positive sign indicates that the covariate of interest is positively associated with the latent variable valued from low to high. For example, the coefficient for age is 0.026, a positive number; thus, one can safely infer that age is positively associated with SRH, the latent variable for which is measured on a continuous scale with low values denoting relatively good health and high values for poor health. One may also attempt to interpret these coefficients on the scale of the latent variable, y^*, as the propensity to have poor health in this case. For example, one can say that for each year increase in age, we would expect the propensity/likelihood to have poor health to increase by 0.026 unit on an unknown scale. Such interpretation, however, is far from appealing in practice for several obvious reasons, for example, that we simply do not know the substantive unit of the response variable.

For ordered regression models using the logit link, it is advisable to use the odds interpretation as one of the most effective options. Based on Eq. 3.1 that $P(y = l|\mathbf{x}, \beta) = F(\tau_l - \mathbf{x}\beta) - F(\tau_{l-1} - \mathbf{x}\beta)$, one can have

$$(y \leq l|\mathbf{x}, \beta) = P(y = 1|\mathbf{x}, \beta) + P(y = 2|\mathbf{x}, \beta) + ... + P(y = l|\mathbf{x}, \beta)$$
$$= F(\tau_l - \mathbf{x}\beta) \tag{3.3}$$

Given that τ_0 is set to $-\infty$, we have $F(\tau_0 - \mathbf{x}\beta) = 0$. When $F(\tau_l - \mathbf{x}\beta)$ is set to be the cumulative density function of a standard logistic distribution, one can have $P(y \leq l|\mathbf{x}, \beta) = \frac{\exp(\tau_l - \mathbf{x}\beta)}{1+\exp(\tau_l - \mathbf{x}\beta)}$, which leads to $\ln\left(\frac{P(y \leq l|\mathbf{x}, \beta)}{1-P(y \leq l|\mathbf{x}, \beta) = P(y > l|\mathbf{x}, \beta)}\right) = \tau_l - \mathbf{x}\beta$. The preceding equation can be confusing for interpretation, since on the right-hand side of the equation, the signs for all β's are reversed (i.e., negative) and on the left-hand side, the odds comparison is the probability of the observed ordered response variable, $y \leq l$, over its complement— the probability of $y > l$. One could interpret the results using this setup, but it can be somewhat counter-intuitive and awkward. To make the interpretation follow usual intuition, one can inverse the odds comparison within the natural log function on the left-hand side and accordingly change the signs on the right-hand side of the equation, $\ln\left(\frac{P(y > l|\mathbf{x}, \beta)}{P(y \leq l|\mathbf{x}, \beta)}\right) = -\tau_l + \mathbf{x}\beta$. With simple algebraic re-arrangement, one can have $\frac{P(y > l|\mathbf{x}, \beta)}{P(y \leq l|\mathbf{x}, \beta)} = \exp(-\tau_l + \mathbf{x}\beta)$. This result looks very similar to the one in the binary logit model, $\frac{P(y=1|\mathbf{x},\beta)}{P(y=0|\mathbf{x},\beta)} = \exp(\mathbf{x}\beta)$, except that the odds comparison changes from $\frac{P(y=1|\mathbf{x},\beta)}{P(y=0|\mathbf{x},\beta)}$ to $\frac{P(y>l|\mathbf{x},\beta)}{P(y\leq l|\mathbf{x},\beta)}$ and the intercept on the right hand side turns into a cut-point. Note that the sign for cut-points is not that important since cut-points are auxiliary parameters. Now if we let the odds, $\Omega = \frac{P(y>l|\mathbf{x},\beta)}{P(y\leq l|\mathbf{x},\beta)}$, and increase the value of a generic covariate x_k by one unit while holding all other covariates constant, then we have $\frac{\Omega_{\mathbf{x}_k+1}}{\Omega_{\mathbf{x}_k}} = \frac{\exp(-\tau_l+x_1\beta_1+...+(x_k+1)\beta_k+...+x_K\beta_K)}{\exp(-\tau_l+x_1\beta_1+...+x_k\beta_k+...+x_K\beta_K)} = \exp(\beta_k)$, thus providing the substantive interpretation of the parameter coefficients from our previous estimation of the ordered cumulative logit model,

```
exp(ordLogit$coefficients)
```

```
##       age    female     white      educ    impinc
## 1.0266097 1.0052273 0.7097431 0.8966593 0.6734442
```

So for example, for each year increase in education, we would expect the odds of having relatively poor health vs. good health to decrease by a factor of $\exp(-0.109) = 0.897$, net of the effects of all other covariates. For dummy variables, for example, gender, we can say that being female (`female = 1`) as opposed to male (`female = 0`), ceteris paribus, increases the odds of having relatively poor health vs. good health by a factor of $\exp(0.005) = 1.005$.

Since odds ratio coefficients are monotonic increasing functions of raw coefficients, one can simply apply the exponential function to the confidence intervals of the raw coefficients and get the confidence intervals for the odds ratio coefficients as follows,

```
exp(confint(ordLogit))
```

```
##              2.5 %    97.5 %
## age     1.0230916 1.0301566
## female  0.8954166 1.1284927
## white   0.6114913 0.8237330
## educ    0.8787977 0.9148115
## impinc  0.6280834 0.7218395
```

Following the framework for post-estimation analysis and interpretation laid out in the previous chapter, we can also calculate predicted probabilities for observed cases in the estimation sample using the `predict` function from the `stats` package that usually comes with the R installation by default,

```
ordlogit.pred <- predict(ordLogit, type="probs")
summary(ordlogit.pred)
```

```
##    1exltHlth         2goodHlth         3fairHlth         4poorHlth
## Min.   :0.01207   Min.   :0.09078   Min.   :0.03374   Min.   :0.00597
## 1st Qu.:0.18609   1st Qu.:0.42305   1st Qu.:0.11221   1st Qu.:0.02202
## Median :0.29842   Median :0.47169   Median :0.16520   Median :0.03511
## Mean   :0.29974   Mean   :0.45122   Mean   :0.19397   Mean   :0.05507
## 3rd Qu.:0.40729   3rd Qu.:0.49919   3rd Qu.:0.25448   3rd Qu.:0.06340
## Max.   :0.72042   Max.   :0.50783   Max.   :0.44810   Max.   :0.55892
```

Since the ordinal response variable, `health`, has four ordinal levels, the `predict` function produces four new variables, whose variable names use their corresponding value labels of `health` previously created, and then places the data into a matrix object `ordlogit.pred`. Frequently, users are interested in calculating predicted probabilities for a hypothetical or specific combination of values of covariates. One can again use the `predict` function to perform such computations. To do that, one needs to feed a vector of covariates with given values, for example, a 35-year old white female with roughly college education (or 16 years of education) and sample mean income, as specified in the data.frame function below

```
x.vector <- data.frame(age=35, female=1, white=1,
                       educ = 16, impinc = mean(mydta$impinc))
x.vector
```

```
##    age female white educ    impinc
## 1   35      1     1   16  9.954116
```

Then we can feed this data frame vector into the `predict` function and calculate the predicted probabilities for the four ordered response levels,

```
(prob = predict(ordLogit, newdata=x.vector, type="probs") )
```

```
##   1exltHlth  2goodHlth  3fairHlth  4poorHlth
## 0.42954102 0.44650026 0.10381960 0.02013912
```

As described in the previous chapter, computing predictions usually is quite straightforward, even in nonlinear models such as binary and ordered regression models; finding their precision estimates (e.g., standard errors and confidence intervals), however, can be cumbersome and frequently convoluted. For the current example, the bootstrap method is selected from several candidates for computing precision estimates. To run bootstrap, we can use various utility functions from the `boot` package (Canty and Ripley, 2019; Davison and Hinkley, 1997). A bootstrap function, `bs`, is first defined to specify details of the re-sampling procedure to re-sample from the target sample with desired sample size (usually full size), run the same ordered regression model (i.e., proportional odds), calculate the predicted probabilities for the ordered response variable for the same hypothetical individual in the preceding prediction example, and then collect results for each iteration.

```
require(boot)
bs <- function(formula, data, selectCase) {
    # allow boot to sample
    d <- data[selectCase,]
    nowfit <- polr(formula, method=c("logistic"), data=d)
    dataF <- data.frame(age = 35, female = 1, white = 1,
                        educ = 16, impinc=mean(mydta$impinc))
    predF = predict(nowfit, dataF, type="probs")
    retnlist = c(predF)
    return(retnlist) }
```

After the bootstrap function is defined, one can set the seed (`set.seed(47306)`) for reproducible results and then call the `boot` function to run the bootstrap procedure off the target sample and model. The `R` argument sets the total number of replications.

```
set.seed(47306)
results <- boot(data=mydta, statistic=bs, R=500,
                formula = health ~ age + female + white + educ + impinc)
```

Once the results are obtained, one can use several ways to present the results. In this case, the percentile method is applied,

```
for (i in 1:4) {
cat("The following is 95 percentile of y = ", i, "\n")
  print(quantile(results$t[, i], c(0.025, 0.975)))
}
```

Sometimes graphical effect displays can be an effective way to learn the relationship between covariates and response variables (Fox and Weisberg, 2019). The following R code chunk turns the race variable, `white`, into a factor variable, re-estimates the same cumulative logit model, and uses the `plot` and the `predictorEffects` functions from the `effects` package (Fox and Weisberg, 2019) to plot predicted probabilities of the four response levels by education (`educ`) and race (`wht`) respectively.

```
mydta$wht <- factor(mydta$white,
                    levels = c(0),
                    labels = c("0nonwht", "1white"))
ord.logit <- polr(health ~ age + female + wht + educ + impinc,
                  data=mydta, method=c("logistic"))
library(effects)
ordlogit.educwht <- predictorEffects(ord.logit, ~ educ + wht)
plot(ordlogit.educwht, axes=list(grid=TRUE))
```

In Fig. 3.4, the left column of the graph panel displays the relationship between education and SRH, while holding all other variables at their sample means. As education increases, it can be shown clearly that the predicted probabilities of having poor (the first graph on the left column) and fair health (the second graph on the left column) decrease. For good health, it appears that its predicted probabilities peak around the medium level of education, roughly around 12 years. It is quite obvious that as education increases, the predicted probabilities of having excellent health grows almost exponentially. On the right column, the four sub-graphs illustrate how race (being white vs. nonwhite) is associated with SRH. The predicted probabilities of having poor health are low for both whites and nonwhite, but the ones of having good health are relatively high for both groups. And the group differences are not that obvious with a simple visual examination. It also appears that the predicted probabilities of having excellent health for whites are higher than those of nonwhites, and this pattern is reversed when the graph of fair health is examined.

3.1.4 Testing the Proportional Odds/Parallel Lines Assumption

The proportional odds model, or what is commonly known as the cumulative logit model among most quantitative researchers in the majority of cases, is based on the proportional odds (PO) assumption, which can be easily generalized to the parallel lines (PL) assumption. These two names are almost interchangeable, except that the PO assumption usually refers to the parallel lines assumption specifically applied to the cumulative logit model, whereas the PL assumption pertains to all major types of ordered regression models. A formal expression of the PO/PL assumption can be found in McCullagh and Nelder (1989, pp. 152-153) that in the cumulative logit model, given $\ln\left(\frac{P(y \leq l|\mathbf{x},\beta)}{1-P(y \leq l|\mathbf{x},\beta)}\right) = \tau_l - \mathbf{x}\beta$, for two different covariates vectors, $\mathbf{x}_i$ and $\mathbf{x}_j$, one has $\frac{\Omega_{\mathbf{x}_i}}{\Omega_{\mathbf{x}_j}} = \frac{P(y \leq l|\mathbf{x}_i,\beta)/(1-P(y \leq l|\mathbf{x}_i,\beta))}{P(y \leq l|\mathbf{x}_j,\beta)/(1-P(y \leq l|\mathbf{x}_j,\beta))} = \exp\left\{-\left(\mathbf{x}_i - \mathbf{x}_j\right)\beta\right\}$. This equality leads to the proportional odds assumption since the two odds are proportional

educ predictor effect plot   wht predictor effect plot

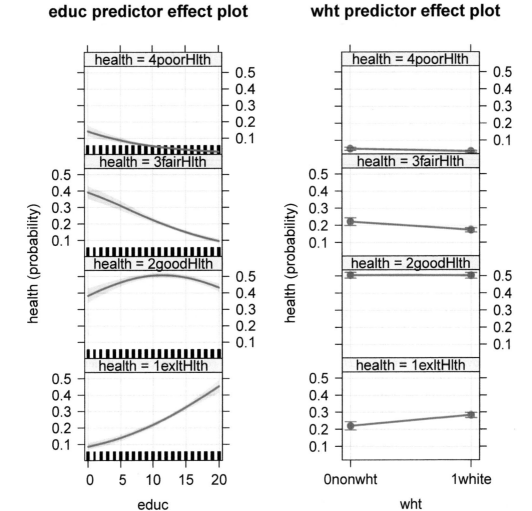

Figure 3.4
Effects Display of Education and Race

to each other by a fixed quantity $\exp\{-(\mathbf{x}_i - \mathbf{x}_j)\beta\}$. In the parameterization of the cumulative ordered logit model, this assumption translates to the invariance property for β across all cut-point equations, regardless of the response level, l. Graphically, the PO/PL assumption can be illustrated as follows. The sub-graph on the left of Fig. 3.5 shows that all the cumulative probability curves are parallel to each other without any rotation, and one curve can be obtained by horizontally shifting another probability curve. In contrast, the middle curve of the graph on the right can only be obtained by first horizontally shifting and then rotating other curves with certain angle.

One can relax the PL assumption by, for example, allowing β to vary across different cut-point equations, such as $\ln\left(\frac{P(y\leq l|x,\beta)}{1-P(y\leq l|x,\beta)}\right) = \tau_l - x\beta_l$ (β_l varies across the $L-1$ cut-point equations), hence the non-parallel cumulative logit model, or generalized ordered logit

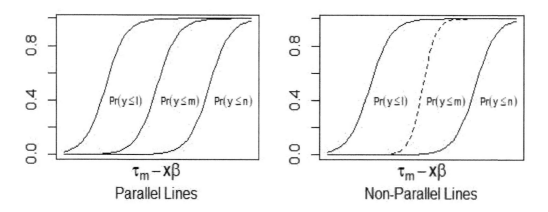

Figure 3.5
Latent and Observed Ordinal Response Variable

model.[2] The PL assumption, as how it is derived using the latent variable approach, does not appear to be that implausible. But when it is carefully considered, the assumption can hardly hold in most empirical circumstances. The orthodox PL assumption requires that for any covariate x_k, we have $H_0 : \beta_{lk} = \beta_k$; or, its corresponding coefficients from all cut-point equations are equal simultaneously. For some of the cut-point equations or some covariates, it might be plausible for this assumption to hold. It is, however, somewhat too strong for all covariates across all cut-point equations to strictly follow this assumption. In practice, quantitative researchers would usually make painstaking efforts to find even a single empirical example for this assumption to hold, given the currently available hypothesis-testing tools. Conversely, quantitative researchers may have to question themselves as to what constitutes a good model. As stated by the venerable Box (1976, p. 792), "all models are wrong," but some are useful. He went further to make an aphoristic note that,

Just as the ability to devise simple but evocative models is the signature of the great scientist so overelaboration and overparameterization is often the mark of mediocrity (p. 792).

This probably provides some wisdom about how to view all other variants and extensions of the cumulative logit model. Unless the substantive theories behind alternatives to the PL assumption or the predictions based on the PL assumption significantly deviate from observed, one may have to invoke Occam's razor or the law of parsimony to use the proportional odds model or parallel lines ordered regression models.

Researchers can use graphical or statistical methods to test the assumption. For one of the graphical methods, Harrell (2015, pp. 315-316) proposes that one can check the estimated logits from a series of binary logit models along different cut-points with the same single predictors: If the estimated logits for $y \geq l, y = 1, ...l..., L$ in the same x category/value range (reasonably chosen, such as quartile points) are arranged in the same order and have roughly same distances from each other, across different categories/value ranges, then one can infer that the PL assumption holds.

Null hypothesis significance testing (NHST) is usually viewed as apparently more legitimate alternatives for adjudicating among statistical assumptions. One can use the Wald, LR, and score tests, or the Brant test. Brant (1990) proposed a weak version of the Wald

---

[2]The PL assumption will be used henforth to denote the PO or PL assumption.

test for the PL assumption. The test statistic is constructed by comparing the coefficients from a series of binary logit models predicting the odds of $y \geq l$, for $l = 1, 2, ..., L - 1$,

$$W_{\mathrm{B}} = (D\hat{\beta}_z)' \left(\mathrm{DV}\left(\hat{\beta}_z\right)\mathrm{D}'\right)^{-1} \left(\mathrm{D}\hat{\beta}_z\right) \sim \chi^2_{(L-2)K} \qquad (3.4)$$

where $\hat{\beta}_z$ contains the full set of ML estimates (except the intercept) from the $L - 1$ binary regression of $Z = \mathrm{I}\,(y > l)$ on the same set of K covariates, and D corresponds to a block design matrix to translate $H_0 : \beta_1 = ... = \beta_{L-1}$ into a matrix format (see Brant 1990:1173). The resultant test statistic W_{B} follows a χ^2 distribution with $(L - 2)K$ degrees of freedom. One can use the `brant` function from the `brant` package (Schlegel and Steenbergen, 2020) to implement the test, for example,

```
library(brant)
brTab <- brant(ordLogit)
```

and then use the `kable` function from the `knitr` package (Xie, 2015) to tabulate the results,

```
library(knitr)
knitr::kable(brTab, digits = 3)
```

	X2	df	probability
Omnibus	41.229	10	0.000
age	8.850	2	0.012
female	0.033	2	0.984
white	3.488	2	0.175
educ	6.061	2	0.048
impinc	18.343	2	0.000

The results show that the omnibus Brant test and single-predictor Brant tests for age, education, and income are statistically significant, suggesting that the PL assumption fails to hold for these three covariates and overall. One can use other NHST procedures, such as the Wald, LR, and score tests to test the PL assumption. Scalar model fit statistics, including AIC, BIC, and pseudo R-squareds, can also be used to adjudicate among models with or without the PL assumption premised to hold. Such examples will be showcased in the following section. Alternative parameterization and model specification are available when the PL assumption is violated, which is quite common in empirical research. Some simple options include using alternative link functions (e.g., complementary log-log) or adding terms (e.g., additional covariates or interaction terms) to the linear predictor of the same model (Agresti, 2010, p. 79). Note that Williams (2006, p. 71), Agresti (2010, pp. 79-80), Fullerton and Xu (2016, pp. 130-131), and Xu et al. (2019) all provide cogent discussion about remedies for when the PL assumption is violated. Please also refer to the guidelines for selecting among different ordered regression models towards the end of our discussion of these models.

3.1.5 Partial, Proportional Constraint, and Non-parallel Models

When sufficient evidence suggests that the PL assumption does not hold, one can explore alternative ordered regression models that relax the assumption in various ways. First, if evidence shows that some of the slope coefficients are not equivalent, then they can be allowed to vary across cut-point equations. Peterson and Harrell (1990) propose partial parallel lines (a.k.a., proportional odds) models, and their estimation also generally falls under the ML framework, except that these models usually take longer and are more difficult to converge. In their paper, Peterson and Harrell (1990, pp. 208-209) discuss two types of partial PL models, including the unconstrained and constrained partial PL models. Our discussion begins with the simpler form of the two, the unconstrained partial PL logit,

$$P(y \leq l|\mathbf{x}, \mathbf{w}) = \Lambda\left(\tau_l - \mathbf{x}\beta - \mathbf{w}\gamma_l\right) \; (1 \leq l \leq L) \tag{3.5}$$

where the initial product matrix of covariates and their slopes in 3.1 is partitioned into two parts, the first is still the product of the covariates and their corresponding slopes that follow the PL assumption (i.e., β's are fixed across all cut-point equations), $\mathbf{x}\beta$, and a second component containing covarites with varying slopes across cut-point equations, $\mathbf{w}\gamma_l$. Note that the subscript l in γ denotes that the parameter is premised to vary across the cut-point equations. This model can be estimated using several R packages with MLE, including the `clm` function in the `ordinal` package (Christensen, 2019) and the workhorse for ordered regression models used in this book, the `vglm` (vector generalized linear models) function and other functions from the `VGAM` package (Yee, 2015). For later comparisons, it is advisable to first estimate the baseline PL model using the `vglm` function. Like most other estimation functions from widely used R packages, the first argument is to specify the formula by beginning with the response variable and then covariates delimited by a tilde sign. What comes next is about the model type. Below, the example requests to have a cumulative logit model with the PL assumption set to true (`parallel = TRUE`). Note that when the `reverse` argument is set to true (`reverse = TRUE`), the `vglm` function estimates a logit model in the form of $\text{logit}(P(y > l|\mathbf{x})) = -\tau_l + \mathbf{x}\beta$ and absorbs (negative) signs in estimating the cut-points (Yee, 2015, pp. 11-13, 388); this model is mathematically equivalent to the proportional odds model (the cumulative logit with the PL assumption), except that it inverts the odds comparisons, thus changing the signs in the linear predictor.

```
library(VGAM)
poLogit <- vglm(as.ordered(health) ~ age + female + white + educ + impinc,
             cumulative(link="logitlink", parallel = TRUE, reverse = TRUE),
                       data = mydta)
```

After the results are obtained, one can conduct any of the variety of post-estimation analyses illustrated with the PL model in the preceding sections. Also recall that results from the Brant test suggest that the slopes for age and income fail to satisfy the PL assumption. Below, a cumulative logit model that allows the slopes of both variables (education is not included since its significance level is close to .05) to vary across cut-point equations are estimated (partial parallel lines or PPL). Notice that for the `parallel` argument, only `female`, `white`, and `educ` are listed as `TRUE` in the `parallel` argument, and "-1" denotes that the PL assumption should discount intercepts/cut-points, since they are allowed to differ across equations according to our model specification.

```
ppoLogit <- vglm(as.ordered(health) ~ age + female + white + educ + impinc,
              cumulative(link="logitlink",
                      parallel = TRUE ~ female + white + educ - 1,
                      reverse = TRUE),
                          data = mydta)
# summary(ppoLogit)
coef(ppoLogit)
```

```
## (Intercept):1 (Intercept):2 (Intercept):3          age:1          age:2
##    4.345475738    3.784295850    2.359479189    0.021387851    0.028788742
##          age:3         female          white           educ       impinc:1
##    0.032987777    0.005348132   -0.342207311   -0.107169406   -0.276727982
##       impinc:2       impinc:3
##   -0.479056041   -0.560111641
```

Note that the results have three values for `age` and `impinc` respectively, corresponding to the slope estimates for the two variables in three cup-point equations for the log odds of $P(y > l|\mathbf{x})$ $l = 1, 2, 3$. We can then compare this PPL model with the PL model previously estimated using the LR test,

```
(modComp01 <- anova(poLogit, ppoLogit, type=1))

## Analysis of Deviance Table
##
## Model 1: as.ordered(health) ~ age + female + white + educ + impinc
## Model 2: as.ordered(health) ~ age + female + white + educ + impinc
##   Resid. Df Resid. Dev Df Deviance Pr(>Chi)
## 1     12724     9391.1
## 2     12720     9357.5  4   33.588 9.05e-07 ***
## ---
## Signif. codes:  0 '***' 0.001 '**' 0.01 '*' 0.05 '.' 0.1 ' ' 1
```

The `anova` function from the base `stats` package can be invoked to conduct the LR test, and the results shows that the two models are statistically different at the conventional significance level with 4 degrees of freedom and a χ^2 of 33.588, with the PPL model outperforming the PL model. Note that the `type` argument in the `anova` function is set to "1" to denote type I (sequential or incremental) sum of squares.

To take a step further, one can also let the slopes of all covariates to vary freely, hence the generalized ordered logit (non-parallel cumulative logit or NP) model,

$$P(y \le l|\mathbf{x}) = \Lambda\left(\tau_l - \mathbf{x}\beta_l\right) (1 \le l \le L) \tag{3.6}$$

with an implicit assumption that $\mathbf{x}\beta_1 \ge \mathbf{x}\beta_2, ..., \ge \mathbf{x}\beta_{L-1}$. The model can also be estimated using the `vglm` function with the `parallel` sub-argument set to `FALSE`, requesting to estimate a non-parallel cumulative logit model.

```
npLogit <- vglm(as.ordered(health) ~ age + female + white + educ + impinc,
          cumulative(link="logitlink", parallel = FALSE, reverse = TRUE),
          # etastart = predict(poLogit), trace = TRUE,
          data = mydta)
coef(npLogit)

## (Intercept):1 (Intercept):2 (Intercept):3          age:1          age:2
##    4.328802357    3.893872979    2.264248117    0.021663066    0.028282871
##          age:3       female:1       female:2       female:3        white:1
##    0.033856938    0.011600019   -0.002904071    0.038328176   -0.419009748
##        white:2        white:3         educ:1         educ:2         educ:3
##   -0.329534884   -0.040085718   -0.100243615   -0.126128757   -0.075310106
##       impinc:1       impinc:2       impinc:3
##   -0.279395370   -0.464796710   -0.618058291
```

```
(modComp02 <- anova(ppoLogit, npLogit, type = 1))

## Analysis of Deviance Table
##
## Model 1: as.ordered(health) ~ age + female + white + educ + impinc
## Model 2: as.ordered(health) ~ age + female + white + educ + impinc
##   Resid. Df Resid. Dev Df Deviance Pr(>Chi)
## 1     12720     9357.5
## 2     12714     9344.9  6   12.596  0.04992 *
## ---
## Signif. codes:  0 '***' 0.001 '**' 0.01 '*' 0.05 '.' 0.1 ' ' 1
```

The results include three different slope coefficients for each covariate. Another LR test using the `anova` function shows the PPL model and the NP model are statistically different with 6 degrees of freedom and a χ^2 of 12.596. After selecting the model, one can conduct post-estimation analysis, including calculating predicted probabilities and computing discrete changes/marginal effects. The R code chunk below calculates the predicted probabilities for the usual hypothetical individual used in this text, based on the estimates from the NP model.

```
whtFem <- data.frame(age = 35, female = 1, white = 1,
                     educ = 16, impinc=mean(mydta$impinc))
(predProb = predict(npLogit, whtFem, type="response"))

##    1exltHlth 2goodHlth  3fairHlth  4poorHlth
## 1  0.4268859 0.4633708 0.09009077 0.0196525
```

The results show that the predicted probabilities for this hypothetical individual having excellent $(y = 1)$ to poor $(y = 4)$ health are 0.427, 0.463, 0.09, and 0.02 respectively.

In between restricting all slopes of each covariate to be equivalent or allowing them to vary freely, one can also straddle somewhere between these two extremes, for example constraining some of the slopes vary by the same scale factor, thus the constrained partial cumulative model (CPPL),

$$P(y \leq l|x, w) = \quad \Lambda \left(\tau_l - \mathbf{x}\beta - \mathbf{w}\gamma_l - \phi_l \mathbf{z}\eta \right) \ (1 \leq l \leq L) \tag{3.7}$$

where ϕ_l is the proportionality constraint factor, $\mathbf{z}$ is a subset of the covariate matrix, and η is the corresponding vector of slopes. So Eq. 3.1 has three components in addition to the cut-point, τ_l, including the covariates that follow the PL assumption, the covariates that relax the PL assumption, and the third part, the covariates that vary by a common factor, ϕ_l, for the lth cut-point equation. For model identification, one needs to constrain $\phi_1 = 1$ and $\phi_L = 0$ and assumes $\phi_1 > \phi_2... > \phi_L$. Peterson and Harrell (1990, p. 209) are probably the first to propose this model, except that they treat ϕ_l as "fixed prespecified scalars" instead of parameters to be estimated, probably due to lack of appropriate computing algorithms and power back then. Note that the stereotype logit model, devised by Anderson (1984) and will be discussed in a later section, only retains the constrained component $(-\phi_l \mathbf{z}\eta)$ and the cut-points (τ_l) in the linear predictor of Eq. 3.7, thus applying the scaled factors to all covariates.

Below, a constrained partial proportional odds model is estimated using the `rrvglm` (reduced rank vector generalized linear model) function from the same VGAM package. In the model, only the slopes for `educ` follows the PL assumption. The slopes for `female` and `white` are allowed to vary freely; and the slopes for what is left in the equation, `age` and impinc, are constrained to vary proportionally across the cut-point equations with the same common factors.

```
###### PARTIAL PROPORTIONAL CONSTRAINTS MODELS ######
# # Setting Up The Model
# 1. Variables that follow Parallelism: formula + parallel
# 2. Variables that vary freely: formula + noRRR
# 3. Variables that have proportional constraints: formula (reduced rank)
ppcLogit <- rrvglm(as.ordered(health) ~ age + female + white + educ + impinc,
                cumulative(link="logitlink",
                parallel = TRUE ~ educ - 1,
                reverse = TRUE), # etastart = predict(poLogit),
                noRRR = ~ female + white, rank = 1, # trace = TRUE,
                data = mydta)
```

In the results summary below, the two values under the Coefficients section corresponding to I(latvar.mat):1 and I(latvar.mat):2 are the estimates of two ϕ_l's. Note that their corresponding standard errors are listed as NAs because the estimated information matrix is not positive definite, and there may not exist effective remedies if the data are ill-conditioned. For more discussion, please refer to Yee (2015, p. 172).

```
summary(ppcLogit)

##
## Call:
## rrvglm(formula = as.ordered(health) ~ age + female + white +
##   educ + impinc, family = cumulative(link = "logitlink", parallel = TRUE ~
##   educ - 1, reverse = TRUE), data = mydta, noRRR = ~female +
##   white, rank = 1)
##
## Pearson residuals:
##                      Min      1Q    Median         3Q     Max
## logitlink(P[Y>=2]) -5.0500 -1.04280  0.32251   0.773243 1.3909
## logitlink(P[Y>=3]) -3.0705 -0.54529 -0.29175  -0.147037 4.6701
## logitlink(P[Y>=4]) -1.2816 -0.21469 -0.12369  -0.092175 8.9167
##
## Coefficients:
##                    Estimate   Std. Error  z value  Pr(>|z|)
## I(latvar.mat):1    1.38498119          NA       NA        NA
## I(latvar.mat):2    1.44686406          NA       NA        NA
## (Intercept):1      4.73155126  0.28374121  16.6756 < 2.2e-16 ***
## (Intercept):2      3.78264732  0.37650271  10.0468 < 2.2e-16 ***
## (Intercept):3      1.71964655  0.40171941   4.2807 9.315e-06 ***
## age                0.02099671  0.00140789  14.9137 < 2.2e-16 ***
## female:1           0.00226224  0.06861862   0.0330    0.4868
## female:2           0.00020513  0.07327701   0.0028    0.4989
## female:3           0.07972674  0.14076762   0.5664    0.2856
## white:1           -0.39776916  0.09062289  -4.3893 5.686e-06 ***
## white:2           -0.33684106  0.06732232  -5.0034 2.816e-07 ***
## white:3           -0.02615149  0.00237419 -11.0149 < 2.2e-16 ***
## educ              -0.08775616  0.00824963 -10.6376 < 2.2e-16 ***
## impinc            -0.33403949  0.02814880 -11.8669 < 2.2e-16 ***
## ---
```

```
## Signif. codes:  0 '***' 0.001 '**' 0.01 '*' 0.05 '.' 0.1 ' ' 1
##
## Names of linear predictors: logitlink(P[Y>=2]), logitlink(P[Y>=3]),
## logitlink(P[Y>=4])
##
## Residual deviance: 9355.908 on 12716 degrees of freedom
##
## Log-likelihood: -4677.954 on 12716 degrees of freedom
##
## Number of Fisher scoring iterations: 6
```

One can use most of the methods described in the previous section to check, compare model fit statistics, and conduct NHST for a single or multiple parameters. Note that the assumption that $\phi_1 > \phi_2... > \phi_L$ needs to hold to ensure the ordinal nature of the response variable and the model. In estimation, however, this assumption cannot be easily imposed under the frequentist framework. Ahn et al. (2009, p. 3145) first shows how this assumption can be explicitly imposed in estimation using the Dirichlet prior for the re-parameterization (i.e., differences between the common factors, ϕ_l) of the common factors under the Bayesian framework, and Fullerton and Xu (2016, pp. 155-157) follow such procedure and estimate a random cut-point multilevel stereotype logit model.

This section has so far covered the full spectrum of modeling variants and post-estimation analyses for the PL model. The next few sections will sample some of these variants and techniques for illustrative purposes, since most of these extensions and post-estimation techniques for the PL model are readily applicable to other major types of polytomous regression models. Below, three additional major types of polytomous regression models (i.e., stopping ratio, adjacent, and multinomial) are introduced. Many are confused about how these models are set up and comparisons are made. Fig. 3.6 illustrates how log odds comparisons are made.[3] From left to right are the cumulative, forward continuation/stopping ratio (CR), backward continuation/stopping ratio, adjacent, and baseline (multinomial) models. The five contiguous small rectangles in the first line of each model panel correspond to the levels of a generic five-response polytomous (first four are ordinal) variable going from left to right. The next four conjoint pairs of circumscribed rectangles correspond to how the binary (logit) comparisons are made. For example, in the cumulative panel, the comparisons are made between $y > 1$ and $y = 1$, $y > 2$ and $y \leq 2$, $y > 3$ and $y \leq 3$, $y > 4$ and $y \leq 4$ respectively. For the forward continuation ratio model, the comparisons are made between $y > 1$ and $y = 1$, $y > 2$ and $y = 2$, $y > 3$ and $y = 3$, and $y > 4$ ($y = 5$) and $y = 4$; in the backward continuation ratio model, the logit comparisons are made between $y = 5$ and $y < 5$, $y = 4$ and $y < 4$, $y = 3$ and $y < 3$, $y = 2$ and $y < 2$; in the adjacent logit model, the comparisons are made between $y = 2$ and $y = 1$, $y = 3$ and $y = 2$, $y = 4$ and $y = 3$, and $y = 5$ and $y = 4$; in the baseline/multinomial logit model, the comparisons are made between the selected reference category ($y = 1$ in this case) and every other category, including $y = 2$ and $y = 1$, $y = 3$ and $y = 1$, $y = 4$ and $y = 1$, and $y = 5$ and $y = 1$. The next few sections will discuss models corresponding to the second to the fifth model panels in Fig. 3.6

3.1.6   Continuation Ratio Regression

The continuation ratio (CR) model is a second major type of ordered regression model. Depending on how the model is parameterized, one can have forward or backward CR models; forward and backward CR models are parameterized differently and should yield different

---

[3]This figure is based and expanded on the graph by Brendan Halpin at http://teaching.sociology. ul.ie/SSS/lugano/node62.html

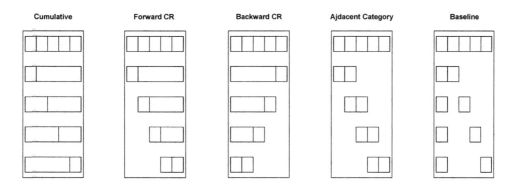

Figure 3.6
Rectangular Illustration of Ordered Regression Models

results, but conceptually they are almost identical. The CR model is also called stopping ratio (SR), stage, sequential, transition model in different fields (Yee, 2015). When it is used generically without specifically referring to forward or backward, a CR/SR usually refers to a forward model. The difference between CR and SR is whether the analytic interest is the continuation or stopping part of the sequence of events. This type of model is frequently used to describe a series of transitions or a sequence of events, the previous (following) event of which is usually subsumed within the following (previous) ones. Educational attainment and academic promotion are two typical examples. For educational attainment, one can start with the odds of obtaining less than high school education vs high school and above (e.g., high school, college, and graduate school). Next, one can be interested in the odds of obtaining college education among those having at least college-level education (e.g., master's and doctoral degree). The process can go on till one is left with the two highest response levels, depending on how the ordinal response variable is coded. Cases terminated at lower levels are taken out of the modeling procedure of later stages, thus yielding progressively shrinking sample sizes at the stage progresses. One can take a look at the second and third panel boxes in Fig. 3.7 and their associated description for a graphical understanding of the forward and backward CR models.

The conceptual idea of the CR model probably first appeared in Cragg and Uhler (1970, p. 399) in analyzing different types of demand for automobiles, and more famously in Cox (1972) in conceiving the Cox proportional hazard regression model that constructs the (partial) likelihood by combining (multiplying) logits of deaths over the risk set at certain time point as time elapses. Early formal treatments of CR models include Fienberg and Mason (1979, pp. 33-36) using log continuation ratios in studying the simultaneous effects of age, period, and cohort and in Maddala (1983, pp. 49-50) while revisiting the automobile demand example first discussed in Cragg and Uhler (1970) and in McCullagh and Nelder (1983, p. 104), just to name a few.

Since the backward CR model only reverses the stages and is very similar to the forward CR model in parameterization (not the results though), the following discussion focuses on the forward CR logit model that has the following setup,

$$\ln\left(\frac{P(y = l|\mathbf{x}, \beta)}{P(y > l|\mathbf{x}, \beta)}\right) = \tau_l - \mathbf{x}\beta \tag{3.8}$$

where $P(y = l|\mathbf{x}, \beta)$ and $P(y > l|\mathbf{x}, \beta)$ are probabilities conditional on $y \geq l$. With simple algebraic rearrangements, one can have $\Omega = \frac{P(y=l|\mathbf{x},\beta)}{P(y>l|\mathbf{x},\beta)} = \exp(\tau_l - \mathbf{x}\beta)$, which can easily lead to the interpretation of β_k for a generic covariate x_k, like in the binary and cumulative logit models, that for each unit increase in x_k, the odds of $y > l$ vs. $y = l$ is expected to change by a factor of $\exp(\beta_k)$ (see how the odds comparison and the sign for the coefficient are changed), holding all other covariates constant. One can also make this interpretation more substantive by stating that the odds of y moving beyond (hence continuation) level l vs. stopping (hence stopping) at l, given that y has reached level l, is expected to change by a factor of $\exp(\beta_k)$. This model and its converse can be readily estimated using the `vglm` function from the `VGAM` package (Yee, 2015). Below, the same ordinal response variable, `health`, and the same set of covariates are used to estimate a forward CR model.

```
crPoLogit <- vglm(as.ordered(health) ~ age + female + white + educ + impinc,
                cratio(link="logitlink", parallel = TRUE, reverse = F),
                data = mydta)
summary(crPoLogit)

##
## Call:
## vglm(formula = as.ordered(health) ~ age + female + white + educ +
##     impinc, family = cratio(link = "logitlink", parallel = TRUE,
##     reverse = F), data = mydta)
##
## Coefficients:
##               Estimate Std. Error z value Pr(>|z|)
## (Intercept):1  4.816799   0.299130  16.103  < 2e-16 ***
## (Intercept):2  3.087980   0.291679  10.587  < 2e-16 ***
## (Intercept):3  2.004940   0.290866   6.893 5.46e-12 ***
## age            0.022510   0.001511  14.897  < 2e-16 ***
## female         0.001063   0.051036   0.021 0.983389
## white         -0.247808   0.065549  -3.781 0.000156 ***
## educ          -0.087386   0.008770  -9.964  < 2e-16 ***
## impinc        -0.361988   0.030296 -11.948  < 2e-16 ***
## ---
## Signif. codes:  0 '***' 0.001 '**' 0.01 '*' 0.05 '.' 0.1 ' ' 1
##
## Names of linear predictors: logitlink(P[Y>1|Y>=1]),logitlink(P[Y>2|Y>=2]),
## logitlink(P[Y>3|Y>=3])
##
## Residual deviance: 9378.565 on 12724 degrees of freedom
##
## Log-likelihood: -4689.282 on 12724 degrees of freedom
##
## Number of Fisher scoring iterations: 6
##
## No Hauck-Donner effect found in any of the estimates

exp(coef(crPoLogit))

## (Intercept):1 (Intercept):2 (Intercept):3         age      female
##    123.5688687    21.9327269     7.4256500   1.0227648   1.0010631
##         white          educ        impinc
##     0.7805096     0.9163231     0.6962906
```

To interpret the results, for example, for the effect of education on the odds of having less (worse) than good health ($y > 2$) vs. having good health ($y = 2$), one can say that for each year increase in education, the odds will decrease by a factor of 0.916, net of the effects of other covariates. Additionally, one can state that being white vs. nonwhites, ceteris paribus, decreases the odds of having less than good health (fair or poor health) vs. good health by a factor of 0.781. Note that the CR model can be appropriate for ordinal response variables that follow data generation processes with a natural progression of events, the later among which are realized only after the preceding ones have occurred. This is one conceptually coherent and dominant account of the (forward) CR model. For this SRH example, the assumption would imply that one usually goes through the excellent health stage first and gradually moves to good, fair, and poor health, and these stages are clearly distinct and do not overlap. In addition, any case can stop at any of these four stages. It appears that this assumption, for the SRH example in particular, does not hold that steadfast. And this example is used mostly for illustration. There could be alternative interpretation of this modeling strategy, with which cases are dropped as one moves from good to poor health.

One can use the `predict.vglm` function from the VGAM package to calculate values for linear predictors or predicted probabilities, with the same R code chunk used for the PL model in previous sections. In the PL CR model, the probability for $y = l$ is calculated as follows,

$$P(y = l|\mathbf{x}, \beta) = F(\tau_l - \mathbf{x}\beta) \prod_{m=1}^{l-1} [1 - F(\tau_m - \mathbf{x}\beta)] \qquad (3.9)$$

One can also take a step further to calculate marginal effects of covariates, for example, the marginal effect of education on the predicted probability of any of the four health status for observed cases. Below, the `margeff` function in the VGAM package is invoked to calculate the marginal effect of education on the predicted probabilities for $y = 1$ (excellent health) to $y = 4$ (poor health) for the first observation in the estimation sample.

```
margeff(crPoLogit, sub=1)["educ", ]
```

```
##    1exltHlth    2goodHlth    3fairHlth    4poorHlth
##   0.021750911 -0.011559074 -0.009065873 -0.001125964
```

If one is interested in some measure of average marginal effects, then the following R code chunk can accomplish that goal.

```
# calculate marginal effects of all covariates for all cases
margAll = margeff(crPoLogit)
# calculate the average
margAvg = rowSums(margAll, dims=2)/dim(margAll)[3]
margAvg
```

```
##                 1exltHlth      2goodHlth     3fairHlth     4poorHlth
## (Intercept) -0.937799636   2.342508e-01   0.5059745144   1.975744e-01
## age         -0.004382471  -4.928013e-05   0.0028020176   1.629733e-03
## female      -0.000206877  -2.326296e-06   0.0001322708   7.693248e-05
## white        0.048246660   5.425254e-04  -0.0308474352  -1.794175e-02
## educ         0.017013532   1.913142e-04  -0.0108779308  -6.326915e-03
## impinc       0.070476764   7.924991e-04  -0.0450606826  -2.620858e-02
```

Here it is fortunate to have the `margeff` function customized for calculating marginal effects. When quantitative scholars do not have such nice R functions, what could be some viable options? One can use the closed form analytic solution, such as taking the first partial derivatives directly and then implementing the resultant formula in R. This process can be cumbersome and error-prone. In practice, numerical solutions can be more efficient and as accurate as closed-form analytical solutions. A common method is to rely on the definition of derivatives, $\frac{dy}{dx} = \lim_{\triangle x \to 0} \frac{f(x+\triangle x)-f(x)}{\triangle x}$; that is, one can increase the value of a generic x_k (from a point of interest) by a sufficiently small amount $(\triangle x_k)$, take a difference of the two functions (e.g., probability) after plugging the values of x_k and $x_k + \triangle x_k$, and divide the difference by $\triangle x_k$. Below is a working example to illustrate how one can produce such marginal effects. The goal is to calculate the marginal effect of education for the same hypothetical individual—a 35-year old white female with 16 years (college) of education and sample average income—that has been used consistently throughout this text,

```
# initial x vector with educ = 16
whtFem01 <- data.frame(age = 35, female = 1, white = 1,
                       educ = 16, impinc=mean(mydta$impinc))
d = 1/100 # delta x for educ
# increase educ by d
whtFem02 <- data.frame(age = 35, female = 1, white = 1,
                       educ = 16+d, impinc=mean(mydta$impinc))
# initial prob
p1 <- predict(crPoLogit, newdata = whtFem01, type = "response")
# prob by adding d to educ
p2 <- predict(crPoLogit, newdata = whtFem02, type = "response")
# approximated marginal effect for educ
(p2-p1)/d

##     1exltHlth    2goodHlth    3fairHlth    4poorHlth
## 1 0.02116873 -0.008600354 -0.01081933 -0.001749042

# combine and list results
margEduc = rbind(p1, p2, (p2-p1)/d)
rownames(margEduc) <- c("femProb01", "femProb02", "diffProb")
margEduc

##             1exltHlth    2goodHlth    3fairHlth    4poorHlth
## femProb01 0.41182243  0.469227331  0.10954806  0.009402180
## femProb02 0.41203412  0.469141328  0.10943987  0.009384690
## diffProb  0.02116873 -0.008600354 -0.01081933 -0.001749042
```

In this case, x_k is education, and $\triangle x_k$ is set to be $1/100$. The last part of the results section lists the marginal effects of education on the corresponding predicted probabilities. It is not surprising to find that education has a positive marginal effect on having excellent health and negative effects on good, fair, and poor health respectively. One can also use the `margeff` function to check the calculation by replacing the first observation with a covariates vector using values for the hypothetical case and tricking `margeff` function to unknowingly calculate these out-of-sample marginal effects.[4]

---

[4]This example is based on email exchanges with Dr. Thomas Yee in March 2021.

```
linePred <- predict(crPoLogit, newdata = whtFem01)  # On the eta scale
# a unit scalar plus the covariate vector to replace
# the values of the first observation in the x matrix
# in crPoLogit estimation results object
crPoLogit@x[1, ] <- c(1, unlist(whtFem01))
# linear predictor
crPoLogit@predictors[1, ] <- linePred
margeff(crPoLogit, sub=1)["educ", ]

##    1exltHlth    2goodHlth    3fairHlth    4poorHlth
##  0.021167100 -0.008593912 -0.010822688 -0.001750500
```

It can be shown that the two sets of results are very similar, identical to the fourth decimal points. One needs to be careful that the first observation in the two name slots, `crPoLogit@x` and `crPoLogit@predictors`, of the CR estimation object, `crPoLogit`, have changed so as to calculate the marginal effects of education for the hypothetical case of our interest. It is advisable to restore these two slots to their original forms if more calculations are to be made. Note that the marginal effects calculated here are specific to this hypothetical individual given the values of its covariates. The marginal effects may well vary with individuals. To get precision estimates for these can be convoluted. One can use either the delta or bootstrap method to derive standard errors and confidence bands, as illustrated in the previous chapter or the section about the PL model in this chapter.

3.1.7 Adjacent Category Regression

Based on the work by Gurland et al. (1960) and especially Theil (1969), several variants and extensions of the binary logit and multinomial logit models have been devised since then, including the adjacent category logit model, among others. The third major type of ordered regression model is the adjacent category logit model, and the name of the model itself suggests that the logit comparisons are made for all pairwise adjacent categories for $y = l$ and $y = l + 1$ with $L - 1$ sufficient sets of logits for $y = 1, ...l, ..., L$. The sources and earliest treatments of the model include Simon (1974), Andrich (1978), and Goodman (1979). More comprehensive discussion of the model under the general ordered regression framework includes Agresti (2010), Clogg and Shihadeh (1994), O'Connell (2006), and Fullerton and Xu (2016). The adjacent category model constructs its linear predictor using the adjacent logit as follows,

$$\ln \left(\frac{P(y = l|x, \beta)}{P(y = l + 1|x, \beta)} \right) = \tau_l - \mathbf{x}\beta \tag{3.10}$$

The fourth box in Fig. 3.7 shows how the adjacent category model compared with other ordered and polytomous regression models. Using the sufficient set of log odds comparisons based on Eq. 3.10, one can derive the probability for $y = l$ as a function of τ_l and $\mathbf{x}\beta$,

$$P(y = l|x) = \frac{\exp \left[\sum_{r=l}^{L} (\tau_r - \mathbf{x}\beta) \right]}{\sum_{q=1}^{L} \left\{ \exp \left[\sum_{r=q}^{L} (\tau_r - \mathbf{x}\beta) \right] \right\}} \quad l = 1, 2, ..., L \tag{3.11}$$

For identification purposes, τ_l and β are set to be zero when $l = L$, and thus $\exp(\tau_L - \mathbf{x}\beta) = 1$. Note that the adjacent logit model is a constrained version of the multinomial logit model. Once the PL assumption is relaxed and the model is turned into a non-parallel (generalized) adjacent logit with all slopes varying freely, one has a model mathematically equivalent to the multinomial logit or sometimes called the base category logit model (Fullerton and Xu, 2016, p. 40).

One can again use the `vglm` function from the VGAM package to estimate the PL adjacent logit model. Below the same structural model of SRH is estimated, as was done in previous sections. Note that by default the `vglm` function estimates $\ln\left(\frac{P(y=l+1|\mathbf{x})}{P(P(y=l|\mathbf{x}))}\right) = \alpha_l + \mathbf{x}\beta$ for the adjacent category model (`acat`). If the `reverse` argument is set to `TRUE`, then it estimates $\ln\left(\frac{P(y=l|\mathbf{x})}{P(P(y=l+1|\mathbf{x}))}\right) = -\alpha_l - \mathbf{x}\beta$ instead. Thus, $\tau_l = -\alpha_l$, based on our previous notation. To make the logit comparisons and accordingly the signs of coefficients consistent across different ordered logit models, we set `reverse = FALSE`. Note that these three seemingly different adjacent logit parameterizations are mathematically equivalent, and one can easily transform one set of results to another. All other arguments of the `vglm` function remain the same as those for other ordered regression models discussed previously, except that the model family type argument needs to be set to `acat`. After the model estimation, 95% confidence intervals of these parameter estimates can be obtained using the `confint` function.

```
acPoLogit <- vglm(as.ordered(health) ~ age + female + white + educ + impinc,
                  acat(link="loglink", parallel = TRUE, reverse = F),
                  data = mydta)
# summary(acPoLogit)
round(cbind(coef(acPoLogit), confint(acPoLogit)), 3)
```

```
##                      2.5 % 97.5 %
## (Intercept):1  3.530  3.055  4.006
## (Intercept):2  2.014  1.554  2.473
## (Intercept):3  1.295  0.840  1.750
## age            0.017  0.015  0.020
## female         0.008 -0.072  0.088
## white         -0.208 -0.309 -0.107
## educ          -0.070 -0.084 -0.056
## impinc        -0.275 -0.322 -0.228
```

One can apply the same approach used in previous sections to interpret odds ratio coefficients,

```
exp(coef(acPoLogit))
```

```
## (Intercept):1 (Intercept):2 (Intercept):3           age        female
##    34.1310013     7.4917564     3.6512885     1.0176306     1.0080105
##         white          educ        impinc
##     0.8122513     0.9323417     0.7596496
```

For example, one can say for each year increase in age, the odds of having good vs. excellent health will increase by a factor of $\exp(0.017) = 1.018$, holding all other variables constant. Similarly, being white vs. nonwhite, ceteris paribus, decreases the odds of having good vs. excellent health by a factor of $\exp(-0.208) = 0.812$. The next step one can take is to calculate predicted probabilities and local marginal effects (marginal effects calculated for individual observations), using the same `predict` (`predict.vlgm`) and `margeff` functions as described in in previous sections. Quantitative researchers, however, are often interested in comparisons, for example, of group differences in the effects of some covariate to address substantive questions. Race, class, and gender differences are among the most common types of such comparisons. The following R code chunk compares the marginal effects of all covariates between males and females,

```
bs <- function(formula, data, selectCase) {
# allow boot to sample
    # create a random sample
    d <- data[selectCase,]
    # estimate the current model
    nowfit <- vglm(formula,
            acat(link="loglink", parallel = TRUE, reverse = F),
            data = d)
    # calculate marginal effects for all cases with all covariates
    margAll = margeff(nowfit)
    # calculate the averaged marginal effect (for illustrative purpose)
    margAvg = rowSums(margAll, dims=2)/dim(margAll)[3]
    # select female cases
    t = 1
    fem01 = mydta$female %in% t
    margFem = margAll[,,fem01]
    # select male cases
    t = 0
    man01 = mydta$female %in% t
    margMan = margAll[,,man01]
    # calculate averaged marginal effects among females
    femAvg = rowSums(margFem, dims=2)/dim(margFem)[3]
    # calculate averaged marginal effects among males
    manAvg = rowSums(margMan, dims=2)/dim(margMan)[3]
    # take a difference between the two averaged marginal effects
    diff = manAvg - femAvg
    # return the matrix array
    return(diff)
}
# set seed for the random number
set.seed(47306)
# run bootstrap with 500 replications
results <- boot(data=mydta, statistic=bs, R=500,
            formula = health ~ age + female + white + educ + impinc)
# show the original statistics
results$t0
# list the bootstrapped statistics, which should have 24 rows.
results$t1
```

Although one can make group comparisons using averaged marginal effects, the results have to be construed with caution because the marginal effects, so calculated, are subject to the distributions of and correlations among covariates within each group. Thus, such group differences in marginal effects should not be viewed as group differences in effects (structural component) solely; instead, the differences can be interpreted as the combined differences in effects and distributions. Possible modifications, such as techniques used to get sample balance in causal inference, can be made to eliminate the confounding issues caused by distributional differences in covariates.

3.1.8   Stereotype Logit

It can be shown from previous sections that ordered regression models can be as parsimonious as the PL (cumulative, continuation ratio, and adjacent) regression model, wherein the slopes of each covariate are constrained to be equal across cut-point (stage/adjacent-category) equations. The other end of the model complexity is non-parallel or sometimes called the generalized ordered (cumulative, continuation ratio, and adjacent) regression model, wherein the slopes of each covariate are allowed to vary freely. One can even add parametric functional forms to the cut-points to further increase the number of parameters (McCullagh and Nelder, 1989, p. 156). Then zero or proportionality constraints can be imposed on any combination of the coefficients to produce a large variety of models standing somewhere in between the two extremes in the spectrum of model complexity. Choosing some of these combinations can appear arbitrary, and ad-hoc decisions are made frequently to align with empirical data, not how researchers are informed by theories.

Anderson (1984), for example, proposes to impose a coherent structure on the slope parameters so as to produce a model that is neither as elaborate as the non-parallel version nor as restrictive as the parallel one: the stereotype logit model. In this model, a base category (usually the highest response level L) is chosen as the reference category for an ordinal response variable $y = 1, ..., l, ..., L$ so as to create a logit (the logarithm of odds) function, which is assumed to be a linear function of covariates as follows,

$$\ln \left(\frac{P(y = l | \mathbf{x}, \beta)}{P(y = L | \mathbf{x}, \beta)} \right) = \tau_l - \phi_l \mathbf{x} \beta \tag{3.12}$$

in which τ_l and ϕ_l are the cut-point and common factor for the lth equation respectively. This equation implies that the slopes for all covariates vary by common factors across cut-point equations. To obtain an ordered regression model and ensure y as an ordinal response variable, ϕ's have to be ordered monotonically increasing from the baseline category (either the lowest or highest response level) to the other end, such as $1 = \phi_1 \geq \phi_2 ... \geq \phi_L = 0$ (Anderson, 1984, p. 6), with the highest response level L being the baseline. This would suggest that as one moves from the baseline further away, the magnitude of the coefficients and accordingly the effects of a predictor on corresponding log of odds also increase. Based on the previous equation, it is straightforward to find the probability for $y = l$, conditional on $\mathbf{x}$, to be

$$P(y = l | \mathbf{x}) = \frac{\exp \left(\tau_l - \phi_l \mathbf{x} \beta \right)}{\sum\limits_{q=1}^{L} \left(\exp \left(\tau_q - \phi_q \mathbf{x} \beta \right) \right)} \tag{3.13}$$

To estimate a stereotype logit model, one can use the same `rrvglm` function as described in the previous section,

```
stereotype <- rrvglm(as.factor(health) ~ age + female + white + educ + impinc,
            multinomial, data = mydta,
            noRRR = ~ 1, Rank = 1)
Coef(stereotype)@C # The C matrix

##              latvar
## age     -0.04774757
## female  -0.02055918
## white    0.57409429
## educ     0.19877142
## impinc   0.76273297
```

```
Coef(stereotype)@A # The A matrix; corner constraints
```

```
##                       latvar
## log(mu[,1]/mu[,4])  1.0000000
## log(mu[,2]/mu[,4])  0.7104234
## log(mu[,3]/mu[,4])  0.2444952
```

Note that in the `rrvglm` function, the `noRRR` argument is set to "~ 1", denoting that all slopes are to be included in the reduced-rank model; this means that all slopes co-vary by the same common factors (sometimes called corner constraints) across different cut-point equations, except the intercepts. The printed C matrix lists all slope estimates and A matrix for corner constrains (common factors). It can be shown from the results that $\phi_2 = 0.71$ and $\phi_3 = 0.244$. And recall that ϕ_1 and ϕ_4 are set to 1 and 0 respectively. Thus, for the slopes in the second cut-point equation for the logit of $P(y = 2|\mathbf{x})/P(y = 4|\mathbf{x})$, their magnitude shrinks by a common factor of 0.71 across the board, and the slopes for the third equation for the logit of $P(y = 3|\mathbf{x})/P(y = 4|\mathbf{x})$ shrink by a common factor of 0.244. For most statistical software applications, or actually all that we are aware of, they cannot impose the ordering constrains for the common factors, thus leaving the estimation essentially appropriate for a categorical response variable. If the common factor estimates turn out to be ordered, then one can safely use ordered regression models, including the stereotype logit model. If the ordering constrains assumption fails to hold, then one needs to make some modification, and the multinomial regression model can be a feasible candidate.

Note that the stereotype logit model, as described previously, is a one-dimensional stereotype logit model. As how it was devised in Anderson (1984, pp. 5-6, 11-12) , the full-blown stereotype logit model can be multidimensional; that is, one can have several common factors to form multiple sets of linear predictors (excluding cut-points). The multi-dimensional stereotype logit model is usually formulated as follows,

$$\ln\left(\frac{P(y = l|\mathbf{x}, \beta)}{P(y = L|\mathbf{x}, \beta)}\right) = \tau_l - \sum_{s=1}^{d} \phi_{sl}\mathbf{x}\beta_s \qquad 1 \leq d \leq D \tag{3.14}$$

Note that the maximum number of dimension D is equal to the minimum of $L - 1$ and K, where L is the highest response level and K is the number of covariates. When d is set to D, its maximum possible value, the corresponding stereotype logit model is equivalent to a multinomial logit model with the same covariates, but additional constraints on ϕ's have to be placed to achieve identifiability in such stereotype logit models with maximum dimensions (Anderson, 1984, p. 6).

3.2 Extentions to Classical Ordered Regression Models

3.2.1 Inflated Ordered Regression

The previous few sections cover ordered regression models that are based on simple data generation processes for typical ordered response variables. Quantitative researchers sometimes, however, might run into circumstances with a large amount of unexplained heterogeneity or unusually poor model fit. To address such issues, one can change parameterization and model specification on the right-hand side—namely, including more covariates and creating polynomial terms. A second possibility is to introduce complexity on the left-hand side, for example, by assuming mixture distributions. It is not uncommon that one observes a disproportionately large number of cases for some response levels of categorical variables, especially on the two extremes and sometimes in the middle. One possibility for having such excess cases is that multiple data generation processes operate simultaneously to mix more than one underlying latent variables into a manifest observed variable. For example, when the number of cases for rating excellent health far exceeds what one otherwise would expect to witness, researchers can assume distributional mixing. One source of these excess cases may come from those subscribing to the idea of positivity, for example, such that they would not only rate their health excellent in all circumstances, but they also rate other areas of their life to be best regardless. This group can be called always-excellent. A second source of excellent health may simply come from the other group that are well informed and rate their health objectively. This second group can be called sometimes-excellent. Note that this is an assumption that needs to be supported either by known theories or compelling empirical observations to begin with. One can surely have alternative assumptions made about the data generation process, and thus have different parameterization of the model given the data.

Based on similar substantive considerations and thought experiments, Harris and Zhao (2007), Bagozzi (2012, 2016), and Bagozzi et al. (2015) have devised a series of inflated polytomous regression models, including the zero/extreme-inflated ordered (probit) regression model, the middle-inflated ordered (probit) regression model, the zero-inflated ordered probit model with correlated errors, and the baseline inflated multinomial logit model. Generally, such models postulate the mixing of usually two processes (switching regimes [5]) to account for excess cases (unobserved heterogeneity) for a particular response category and then make an assumption about the correlation between the error terms in the two structural models.

Assuming the preceding assumption holds (always-excellent and sometimes-excellent), we can premise that the data generation process of SRH follows a two-stage (switching regimes) process. The first step is to separate the always-excellent from the sometimes-excellent, for example, by assuming $w = 0$ for the always-excellent group and $w = 1$ for the sometimes-excellent group, where w is an *unobserved* binary group indicator variable. Sometimes such separation or grouping is also called regime in econometric literature. One can then set up a binary probit regression model, wherein $w^* = \mathbf{z}\gamma + u$ such that $w^* > 0 \iff w = 1$ and $w^* \leq 0 \iff w = 0$; in the structural model predicting w^*, $\mathbf{z}$ is a covariate matrix, γ is a parameter vector, and u is the disturbance term that follows a standard normal distribution. Note that neither w^* nor w is observable. Following the derivation in the previous chapter, one can get $P(w = 1|\mathbf{z}) = \Phi(\mathbf{z}\gamma)$. Then in the sometimes-excellent group ($w = 1$), people can rate their health to be from excellent to poor. In the second step for the sometimes-excellent group only, one can have a typical ordered regression model, usually the PL ordered probit model. In this step, the latent variable approach can be applied to derive the model by assuming a latent continuous variable, y^*, such that $P(y = l|\mathbf{x}, \beta) = P(\tau_{l-1} < y^* = \mathbf{x}\beta + \varepsilon \leq \tau_l|\mathbf{x}, \beta)$ for an observed ordinal response variable $y = 1, ..., l, ..., L$ (see Eq. 3.1). After combining the two steps, one can set up the likelihood function using the following sets of equations,

$$y = \begin{cases} 0 & \text{if } (w^* \leq 0) \text{ or } (w^* > 0 \text{ and } y^* \leq 0) \\ l & \text{if } (w^* > 0 \text{ and } \tau_{l-1} \leq y^* \leq \tau_l) \\ L & \text{if } (w^* > 0 \text{ and } \tau_{L-1} \leq y^*) \end{cases} \tag{3.15}$$

---

[5]Regime is a term used frequently in economics to denote a set of rules or institutions distinct from other regimes.

$$P(y) = \begin{cases} P(y=0|\mathbf{x},\mathbf{z}) = P(w=0|\mathbf{z}) + P(w=1|\mathbf{x})P(y=0|\mathbf{z},w=1) \\ P(y=l|\mathbf{x},\mathbf{z}) = P(w=1|\mathbf{x})P(y=l|\mathbf{z},w=1) \quad \text{for } l=1,...,L \end{cases}$$

$$= \begin{cases} P(y=0|\mathbf{x},\mathbf{z}) = (1-\Phi(\mathbf{z}\gamma)) + (\Phi(\mathbf{z}\gamma)\,\Phi(-\mathbf{x}\beta)) \\ P(y=l|\mathbf{x},\mathbf{z}) = \Phi(\mathbf{z}\gamma)\,(\Phi(\tau_l-\mathbf{x}\beta)-\Phi(\tau_{l-1}-\mathbf{x}\beta)) \quad \text{for } l=1,...,L-1 \\ P(y=L|\mathbf{x},\mathbf{z}) = \Phi(\mathbf{z}\gamma)\,(1-\Phi(\tau_{L-1}-\mathbf{x}\beta)) \end{cases}$$

$$(3.16)$$

Note that in the model specification, the inflated response level is set to be a generic zero, thus leading to a zero-inflated ordered probit model. Usually, a constraint on the cut-points (e.g., zero for the first cut-point) only parallelly shifts the whole set of cut-points. In practice, the lowest response level can be any other numerical value, such as 1 in the case of SRH. In addition, the $\mathbf{z}$ and $\mathbf{x}$ covariate matrices can have overlap, but a large overlap (e.g., multi-collinearity) may lead to a singular Hessian matrix, thus making the resultant variance-covariance matrix not positive definite. So it is helpful to have at least one different covariate in either $\mathbf{z}$ or $\mathbf{x}$. With the likelihood function specified above, the ML estimation ensues, and nothing appears surprising. The challenge is that a custom-made `glm`-like function has not been made available yet. Bagozzi et al. (2015) programmed a function for estimating zero-inflated ordered probit model (ZiOP), and that function is lightly revised to estimate a ZiOP regression of SRH below. In this model, age, gender (`female` = 1), and education are used to differentiate the two regimes of always-group and the sometimes-group, and age, race (`white` = 1), and income are included as the covariates for the ordered probit model,

```
source("prgm/ZIOP.R")
# set starting values
est <- rbind(0.1, 0.1, 0.1, 0.1, 0.1, 0.1, 0.1, 0.1, 0.1, 0.1)
# set data, Y, X, and Z
data <- mydta
Y <- data$health
X <- cbind(data$age, data$white, data$impinc)
Z <- cbind(1, data$age, data$female, data$educ)
# optimize p=est
output.ZIOP <- nlm(f = ZIOP, p = est, Y = Y, X = X, Z = Z, iterlim = 500,
    data = data, hessian = TRUE)
names(output.ZIOP$estimate) <- c("ord:tau1", "ord:tau2", "ord:tau3",
    "binInfl:_cons", "binInfl:age", "binInfl:female", "binInfl:educ",
    "ord:age", "ord:white", "ord:impinc")
print(output.ZIOP$estimate)
```

```
##      ord:tau1      ord:tau2      ord:tau3 binInfl:_cons  binInfl:age
##    -3.03509156    0.27724847    0.02712275    0.10420142   0.18446592
## binInfl:female  binInfl:educ       ord:age     ord:white   ord:impinc
##     0.10276901    0.14229768    0.01702728   -0.20277122  -0.30886371
```

```
vcv <- solve(output.ZIOP$hessian)
zval = output.ZIOP$estimate/sqrt(diag(vcv))
pval <- 2 * pnorm(abs(zval), lower.tail = FALSE)
pval
```

```
##      ord:tau1      ord:tau2      ord:tau3 binInfl:_cons  binInfl:age
##   3.165812e-55  4.300453e-47  4.001683e-01  9.996744e-01 9.954301e-01
## binInfl:female  binInfl:educ       ord:age     ord:white   ord:impinc
##            NaN  9.980495e-01  3.639528e-66  4.483371e-06 4.362005e-60
```

It can be shown from the results that none of the four parameter estimates for the binary inflation equation are statistically significant, suggesting a lack of evidence for such inflation. Note that inflation or mixture may happen to any or multiple response levels.

Harris and Zhao (2007) also generalized the ZiOP model to the zero-inflated ordered probit model with error correlation, by allowing for the two error/disturbance terms in the binary inflation equation (u) and (main) ordered regression equation (ε) to be correlated. According to Harris and Zhao (2007, p. 1077), we have

$$
y = \begin{cases} 0 \text{ if } (w^* \leq 0) \text{ or } (w^* > 0 \text{ and } y^* \leq 0) \\ l \text{ if } (w^* > 0 \text{ and } \tau_{l-1} \leq y^* \leq \tau_l) \\ L \text{ if } (w^* > 0 \text{ and } \tau_{L-1} \leq y^*) \end{cases} \tag{3.17}
$$

$$
P(y) = \begin{cases} P(y = 0|\mathbf{x}, \mathbf{z}) = (1 - \Phi(\mathbf{z}\gamma)) + \Phi_2(\mathbf{z}\gamma, -\mathbf{x}\beta; -\rho) \\ P(y = l|\mathbf{x}, \mathbf{z}) = \Phi_2(\mathbf{z}\gamma, \tau_l - \mathbf{x}\beta; -\rho) - \Phi_2(\mathbf{z}\gamma, \tau_{l-1} - \mathbf{x}\beta; -\rho) \\ P(y = L|\mathbf{x}, \mathbf{z}) = \Phi_2(\mathbf{z}\gamma, \mathbf{x}\beta - \tau_{L-1}; -\rho) \end{cases} \tag{3.18}
$$

wherein $\Phi_2(v_1, v_2; \rho)$ denotes the cumulative density function of the standardized bivariate (v_1, v_2) normal distribution, and ρ corresponds to the correlation between the two error terms. Although this single relaxation seems trivial, the level of mathematical complexity and computational cost increase exponentially. One can use the ZIOPC function by Bagozzi et al. (2015) to estimate the zero-inflated ordered probit regression model with correlated errors. Note that the sourced function (ZIOPC.R) is different, and we have an additional value in the est vector for ρ.

```
source("prgm/ZIOP.R")
est<-rbind(.1,.1,.1,.1,.1,.1,.1,.1,.1,.1,.1)
#set data, Y, X, and Z
data <- mydta
Y<- data$health
X<-cbind(data$age, data$white, data$impinc)
Z<-cbind(1, data$age, data$female, data$educ)
output.ZIOPC<-nlm(f=ZIOPC, p=est, Y=Y,X=X,Z=Z,
iterlim=500, data=data, hessian=TRUE)
> print(output.ZIOPC$estimate)
      ord:tau1         ord:tau2         ord:tau3
        -2.879            0.270            0.003
 binInfl:_cons    binInfl:age binInfl:female
         0.105            0.195            0.102
 binInfl:educ          ord:age        ord:white
         0.139            0.015           -0.140
   ord:impinc              rho
        -0.290            0.106
```

Since the interpretation of the results from inflated ordered regression models is almost identical to that from non-inflated ordered regression models, we do not intend to belabor the obvious except that the coefficients need to be interpreted conditionally on the always- or sometimes-regime.

Note that having excess cases for the lowest/highest level is one source of inflation. Sometimes, such inflation can occur in the middle category, especially when such middle category provides an opportunity for avoiding direct answers, for example, neutral or independent. As human beings have this tendency to conform, to please others, and most importantly, to avoid embarrassment, a careful exploration of such response inflation is advised to be part of routine in descriptive data analysis. Empirical analysis of such middle-category inflated polytomous regression models can be conducted using methods similar to the extreme-category inflated models, except that the inflation occurs for the middle category and the two-step process involves the inflation of the middle instead of extreme categories. Hence the first step is to use a binary regression model to predict whether one is in the always-middle vs. sometimes-middle group, and the second step is to model the ordinal response variable, conditional on the sometimes-middle group. Up until now, we have only considered a two-step process, wherein the first step is binary and the inflation occurs only to one response category. More complicated processes possibly transpire in practice, for example, a multi-stage event with a multinomial data-generation process in the first step, the inflation or even deflation occurs for multiple response categories that are discussed in Brown et al. (2020) and Cai et al. (2021) in the second step, and possibly a few more stages. To summarize, inflation models generally fall under the umbrella of finite mixture modeling (McLachlan and Peel, 2000) to explain unobserved heterogeneity that cannot be directly measured in empirical data, and it is advised that quantitative researchers can gain significant explanatory power if the inflation process is relatively simple and obvious, thus not taxing too many degrees of freedom or much computational power (e.g., the ZiOPC model with correlated errors). The key is to strike a balance between model simplicity and accuracy for a reasonable data generation process.

3.2.2 Heterogeneous Choice Models

Up to this point, all variants and extensions of the prototypical PL model still subscribe to a fundamental assumption; that is, the error term in the structural model for the latent continuous variable, based on which a polytomous response variable is observed on an ordinal scale, is homoscedastic. Like in the classic OLS regression models, one can relax this assumption by having the following regression model,

$$y^* = \mathbf{x}\beta + \sigma\varepsilon \tag{3.19}$$

We can still have the same conceptual relationship linking the observed ordinal response variable, y, with its corresponding latent variable, y^*, as well as $\mathbf{x}\beta$, by having $\tau_{l-1} < y^* \leq \tau_l \iff y = l \ \forall \ y = 1, ..., l, ..., L$. With simple algebraic rearrangements after substituting y^* with $\mathbf{x}\beta + \sigma\varepsilon$, we can have $\frac{\tau_{l-1} - \mathbf{x}\beta}{\sigma} < \varepsilon \leq \frac{\tau_l - \mathbf{x}\beta}{\sigma}$. Then by making a probability distributional assumption about ε, we get

$$P(y = l|\mathbf{x}, \beta, \tau) = P(\tau_{l-1} < y^* \leq \tau_l) \tag{3.20}$$
$$= F\left(\frac{\tau_l - \mathbf{x}\beta}{\sigma}\right) - F\left(\frac{\tau_{l-1} - \mathbf{x}\beta}{\sigma}\right)$$

So when σ is equal to one, the above probability equation collapses to the one for a typical PL regression model. If σ is a constant, then one can cogently argue that all slope coefficients can only be identified up to a scale. If we allow σ_i, the scale factor, to vary across cases, then we have heteroscedastic errors, $u_i = \sigma_i\varepsilon$. Since σ_i is positive, the natural candidate for parameterizing it is the natural exponential function, $\sigma_i = \exp(\mathbf{z}_i\gamma)$, where $\mathbf{z}_i$ is a vector of covariates, and γ, the associated auxiliary parameter vector to be estimated. Thus, we have the heterogeneous choice model (McCullagh and Nelder, 1989, p. 155; Williams, 2009). It is interesting to note that the scale factor, σ, appears frequently in early literature about polytomous regression models (McKelvey and Zavoina, 1975; Andrich, 1978), and somehow it gets ignored in practice as the empirical research advances. The heterogeneous choice model can be estimated using the `oglmx` function from the `oglmx` package by Carroll (2018). Below, we estimate a heterogeneous choice ordered logit model with the same ordinal response variable and covariates as those in previous sections used for the mean equation $(\tau_l - \mathbf{x}\beta)$; an additional equation is included to estimate the scale factor σ, or the SD equation, $\sigma_i = \exp(\mathbf{z}_i \gamma)$. In this example, we first run a heterogeneous PL regression by requesting to have a constant-only model for the scale factor (~ 1,..., `constantSD=TRUE`) in the SD equation,

```
# install.packages("oglmx")
library(oglmx)
hetOlogit01<-oglmx(health ~ age + female + white + educ + impinc,
                   ~ 1, data=mydta, link="logit",
                   constantMEAN = FALSE, constantSD = TRUE)
educRawCoef = coef(hetOlogit01)["educ"]/exp(coef(hetOlogit01)["(Intercept)"])
educRawCoef
```

```
##       educ
## -0.109079
```

If the SD equation is a constant-only equation for a general PL model using the logit link, then one can simply use $\exp\left(\frac{\beta_k}{\sigma}\right)$ to provide odds ratio interpretation. For example, for each year in increase in education, we would expect the odds of having relative poor health vs. good health to decrease by a factor of $\exp(-0.109) = 0.897$, holding all covariates constant. One can also compute predicted probabilities and marginal effects to provide different types of substantive interpretation.

Next, we estimate a second heterogeneous PL model by including covariates in the SD equation,

```
hetOlogit02<-oglmx(health ~ age + female + white + educ + impinc,
                   ~ female + white, data=mydta, link="logit",
                   constantMEAN = FALSE, constantSD = FALSE)
summary(hetOlogit02)
```

```
## Heteroskedastic Ordered Logit Regression
## Log-Likelihood: -4694.739
## No. Iterations: 8
## McFadden's R2: 0.07492233
## AIC: 9409.478
## ----- Mean Equation ------
##           Estimate Std. error  t value   Pr(>|t|)
## age       0.0273609  0.0021765  12.5712  < 2.2e-16 ***
## female    0.0091857  0.0617053   0.1489     0.8817
## white    -0.3630492  0.0794078  -4.5720  4.832e-06 ***
## educ     -0.1129409  0.0112534 -10.0362  < 2.2e-16 ***
## impinc   -0.4103391  0.0402384 -10.1977  < 2.2e-16 ***
## ----- SD Equation ------
##           Estimate Std. error  t value Pr(>|t|)
## female   -0.0070848  0.0346664  -0.2044   0.8381
## white     0.0524845  0.0424907   1.2352   0.2168
## ----- Threshold Parameters -----
##                                   Estimate Std. error   t value   Pr(>|t|)
## Threshold (1exltHlth->2goodHlth) -5.58270    0.41564  -13.4316  < 2.2e-16 ***
```

```
## Threshold (2goodHlth->3fairHlth) -3.25565  0.37058  -8.7854 < 2.2e-16 ***
## Threshold (3fairHlth->4poorHlth) -1.25500  0.35349  -3.5503 0.0003848 ***
## ---
## Signif. codes:  0 '***' 0.001 '**' 0.01 '*' 0.05 '.' 0.1 ' ' 1
```

The results from the mean equation show that, compared with those in the PL model, the magnitude of the coefficients in the mean equation is generally unleashed, and the results from the SD equation indicates that whites and nonwhites differ (not statistically significant though) in their scale factor, thus contributing to the heteroscedasticity of the error term. Note that a positive coefficient corresponds to a positive association between the covariate and the error variance.

3.2.3 General Guidelines for Model Selection

Having many choices can be a good thing insomuch as one is well informed about the advantages and limitations of these choices and that these choices can be compared on a single dominant dimension, and there are perceptible differences among them. As illustrated in the previous sections, the variegated ordered regression models that are extended along different dimensions of the prototypical PL model can almost surely pose challenges to users to choose. Below, a few guidelines are provided to help quantitative researchers with their model selection process.

First, one needs to carefully examine the response variable of interest to see if it is a typical ordinal variable as described in the first section of this chapter. Variables measured on a Likert scale are usually typical ordinal variables, although complications (e.g., extreme- or middle-category inflation) can exist. There are variables that are usually viewed as regular ordinal variables but turn out to be atypical. For example, the measurement of tobacco consumption, which has led to the development of zero-inflated ordered probit model (Harris and Zhao, 2007) and excess zeros on "social bads" (e.g., illegal drug consumption) (Greene et al., 2018) , can result from mixing different data generation processes. The lowest level of tobacco consumption is usually coded as zero, denoting non-current smoker. The response levels then increase from one to three or higher, depending on how the consumption frequency is defined (e.g., 1 for smoking weekly or less, 2 for daily with fewer than 20 cigarettes per day, and etc.). It is usually observed that there is an unusually high concentration of zeros for this variable, and quantitative researchers have surmised a plausible decision process, in which one first decides to smoke or not; then among those deciding to partake, smokers may end up in different levels, including zero consumption. A second example is the measurement of blood pressure from hypotension to stage 1 and stage 2 hypertension. If the causes of hypotension and hypertension are carefully explored, then one can reasonably argue that the two seemingly related health conditions may follow different processes, possibly involving mixing distributions like that in tobacco consumption (inflation models), or involving truncation (two-part hurdle models) if the causes of hypotension is uni-dimensional and distinct from those for hypertension.

Another consideration is about whether the ordinal response variable follows a sequential/continuation process, wherein lower response levels are requisite for higher levels. If so, then one can go with the CR model if the transition in each stage is of research interest. If however the interest is in the comparisons of adjacent categories or how the response level moves up/down to the next level, then adjacent category logit is probably a good candidate. One can imagine that, for example, medical researchers might be interested in learning about the predictors that move a patient from one stage of a disease to its next more advanced stage (e.g., the four stages of cancer, different levels of blood glucose as well as blood pressure).

One can draw a diagram of decision tree to lay out all the possible decision points and paths in model selection. Once the model type is settled on the commonly used ordered regression, for example, the next step recommended is to run a stereotype logit model. If the cut-points and common proportionality factors satisfy the ordinality assumption (i.e., arranged in the order corresponding to their response levels), then one can safely go with ordered regression models given that all major demographics are included. Otherwise, important omitted variables, if available, should be added back to the model, or, more complicated models, such as multinomial logit, can be considered. A routine for users of ordered regression models is to test the PL assumption, which is frequently violated. Agresti (2010, pp. 75-80) suggested roughly six options for remedy when the proportional odds fits poorly, including 1) using non-parallel or partial cumulative logit (separate effects for each logit), 2) using a link function for which the response curve is asymmetric (e.g., complementary log-log), 3) adding additional terms (e.g., interaction terms), 4) adding dispersion parameters (e.g., heterogeneous choice model), 5) fitting models using separate parameters/covariates for each logit, and 6) parameterizing cut-points as a linear function of covariates. We can also use models with proportionality constraints or other forms of reduced rank vector generalized ordered regression models (Yee, 2015).

Although these options can improve the model fit, the theoretical motivations behind such changes is often quite lacking. For example, if one opts for a partial parallel model after testing the PL assumption, then a follow-up question would be why some covariates follow the PL assumption whereas other do not. Thus, this text would not suggest that quantitative researchers make post-hoc decisions to go with some partial (or partial proportional constraint) version of the ordered regression models simply based on results from statistical testing. When there are no strong theories guiding such decisions, it is advised that one probably choose among PL, stereotype, generalized (NP), or multinomial logit. In adjudicating among models, unless the fit is obviously poor, the principle of Occam's razor or the law of parsimony should be followed.

It is of note that the PL assumption (i.e., slopes for same covariates are equal amongst themselves across all cut-point equations for all covariates) is violated in an overwhelming majority of the cases with the usual NHST (e.g., the Wald and Brant, LR, and score tests) or similar procedures. Echoing Meehl (1967), Edwards and Berry (2010), and Kruschke and Liddell (2018), Xu et al. (2019) cogently argue that the global PL assumption is an unrealistic assumption, too strong to hold. There can be several alternatives. One is that we can largely ignore the NHST results and focus on the predictive power or fit statistics instead. If the benchmark models, such as the PL, stereotype logit, or multinomial logit model, turn out to have sufficient predictive power, then we can go with such models regardless. An alternative is to confront the unrealism of the PL assumption, and tweak it to become a more feasible one, for example, the hyper-parameter cumulative logit model wherein slopes for the same covariates come out of the same or similar distributions (Xu et al., 2019).

3.3 Multinomial Regression

3.3.1 Multinomial Logit Regression

According to McFadden (1973), Gurland et al. (1960), Bloch and Watson (1967), McFadden (1968), Rassam et al. (1971), and Theil (1969) and Theil (1971) are among the first few to develop the multinomial regression model. McFadden (1973) probably provides the first rigorous treatment of the multinomial logit model, which is categorized as a special case of conditional logit model (alternative-specific logit). In setting up the model, McFadden (1973, p. 108) supposes that individuals make their discrete choices among L alternatives, indexed by $l = 1, ..., L$ using the economic behavioral principle of utility-maximization given the attributes vector $\mathbf{x}$, then one can have

$$U = S(\mathbf{x}) + \varepsilon(\mathbf{x}) \tag{3.21}$$

where U is the utility function, S is the systematic component of the function and "represents the tastes of the population," and ε is the stochastic (random) component of the function, representing some individual-specific attributes that are not included in measurement. Then a plausible assumption can be made that the probability of choosing l, $P(y = l) = P(U_l > U_m \forall\ l \neq m)$; that is, when the choice l provides the greatest utility for an individual, then that person will choose l. If the stochastic term in Eq. 3.21 is assumed to follow a type I extreme-value distribution, then one can derive the formula for the probability of choosing l among a set of alternatives given an individual attributes vector to be

$$P(y = l | \mathbf{x}, \beta) = \frac{\exp(\mathbf{x}\beta_l)}{\sum\limits_{m=1}^{L} \exp(\mathbf{x}\beta_m)} \tag{3.22}$$

This probability function can be used to construct the likelihood function for ML estimation. The multinomial logit regression can be viewed, to some extent, as a series of binary logit regressions for a sufficient set of $L - 1$ equations. Odds ratio coefficient is usually the most popular quantity of interest for interpretation in multinomial logit models. Based on Eq. 3.22, one can have $\ln\left(\frac{P(y=l|\mathbf{x})}{P(y=m|\mathbf{x})}\right) = \mathbf{x}\beta_{l|m}$ for the odds comparison of l vs m (base category), $\frac{P(y=l|\mathbf{x})}{P(y=m|\mathbf{x})} = \Omega_{l|m}$. For interpreting the parameter β_k associated with a generic covariate x_k, like the odds ratio interpretation in binary logit and cumulative ordered logit models, one can state that for each unit increase in x_k, one would expect the odds of l vs m to vary by a factor of $\exp(\beta_k)$, while holding all other covariates constant. Below we use the `multinom` function from the **nnet** package (Venables and Ripley, 2002) to estimate a multinomial regression of the same model used throughout this chapter. After invoking the **nnet** package, one can use the `relevel` function to set the reference category first, and then estimate the model with the `multinom` function. In this example, $y = 1$ (`ref = 1`) is set to be the reference category,

```
library(nnet)
mydta$hlthFac <- relevel(mydta$hlthFac, ref= 1)
mLogit <- multinom(hlthFac ~ age + female + white + educ + impinc,
                data=mydta, model=T)


## # weights:  28 (18 variable)
## initial  value 5883.433269
## iter  10 value 4814.636916
## iter  20 value 4687.566776
## final  value 4675.893964
## converged
```

After estimation, one can use the `summary` function to present the results,

```
summary(mLogit, digits=3)

## Call:
## multinom(formula = hlthFac ~ age + female + white + educ + impinc,
##     data = mydta, model = T)
##
## Coefficients:
##           (Intercept)    age  female  white   educ impinc
## 2goodHlth        2.56 0.0138 0.00602 -0.360 -0.070 -0.150
## 3fairHlth        5.63 0.0345 0.00845 -0.573 -0.170 -0.512
## 4poorHlth        6.37 0.0506 0.02445 -0.451 -0.172 -0.827
##
## Std. Errors:
##           (Intercept)     age female white   educ impinc
## 2goodHlth       0.457 0.00235 0.0738 0.103 0.0135 0.0472
## 3fairHlth       0.565 0.00288 0.0978 0.126 0.0171 0.0582
## 4poorHlth       0.799 0.00439 0.1562 0.194 0.0251 0.0824
##
## Residual Deviance: 9351.788
## AIC: 9387.788
```

Since the summary function does not produce precision estimates automatically, one needs to add a few lines to compute the z-statistics and p-values associated with the parameter estimates.

```
z <- summary(mLogit)$coefficients/summary(mLogit)$standard.errors
pval <- (1 - pnorm(abs(z), 0, 1)) * 2
print(pval, digit=3)

##           (Intercept)      age female    white     educ  impinc
## 2goodHlth    2.26e-08 4.87e-09  0.935 4.63e-04 2.18e-07 0.00147
## 3fairHlth    0.00e+00 0.00e+00  0.931 5.72e-06 0.00e+00 0.00000
## 4poorHlth    1.55e-15 0.00e+00  0.876 2.02e-02 7.66e-12 0.00000
```

We can also compute odds ratio coefficients, which sometimes (e.g., when the proportion for either level involved in comparisons is small, for example, 10%) approximates relative risk (ratio), as follows,

```
exp(coef(mLogit))

##           (Intercept)      age   female     white      educ    impinc
## 2goodHlth    12.89452 1.013866 1.006043 0.6975227 0.9323760 0.8606509
## 3fairHlth   278.26296 1.035081 1.008484 0.5640552 0.8433095 0.5995401
## 4poorHlth   586.64275 1.051937 1.024751 0.6368601 0.8420557 0.4371904
```

For interpretation, one can state, for example, for each year increase in education, ceteris paribus, we would expect that the odds of having good ($y = 2$) vs excellent ($y = 1$) health will decrease by a factor of $\exp(-0.07) = 0.932$. Similarly, for the same amount of change in education, we would expect the odds of fair ($y = 3$) vs excellent health and the odds of poor ($y = 4$) vs excellent health will decrease by a factor of $\exp(-0.17) = 0.843$ and $\exp(-0.172) = 0.842$ respectively.

One can also calculate predicted probabilities using the `predict.multinom` function from the `nnet` package.

```
whtFem <- data.frame(age = 35, female = 1, white = 1,
                     educ = 16, impinc=mean(mydta$impinc))
pred.whtFem <- predict(mLogit, newdata = whtFem, type="probs")
pred.whtFem
```

```
##   1exltHlth  2goodHlth  3fairHlth  4poorHlth
## 0.43032135 0.46175013 0.09151841 0.01641012
```

Note that the large number of coefficients in the multinomial logit model can make it cumbersome and almost infeasible to provide meaningful interpretation. The primary challenge is to sort out a pattern. Scholars have devised various visualization tools for easy interpretations, for example, the discrete change and odds ratio plots (`mlogitview` command in SPost) by Long (1997). In R, one can graph effect star plots using the `star.nominal` function from the `EffectStars` package (Schauberger, 2019). To read these star plots, using odds ratios for example, if the vertex is close to the circumference, then its corresponding odds ratio coefficient is close to one. If the vertex is outside the circumference and the further it goes, the greater the value is away from one. If the vertex is within the circle and the further away it is from the circumference and thus closer to the center, then the odds ratio coefficient is closer to zero, indicative of a negative effect.

```
library(EffectStars)
star.nominal(hlthFac ~ age + female + white + educ + impinc,
    refLevel = 1, # reference category
    symmetric = F, # TRUE for symmetric side constraints, F for refLevel category
    dist.x = 1.0, # vary dist bw star center on x
    dist.y = 1.0, #  vary dist bw star center on y
    dist.cat = 0.15, # vary dist bw star and category labels
    dist.cov = 1.1,  #vary dist bw stars and covariates labels
    cex.labels = 0.8, # size of covariate label
    cex.cat = 0.8, # size of cat label
    select = 1:6, # select variables
    data=mydta)
```

```
## $odds
##            (Intercept)      age    female      white      educ    impinc
## 1exltHlth      1.00000 1.000000 1.000000 1.0000000 1.0000000 1.0000000
## 2goodHlth     12.89351 1.013864 1.006045 0.6975260 0.9323794 0.8606583
## 3fairHlth    278.24242 1.035079 1.008476 0.5640598 0.8433135 0.5995446
## 4poorHlth    586.76488 1.051935 1.024718 0.6368719 0.8420628 0.4371793
##
## $coefficients
##            (Intercept)        age       female       white       educ      impinc
## 1exltHlth     0.000000 0.00000000 0.000000000  0.0000000  0.0000000  0.0000000
## 2goodHlth     2.556724 0.01376858 0.006026363 -0.3602155 -0.0700155 -0.1500577
## 3fairHlth     5.628493 0.03447778 0.008439812 -0.5725950 -0.1704166 -0.5115849
## 4poorHlth     6.374624 0.05063134 0.024417428 -0.4511868 -0.1719007 -0.8274119
##
## $se
##            (Intercept)        age       female       white       educ      impinc
## 1exltHlth    0.0000000 0.000000000 0.00000000 0.0000000 0.00000000 0.00000000
```

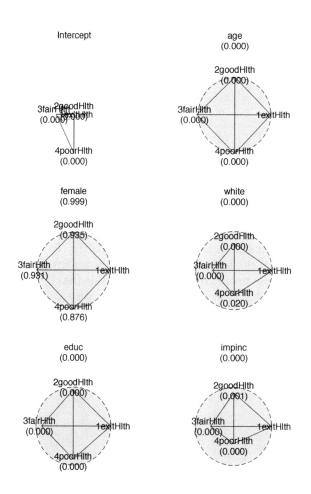

Figure 3.7
Multinomial Logit Effect Star Plot

```
## 2goodHlth    0.4573177 0.002353307 0.07382168 0.1028788 0.01350922 0.04719321
## 3fairHlth    0.5648582 0.002884661 0.09781724 0.1262241 0.01712273 0.05824135
## 4poorHlth    0.7992356 0.004391532 0.15622015 0.1942737 0.02511501 0.08239262
##
## $pvalues
##              (Intercept)         age      female        white         educ
## 1exltHlth            NaN         NaN         NaN          NaN          NaN
## 2goodHlth   2.261609e-08 4.89398e-09   0.9349377 4.628908e-04 2.185859e-07
## 3fairHlth   0.000000e+00 0.00000e+00   0.9312427 5.723983e-06 0.000000e+00
## 4poorHlth   1.554312e-15 0.00000e+00   0.8757955 2.020986e-02 7.672307e-12
##                  impinc
## 1exltHlth           NaN
## 2goodHlth   0.001474554
```

```
## 3fairHlth 0.000000000
## 4poorHlth 0.000000000
##
## $p_rel
##        age    female      white educ impinc
## [1,]     0 0.9988745 5.002436e-05    0      0
##
## $xlim
## [1] 1525.589 5867.649
##
## $ylim
## [1] 1461.045 8197.105
```

The multinomial logit model (along with the conditional multinomial logit model), as formulated in McFadden (1973), heavily relies on the assumption of independence of irrelevant alternatives (IIA), a primitive form of which was introduced by Arrow (1951) as part of the Arrow's Theorem and made widely known by Arrow (1959) in his *Axiom of Choice*. According to McFadden (1973, p. 109), the IIA assumption presupposes that given measured attributes vector $\mathbf{x}$, the odds of choice l over m out of the initial alternative set C remain the same even after a new alternative, k, is introduced,

$$\frac{P(l|\mathbf{x}, C)}{P(m|\mathbf{x}, C)} = \frac{P(l|\mathbf{x}, \{C, k\})}{P(m|\mathbf{x}, \{C, k\})} \tag{3.23}$$

Election is a commonly used example for the IIA assumption. Assuming we have two candidates A and B, and the voters prefer A over B, for IIA to hold, the same results should stay even after we introduce a third candidate D. A strong version of the IIA assumption would require that even the odds of choosing A over B remain the same after new alternatives are introduced. Although several statistics are available for testing the IIA assumption, they are not yet considered to be very effective based on a limited number of studies in this area that are almost exclusively based on choice-set partitioning tests (Fry and Harris, 1998; Cheng and Long, 2007). Among these non-optimal alternatives, the Hausman-McFadden test is probably the most well-received and practiced in empirical research. Based on Hausman and McFadden (1984), the test statistic, can be constructed by the following formula,

$$\text{HM} = \left(\widehat{\beta}_S - \widehat{\beta}_F\right)' \widehat{Q}_g \left(\widehat{\beta}_S - \widehat{\beta}_F\right) \tag{3.24}$$

wherein $\widehat{\beta}_F$ is the estimated parameter vector from running a multinomial logit with the full set of choices and $\widehat{\beta}_S$ is the estimated parameter vector from running a multinomial logit with a subset of alternative choices, and $\widehat{Q}_g$ is the generalized inverse of $\widehat{Q} = \widehat{V}\left(\widehat{\beta}_S\right) - \widehat{V}\left(\widehat{\beta}_F\right)$. It turns out that this test statistic is asymptotically distributed as a χ^2 with its degrees of freedom equal to the rank of Q (the number of elements in $\widehat{\beta}_S$). Note that when the test statistic is small, that would imply that there is statistically insignificant difference between the two sets of parameters, and thus the IIA assumption (i.e., the null hypothesis) holds, or in other words, we fail to reject the assumption; otherwise, the assumption can be rejected. When the subset of choices is properly chosen, $\hat{\beta}_S$ is considered to be consistent under both the null and alternative hypotheses but less efficient under the alternative, while $\hat{\beta}_F$ is consistent under both null and alternative hypotheses, but inefficient under the alternative.

To run the HM test, below we first invoke the `mlogit` package (Croissant, 2020), which uses the (simulated) ML method to estimate random utility discrete choice models described in Train (2009). The results should be identical to those produced by the `multinom` function. Note that in the R code chunk below, the vertical sign, "|", separates the two components of a typical conditional logit model (McFadden, 1973); the part before the "|" sign denotes the covariates that are alternative-specific (i.e., values are specific and supposedly different across at least some alternative choices of the response variable), and the part after lists those that are uniform across all alternative choices.

```
library(mlogit)
data <- mlogit.data(mydta, varying=NULL, choice="hlthFac", shape="wide")
mymlogit <- mlogit(hlthFac ~ 1 | age + educ + impinc + female + white, data=data)
```

Below one can estimate a second multinomial model by restricting the alternative choices to be, for example, from having excellent to fair health, excluding poor health,

```
mlogitSub <- mlogit(hlthFac ~ 1 | age + educ + impinc + female + white,
                    alt.subset = c("1exltHlth", "2goodHlth", "3fairHlth"),
                    data=data)
coef(mlogitSub)
```

```
## (Intercept):2goodHlth (Intercept):3fairHlth          age:2goodHlth
##            2.553658588            5.637337997            0.013628420
##            age:3fairHlth          educ:2goodHlth          educ:3fairHlth
##            0.034228845           -0.070136354           -0.172998690
##         impinc:2goodHlth       impinc:3fairHlth        female:2goodHlth
##           -0.149112221           -0.507063541            0.005900171
##         female:3fairHlth        white:2goodHlth         white:3fairHlth
##            0.010887899           -0.358928782           -0.589582344
```

Then we can use the `hmftest` function from the `mlogit` package to conduct the Hausman and McFadden test of the IIA assumption,

```
hmftest(mymlogit,mlogitSub)
```

```
##
##  Hausman-McFadden test
##
## data:  data
## chisq = 0.66644, df = 12, p-value = 1
## alternative hypothesis: IIA is rejected
```

The results show that the test statistic is statistically significant, and we can safely reject the the null hypothesis and accordingly the IIA assumption. Like how the PL assumption is discussed in previous sections, the IIA assumption is a rather stringent and probably unrealistic assumption that requires all alternative choices are conceptually distinct enough. In the HM test of the IIA assumption, the construction of the test statistic requires that parameters from the full-set multinomial logit and their counterparts from a subset multinomial logit (can be a binary logit) are simultaneously equal to each other; such likelihood is low, especially when the number of alternative choices or covariates increases. Without turning to any reliable and valid statistical testing, one can reasonably surmise that the IIA assumption would not hold in many cases. Note that ideally, the dissimilarity among alternative choices has to be both distinctive and equidistant, the second of which rarely exists in practice. So why should we hang on to this assumption? In our view, it is normative for polytomous response variables to violate the IIA assumption, and the adherence to it is more of an exception than the rule. Like the PL assumption, the IIA assumption is an important concept to know, but the violation of this assumption does not necessarily lead to an overhaul, or even a minor modification, of the model. As of the time when this text is written, we are not aware of any choice-set partitioning test that can effectively test the IIA assumption. So we offer two tentative solutions. First, if the polytomous response variable has reasonably distinctive alternative choices and the multinomial logit model provides sufficient predictive power, then one can safely use the model without modification; second, one can turn to the multinomial probit model, which is computationally expensive.

3.3.2 Multinomial Probit Regression

When $\varepsilon \sim \mathcal{N}(0, \Sigma)$ in 3.21 and everything else stays the same, we have the multinomial probit (MNP) model that allows for non-zero correlations among error terms and thereby the violation of IIA. One can use functions from several packages to estimate the MNP model, for example the `mlogit` function from the `mlogit` package by setting the `probit` argument to true. But the downside about using this function is that it can take very long time to converge even for reasonably-sized model, such as the following.

```
mymprobit <- mlogit(hlthFac <- as.factor(mydta$health) ~ 1 |
    age + educ + impinc + female + white, data = data, probit = TRUE)
```

A good candidate for running a multinomial probit model is the `mnp` function from the MNP package (Imai and van Dyk, 2021), which can be used to estimate both alternative-specific (conditional multinomial probit) and the usual multinomial probit model. The `mnp` function uses MCMC for fitting the model and computing associated model fit statistics. The `n.draw` decides the number of MCMC draws.

```
library(MNP)
mnProbit <- mnp(health ~ age + female + white + educ + impinc,
    data = mydta, n.draws = 10000, verbose = TRUE)
summary(mnProbit)

Call: mnp(formula = health ~ age + female + white + educ + impinc,
data = mydta, n.draws = 10000, verbose = TRUE)
Coefficients:
```

	mean	std.dev.	2.5%	97.5%
(Intercept):2gooddHlth	1.151807	0.443698	0.464538	2.070
(Intercept):3fairHlth	1.951769	0.396737	1.185151	2.720
(Intercept):4poorHlth	1.994800	0.554738	0.876537	3.004
age:2gooddHlth	0.003214	0.001337	0.001078	0.006
age:3fairHlth	0.012968	0.002137	0.009080	0.017
age:4poorHlth	0.020974	0.003262	0.014806	0.027
female:2gooddHlth	0.024015	0.028585	-0.024404	0.093
female:3fairHlth	0.028886	0.051739	-0.076043	0.132
female:4poorHlth	0.061198	0.074915	-0.086139	0.207
white:2gooddHlth	-0.007870	0.036169	-0.078132	0.075
white:3fairHlth	-0.150001	0.072250	-0.299908	-0.016
white:4poorHlth	-0.135502	0.095207	-0.323676	0.049
educ:2gooddHlth	-0.032040	0.012616	-0.058756	-0.013
educ:3fairHlth	-0.082661	0.012733	-0.107254	-0.057
educ:4poorHlth	-0.067355	0.016440	-0.098427	-0.035
impinc:2gooddHlth	-0.070374	0.030113	-0.139171	-0.025

```
impinc:3fairHlth          -0.199322  0.034356 -0.267521 -0.135
impinc:4poorHlth          -0.349577  0.053602 -0.456774 -0.248

Covariances:
                          mean std.dev.       2.5% 97.5%
2gooddHlth:2gooddHlth   0.38870  0.38595  0.03916 1.575
2gooddHlth:3fairHlth    0.33924  0.19498 -0.04018 0.765
2gooddHlth:4poorHlth    0.29670  0.23825 -0.11184 0.841
3fairHlth:3fairHlth     1.31732  0.45963  0.58599 2.125
3fairHlth:4poorHlth     0.11888  0.39385 -0.70816 0.851
4poorHlth:4poorHlth     1.29398  0.42727  0.58774 2.265

Base category: 1exltHlth
Number of alternatives: 4
Number of observations: 4260
Number of estimated parameters: 23
Number of stored MCMC draws: 10000
```

With these raw coefficients, one can roughly know the direction of the associations between covariates and choice comparisons. One can also take a step further to make prediction as follows, by using the `predict.mnp` function from the MNP package.

```
mydtaEg <- mydta[689, ]
predict(mnProbit, newdata = mydtaEg, type="prob")$p
     1exltHlth 2gooddHlth 3fairHlth 4poorHlth
[1,]    0.2488      0.5396     0.1882     0.0234
```

It can be shown that for the randomly selected case 689 from the `mydta` data, the predicted probabilities for having excellent, good, fair, and poor health are 0.249, 0.540, 0.188, and 0.023 respectively, given corresponding covariates' values (which are not shown here). Note that the new data frame for prediction after using the `mnp` function has to have the same set of variables as those in the estimation data.

3.4   Bayesian Polytomous Regression

3.4.1   Bayesian Estimation

As discussed in the first two chapters, the goal of Bayesian estimation is to derive the posterior distribution of the parameters of interest, and the posterior is proportional to the product of the likelihood and priors. One can use conjugate priors to analytically derive posteriors. But given the dramatic improvement in the computing power nowadays, the use of the MCMC methods become accessible and convenient. Several R packages, including arm (using approximate EM algorithm) (Gelman and Su, 2020a), bayesm (Rossi, 2019), brms (Bürkner, 2017), MCMCpack (Martin et al., 2011), nimble (de Valpine et al., 2017), rstanarm (Goodrich et al., 2018), and UPG (Zens et al., 2020) are available for estimating a large variety of polytomous response variable models.

3.4.1.1 Bayesian Parallel Cumulative Ordered Regression

The classical PL model can be relatively easily estimated with Bayesian methods. One can use, for example, the `stan_polr` function from the `rstanarm` package to estimate the model. The first few arguments of the `stan_polr` function, including `formula` and `data`, are almost identical to those of other commonly used estimation functions, such as `glm` or `lm`. The `prior` argument is to specify the priors for parameters by users, who usually have two options; one is to specify a beta prior for R^2, or the proportion of variance in the response variable explained by the predictors; if the `prior` argument is omitted, then uniform priors are used for coefficients.[6]

To ensure reproducibility, we set the seed to 47306, the seed number used throughout this text. The number of MCMC chains is set to 4 by default. we also assign the Bayesian estimation results to an object named `bayesPO`, and then issue the `summary` function to print a summary of the results.

```
library(rstanarm)
bayesPO <- stan_polr(as.ordered(health) ~ age + female + white + educ + impinc,
 data = mydta, prior = R2(0.2), seed = 47306)
summary(bayesPO, probs = c(0.1, 0.5, 0.9), digits=3)
```

After the code chunk is executed, we get the following results with four sections, including model information, estimates, fit diagnostics, and MCMC diagnostics. Note that the `mean`, `sd`, `10%`, `50%`, and `90%` columns correspond to the means, standard deviations, 10 percentiles, 50 percentiles (median), and 90 percentiles of the parameters from the simulated posterior distribution. With the 10- and 90-percentile columns, we can construct the 80% credible intervals of the parameters. One can also set the `probs` arguments of the summary function in the preceding R code chunk to be `c(0.025, 0.5, 0.975)` so that 95% credible intervals can be obtained.

```
Model Info:
function:     stan_polr
family:       ordered [logistic]
formula:      as.ordered(health) ~ age + female + white + educ + impinc
algorithm:    sampling
sample:       4000 (posterior sample size)
priors:       see help('prior_summary')
observations: 4260

Estimates:
          mean    sd     10%     50%     90%
age       0.021  0.002   0.019   0.021   0.023
female    0.049  0.058  -0.026   0.050   0.122
white    -0.181  0.078  -0.280  -0.180  -0.082
educ     -0.116  0.010  -0.129  -0.116  -0.103
impinc   -0.342  0.034  -0.386  -0.342  -0.298
1|2      -4.965  0.336  -5.394  -4.965  -4.532
2|3      -2.750  0.331  -3.175  -2.749  -2.320
3|4      -0.865  0.332  -1.287  -0.864  -0.434
```

---

[6]For more details, please see `https://mc-stan.org/rstanarm/articles/polr.html` and `https://rdrr.io/cran/rstanarm/man/stan_polr.html`

```
Fit Diagnostics:
              mean   sd     10%    50%    90%
mean_PPD:1  0.305  0.010  0.293  0.305  0.317
mean_PPD:2  0.453  0.011  0.439  0.454  0.467
mean_PPD:3  0.188  0.008  0.177  0.188  0.199
mean_PPD:4  0.054  0.005  0.048  0.054  0.060
```

The mean_ppd is the sample average posterior predictive
distribution of the outcome variable (for details see
help('summary.stanreg')).

```
MCMC diagnostics
                  mcse   Rhat    n_eff
age             0.000  0.999   3622
female          0.001  1.000   4357
white           0.001  0.999   5038
educ            0.000  1.000   3718
impinc          0.001  0.999   3950
1|2             0.006  0.999   3362
2|3             0.006  0.999   3194
3|4             0.006  0.999   3118
mean_PPD:1      0.000  1.000   3489
mean_PPD:2      0.000  1.000   3483
mean_PPD:3      0.000  1.000   3327
mean_PPD:4      0.000  1.000   3364
log-posterior   0.084  1.002    853
```

For each parameter, mcse is Monte Carlo standard error,
n_eff is a crude measure of effective sample size, and
Rhat is the potential scale reduction factor on
split chains (at convergence Rhat=1).

Under the section of MCMC diagnostics, the **Rhat** column presents the potential scale reduction factor (PSRF) statistic, $\hat{R}$, which measures the ratio of the sum of the average within-chain and between-chain variances over the average within-chain variance of simulated samples (Gelman et al., 2014, pp. 284-285). If this statistic is close to one, then that should provide strong evidence for convergence. It can be shown from the results that all estimates of the PSRF statistic have an absolute difference from one to be smaller than one hundredth (usually smaller than 1.1), so we can be confident that the chain convergence is achieved.

3.4.1.2   Bayesian Non-Parallel Cumulative Ordered Regression

When we go beyond commonly used polytomous regression models, rarely does any R package have custom-made function meeting such estimation needs. Consequently, one may have to turn to low-level utility functions. Below we use **Stan** codes to illustrate how to estimate a Bayesian non-parallel (NP) cumulative ordered logit model, or what is sometimes called the generalized ordered logit model. The first code block, **data**, synchronizes the data information (e.g., response variable and predictor matrix names, response levels, number of cases, and number of predictors) between the empirical data to be processed and the Stan environment; the **parameters** code block declares the parameters to be estimated.

```
data {
int<lower=2> K; // response level
int<lower=0> N; // number of cases
int<lower=1> D; // number of predictors
int<lower=0, upper=K> y[N]; // response variable
row_vector[D] x[N];  // predictors
}


parameters {
vector[D] b1; // parameters in cut-point eq. 1
vector[D] b2; // parameters in cut-point eq. 2
vector[D] b3; // parameters in cut-point eq. 3
vector[K-1] thresh; // ordered[K-1] thresh
}
```

The following `model` code block supplies priors and sets up the likelihood function. Since the NP cumulative logit model assumes that the slopes are free to vary across cut-point equations, we have three different slopes vectors, b1, b2, and b3, and weakly informative priors are used for all parameters. `eta`'s are linear predictors and `prob` is the vector containing the four probabilities corresponding to $y = 1, 2, 3, 4$ respectively.

```
model {
vector[K] prob;  // predicted probabilities
vector[K-1] eta; // linear predictors
for (d in 1:D) { // assign priors
b1[d] ~ normal(0, 10);
b2[d] ~ normal(0, 10);
b3[d] ~ normal(0, 10);
}
for ( k in 1:K-1)
   thresh[k] ~ normal( k+0.5, 10);
for (n in 1:N) { // set up the likelihood
eta[1] <- x[n]*b1;
eta[2] <- x[n]*b2;
eta[3] <- x[n]*b3;
prob[1] <- inv_logit(thresh[1] - eta[1]); // prob y = 1
for (k in 2:(K-1)) { // prob y = 2...K-1
// fmax(real x, real y): step-wise function returning the maximum
prob[k] <- fmax(inv_logit(thresh[k] - eta[k])
                    - inv_logit(thresh[k-1] - eta[k-1]), log(1));
}
prob[K] <- 1 - inv_logit(thresh[K-1] - eta[K-1]); // prob y = K
y[n] ~ categorical(prob);
}
}
```

The last code block `generated quantities` is to calculate quantities of interest using the posterior distribution. Generated quantities can include statistics such as predicted probabilities, rates, differences in probabilities and rates, and log likelihoods, the last of which can be used for computing Bayesian model fit statistics such as WAIC (widely applicable information criterion) and LOO (leave-one-out) using the `waic` and `loo` functions from the `loo` package (Vehtari et al., 2017).

```
generated quantities {
vector[K] probG;
vector[K-1] etaG;
vector[N] log_lik;
for (n in 1:N) {
etaG[1] <- x[n]*b1;
etaG[2] <- x[n]*b2;
etaG[3] <- x[n]*b3;
probG[1] <- inv_logit(thresh[1] - etaG[1]);
for (k in 2:(K-1)) {
probG[k] <- fmax(inv_logit(thresh[k] - etaG[k])
                         - inv_logit(thresh[k-1] - etaG[k-1]), log(1));
}
probG[K] <- 1 - inv_logit(thresh[K-1] - etaG[K-1]);
log_lik[n] <- categorical_log(y[n], probG);
}
}
```

Quantitative researchers can conduct the usual Bayesian analysis of the posterior distribution, for example, Bayesian statistical inferences and predictions. In Fig. 3.8, we use the posterior to graph the predicted probabilities of the four response levels of SRH and their corresponding credible intervals for different values of education. While computing the predictions, we hold all other predictors, except for education, to their sample means.

The pattern presented in this graph is very similar to that in Fig. 3.4 that as education increases, the predicted probabilities of having fair and poor health decrease, whereas the predicted prbabilities of having excellent health rise up. The predicted probabilities of having good health have a curvalinear relationship with education, peaking around 12 years and tapering off on both ends.

3.4.1.3 Bayesian Stereotype Logit Model

The stereotype logit model straddles somewhere between ordered and multinomial regression models. It is only when the common scale factors in the model satisfy the inequality and the value constraint condition that $0 = \phi_1 \leq \phi_2... \leq \phi_L = 1$, a stereotype logit model is an ordered regression model and the response variable y can be viewed as stochastically ordinal. Note that here we use $y = 1$ as the baseline category, so the inequality and value constraint condition specified in the previous section of the frequentist ML estimation of the stereotype logit is reversed to reflect this change. Commands or functions that are programmed with the frequentist approach and ML estimation can hardly incorporate that condition. Bayesian estimation nonetheless can approach this challenge with remarkable ease. For the inequality and value constraint condition, one needs to re-parameterize the common scale factors by having $\gamma_l = \phi_{l+1} - \phi_l$ (Ahn et al., 2009, p. 3145). Here if γ's are assumed to follow a Dirichlet(1, . . . ,1) distribution, then the inequality and value constraint condition is met. The Dirichlet distribution is a multivariate generalization of the beta distribution that for a vector of K positive real random variables $x_1, ..., x_k, ..., x_K$, where $x_k \in (0, 1)$, we have $\sum_{k=1}^{K} x_k = 1$. Just like the beta distribution is a conjugate of binomial, the Dirichlet distribution is the conjugate of categorical or multinomial distribution. When the Dirichlet distribution is assigned to have its concentration parameter $\alpha_k = 1$, one holds the prior belief that the scores are uniformly distributed (i.e., equal spacing of scores). Note that in the **data** block, we add a line for **alpha** (α), which is assigned a real vector of (1, 1, 1) outside this Stan syntax file. In the **parameters** block, we assign **gamma** (γ) to

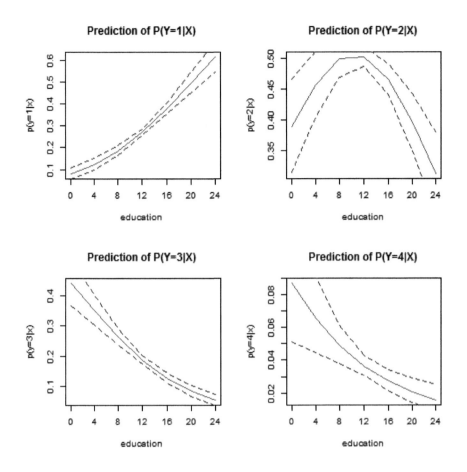

Figure 3.8
Credible Intervals of Predicted Probability

be a simplex such that $\sum_{l=1}^{L-1} \gamma_l = 1$, where L is the total number of response levels. In the `transformed parameters` block, we set up the equality equations between γ and ϕ as described previously. Then in the `model` block, we include the prior for gamma (γ) as $\mathrm{Dir}(\alpha)$, in which $\alpha_1 = \alpha_2 = \alpha_3 = 1$.

```
data {
int<lower=2> L; // number of response levels
int<lower=1> N; // number of observations
int<lower=1> K; // number of predictors
int<lower=1, upper=L> y[N]; // y variables
row_vector[K] x[N]; // x vectors
vector<lower=0>[L-1] alpha;
}

parameters {
vector[K] b;
vector[L-1] theta;
```

```
simplex[L-1] gamma;
}
transformed parameters{
vector<lower=0, upper=1>[L] phi;
phi[1] <- 0;
phi[2] <- gamma[1];
phi[3] <- gamma[1] + gamma[2];
phi[4] <- gamma[1] + gamma[2] + gamma[3];
}

model {
vector[L] prob;
vector[L] emu;
// specify priors
gamma ~ dirichlet(alpha);
for (k in 1:K)
b[k] ~ normal(0, 100);
for (l  in 1:L-1)
theta[l] ~ normal(0, 100);
for (n in 1:N) {
emu[1] <- exp(phi[1]*x[n]*b);
for (l in 2:L)
emu[l] <- exp(theta[l-1] + phi[l]*x[n]*b);
for (l in 1:L)
prob[l] <- emu[l]/sum(emu);
y[n] ~ categorical(prob);
}
}
```

Once the ϕ's are set up in the **transformed parameters** code block, one can use the likelihood function based on Eq. 3.13 to set up the Bayesian model for MCMC simulation in the **model** code block.

4

<hr>

Count Regression

<hr>

> Those seasoned with steering leave no
> impression, those versed in edification
> commits no misdeed, and those adept
> at counting use no tally sticks.

<hr>

in Chapter 27 of *Dao De Jing* by Laozi

<hr>

4.1 Poisson Distribution

Like many other scientific discoveries, there was an anteriority dispute over who discovered the Poisson distribution. French mathematician Siméon Denis Poisson worked out the distribution as an approximation to the negative binomial cumulative distribution to examine the number of wrongful convictions in his *Recherches sur la Probabilité des Jugements en Matière Criminelle et en Matiére Civile, Précédées des Règles Générales du Calcul des Probabilités* (*Research on the Probability of Judgments in Criminal and Civil Matters: Preceded by the General Rules of the Calculation of Probabilities*), published in 1837 (Hald, 2003, pp. 216-217). Although the name of the distribution clearly credits its discovery to Poisson, some argue that because most of the important results were already obtained by Abraham de Moivre in 1711 (Johnson et al., 2005), the Poisson distribution probably presents itself as another example of Stilger's law of eponymy—"No scientific discovery is named after its original discoverer" (Stigler, 1982, p. 33). Setting aside all these interesting historical accounts and disputes and based on classical theories in probability and statistics, the Poisson distribution is a discrete probability distribution to describe random variables for counting the number of events of the same (or similar) nature occurring in a specified time frame, and the process of such event recurring is assumed to be memoryless; that is, the probability of the event recurring is not affected by its previous occurrences. So for a random variable, $Y = 0, 1, 2, ...$, that follows a Poisson distribution, the probability of having y counts of such event occurring is

$$P(Y = y) = \exp(-\mu) \frac{\mu^y}{y!} \tag{4.1}$$

where y and $\mu > 0$ correspond to the realized and expected event count per time frame, respectively. The Poisson distribution is closely related to the binomial distribution, which is the sum of a series of independent, identically distributed Bernoulli random variables. If we have a binomial random variable, Y, with parameters n, the total number of trials and p, the probability of success for each identical trial, and let $\mu = np$, then

DOI: 10.1201/9780429056468-4

Table 4.1

A Frequency Table of Bortkiewicz's Prussian Horse-Kick Data

Death Count	Observed Frequency	Expected Frequency
0	109	108.670
1	65	66.289
2	22	20.218
3	3	4.111
4	1	0.627

$$P(Y = y) = \frac{n!}{(n-y)!y!}p^y(1-p)^{n-y}$$
$$= \frac{n!}{(n-y)!y!}\left(\frac{\mu}{n}\right)^y\left(1-\frac{u}{n}\right)^{n-y} \quad (4.2)$$
$$= \frac{n(n-1)...(n-y+1)}{n^y}\frac{\mu^y}{y!}\frac{(1-\frac{\mu}{n})^n}{(1-\frac{\mu}{n})^y}$$

When $n \to \infty$ and μ remains small, $\frac{n(n-1)...(n-y+1)}{n^y} \to 1$, $(1-\frac{\mu}{n})^y \to 1$, and $(1-\frac{\mu}{n})^n \to \exp(-\mu)$ (Ross, 1998). Thus we have Eq. 4.1. Neither de Moivre nor Poisson made substantial efforts to apply the Poisson distribution to empirical data. It is not until 1898 when Ladislaus von Bortkiewicz first described how the Poisson distribution can be used in various settings in his *Das Gesetz der kleinen Zahlen* [The Law of. Small Numbers], that this distribution became well known within the scientific community and without (von Bortkiewicz, 1898). In this book, von Bortkiewicz used the Poisson distribution to fit the number of Prussian cavalry soldiers dying of horse-kicking in 15 corpse spanning over 20 years from 1875 to 1894, with a total of 300 observations. The data were then trimmed by R. A. Fisher, due to some inconsistency in the data collection process. The final subset used in most statistical texts usually contains data from 10 corps, totaling 200 observations. Table 4.1 is a frequency distribution table of the horse-kick data reconstructed from the raw tallies by corps and year.

Below we use R to read in the horse-kick data refined by Fisher, and then compare the observed counts and proportions with their expected counterparts calculated from a theoretical Poisson distribution using the observed sample mean as the parameter estimate.

```
load("data/hkFisher.Rdata")
summary(fisher$deaths, digits=3)

##    Min. 1st Qu.  Median    Mean 3rd Qu.    Max.
##    0.00    0.00    0.00    0.61    1.00    4.00

xtabs(~ deaths, data = fisher) # freq dist table

## deaths
##   0   1   2   3   4
## 109  65  22   3   1

N = length(fisher$deaths) # total observation
print(mu <- mean(fisher$deaths)) # observed sample mean
```

```
## [1] 0.61

freq = table(fisher$deaths)
obs.prob = freq / NROW(fisher$deaths) # observed relative freq
obs.count = obs.prob*N # observed count
# create a function to calculate P(Y=y) for Y~Pois
poisp = function (y, u) {p=exp(-u)*(u^y)/factorial(y); return(p)}
exp.prob <- vector("numeric", 5)
# collect results for P(Y=0-4)
for (i in 0:4) {
        this = poisp(i, mu)
        k = i + 1
        exp.prob[k] = this }
exp.count = exp.prob*N
round(cbind(freq, exp.count, obs.prob, exp.prob), 3)      # tabulate

##    freq exp.count obs.prob exp.prob
## 0  109    108.670    0.545    0.543
## 1   65     66.289    0.325    0.331
## 2   22     20.218    0.110    0.101
## 3    3      4.111    0.015    0.021
## 4    1      0.627    0.005    0.003
```

Based on our calculation, the average death number per corps per year is 0.61. The first to fourth *named* columns of the frequency distribution table at the end of the results section correspond to the observed and expected counts, and observed and expected proportions for each death number category (the first, unnamed column) respectively. It can be shown from the results that the Poisson distribution with an average of 0.61 death count per corps per year fits the empirical data quite well since the differences between the observed (the first column) and expected counts (the second column) are relatively small. We can take a step further to test if the Poisson distribution fits the data well in a statistical sense. Below we use the `chisq.test` function to test the difference,

```
chisq.test(x = freq, p=exp.prob, rescale.p = TRUE)

##
##  Chi-squared test for given probabilities
##
## data:  freq
## X-squared = 0.70503, df = 4, p-value = 0.9507
```

And the results indicate that the hypothesis that there is no statistically significant difference between the observed and expected counts is not rejected with a p-value way above .05; thus we can safely infer that the data probably follow a Poisson distribution.

4.2 Basic Count Regression Models

The previous section illustrates how the count data can be fit using a single parameter Poisson distribution. In empirical studies, analysts usually introduce covariates and explore how these covariates are associated with distributional parameters and accordingly the count responses; hence we turn to the classical regression framework. There are a number of count regression models that can be used to fit count data, presumably generated through slightly different processes. The most commonly used ones include the Poisson and the negative binomial regression models.

4.2.1 Explore the Count Response Variable

Before directly plunging into multivariate regressions, one is usually advised to take a close look at the descriptive statistics of the count response variable of interest. Below we use data adapted from the cumulative General Social Survey 1972–2012 to examine one response variable, physlth (Smith et al., 2019). This variable, an approximate count response, records the number of days of having poor physical health over the past 30 days. Strictly speaking, this count response variable is an upper-truncated since its upper limit is 30.

```
# load libraries
require(foreign)
require(MASS)
# read in recoded cumulative GSS data
readin <- read.dta("data/gssCum7212Teach.dta", convert.factor=F)
# create a list of variable names used for variable selection afterwards
usevar <- c("physlth", "age", "educ", "impinc", "female")
# subset the data (select variables)
mydta <- subset(readin[complete.cases(readin[usevar]),],
        select=c(usevar))
# summary(mydta)
# calculate mean and variance
hlthM = mean(mydta$physlth)
hlthV = var(mydta$physlth)
# as.matrix(table(mydta$physlth, useNA=c("ifany")))
myTab <- transform(table(mydta$physlth, useNA=c("ifany")))
colnames(myTab) <- c("physlth", "freq")
myTab
```

```
##    physlth freq
## 1        0  279
## 2        1   43
## 3        2   33
## 4        3   24
## 5        4    9
## 6        5   14
## 7        6    6
## 8        7   10
## 9        8    2
## 10      10    9
## 11      11    1
## 12      14    4
## 13      15    6
## 14      19    1
## 15      20    8
## 16      21    1
```

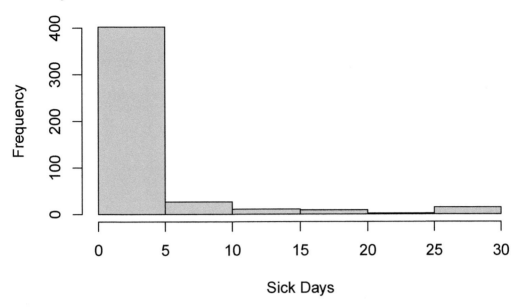

Figure 4.1
Histogram of Sick Days

```
## 17        24      1
## 18        30     15
```

```
hist(mydta$physhlth, main="", xlab="Sick Days")
```

It can be shown from the frequency distribution table above (`myTab`) that there is a total of 17 count categories, from zero to 30, with zero greatly outnumbering other categories. Fig. 4.1, a histogram graph of the raw counts, also clearly shows that there are excess zeros. Thus it is not surprising that the mean (2.854) is much smaller than the variance (41.196) of `physhlth`, a common phenomenon among count response variables. This is usually a red flag that the Poisson distribution or regression might not be the best tool for analyzing the count data at hand.

4.2.2 Plot Observed vs. Predicted Count Proportions

A more careful descriptive analysis of the count response data involves plotting the observed vs. predicted count proportions to see if there is any significant discrepancy between the two sets. Below, we use `count`, `obsProb`, and `prdProb` to create matrices containing observed count categories, proportions (relative frequencies), and the predicted proportions (probabilities) for observed count categories respectively. While computing the predicted proportions, we use the sample mean for `physhlth` and the probability mass function from the Poisson distribution in Eq. 4.1.

```
# compute the number of observed count categories
nrow(mytab <- as.matrix(table(mydta$physhlth, useNA="ifany")))
```

```
## [1] 18
```

```
# compute observed proportion
```

```
obsProp <- mytab/sum(mytab)
# create an empty column vector with dim nrow and 1
count <- matrix(, ncol = 1, nrow=length(mytab))
# create an empty column vector with dim nrow and 1
prdProp<- matrix(, ncol = 1, nrow=length(mytab))
# loop to plug in values for the count and prdProb column matrices
for (i in 1:length(mytab)) {
        # extract count values
        crtcnt <- as.numeric(rownames(mytab)[i])
        # compute expected count
        p <- dpois(crtcnt, mean(mydta$physhlth))
        count[i,1] <- crtcnt
        prdProp[i,1] <- p
}
plot(count, obsProp, type="o")
lines(count, prdProp, type="o", lty=2, pch=22)
legend("topright", c("observed", "predicted"), pch=21:22)
```

It can be shown from Fig. 4.2 that the predicted proportions are quite different from the observed ones for counts below seven and especially zero. Although we could improve the predictive power by adding covariates, having such excess zeros poses enormous challenges to simply using the Poisson regression model without additional consideration. After we cover Poisson regression, we will revisit this issue and discuss remedies.

4.2.3 Poisson Regression

For a discrete random variable $Y = 0, 1, 2, \ldots$ to follow a Poisson distribution, we have its probability mass function (PMF) as $P(Y = y) = \exp(-\mu) \frac{\mu^y}{y!}$, where y is any possible count,

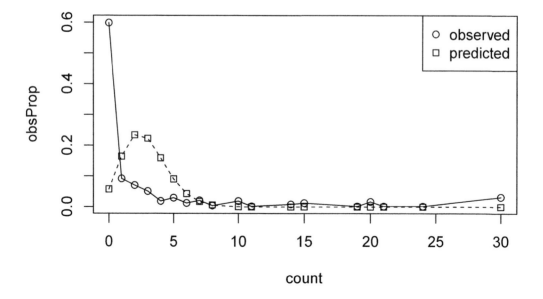

Figure 4.2
Predicted vs. Observed Count Proportions for Sick Days

presumably a non-negative integer, and $E(Y) = \mu$ is the only parameter. In addition to its memorylessness aforementioned, the Poisson distribution is characterized by the property of equi-dispersion; that is, its mean is equivalent of its variance, or $E(Y) = V(Y) = \mu$. This assumption, however, is often violated because of a considerable amount of heterogeneity often concealed in empirical data; thus, the Poisson distribution and regression is commonly used as a baseline in empirical analyses. The main idea about Poisson regression centers around modeling the only parameter μ. Since μ has to be a non-negative number, a natural option is to parameterize $\mu = \exp(\mathbf{x}\beta)$. Then we can combine the few terms aforementioned to form the likelihood function for the Poisson regression,

$$L = \prod \exp\left(-\exp\left(\mathbf{x}\beta\right)\right)\left(\exp\left(\mathbf{x}\beta\right)\right)^y / y! \qquad (4.3)$$

Constant terms from the data, such as y, can be dropped from the likelihood function to simplify the computation. And the estimation can be easily done through MLE, as discussed in Chapter 2. There are a variety of functions from several R packages to estimate count regression models, including the `glm` function from the `stats` package (R Core Team, 2020) to estimate Poisson regression models, the `glm.nb` function from the `MASS` package (Venables and Ripley, 2002) to estimate negative binomial regression models, and a few functions from the `pscl` and `countreg` packages to estimate hurdle and zero-modified models (Zeileis et al., 2008). Below, we use the `glm` function to run a Poisson regression with `physhlth` as the count response variable, and `age`, `education`, `income`, and `female` as covariates. Since `glm` was written to estimate a wide variety of generalized linear models, we need to set the `family` option to `poisson` for a Poisson regression.

```
summary(mypoisson <- glm(physhlth ~ age + educ + impinc + female,
    family = "poisson", data = mydta))

##
## Call:
## glm(formula = physhlth ~ age + educ + impinc + female, family = "poisson",
##     data = mydta)
##
## Deviance Residuals:
##     Min       1Q   Median       3Q      Max
## -4.2204  -2.3559  -1.8861  -0.2083  10.1671
##
## Coefficients:
##             Estimate Std. Error z value Pr(>|z|)
## (Intercept)  2.876241   0.254223  11.314  < 2e-16 ***
## age          0.012141   0.001989   6.104 1.03e-09 ***
## educ        -0.044085   0.009435  -4.673 2.97e-06 ***
## impinc      -0.202019   0.025531  -7.913 2.52e-15 ***
## female       0.429398   0.059086   7.267 3.67e-13 ***
## ---
## Signif. codes:  0 '***' 0.001 '**' 0.01 '*' 0.05 '.' 0.1 ' ' 1
##
## (Dispersion parameter for poisson family taken to be 1)
##
##     Null deviance: 3915.4  on 465  degrees of freedom
## Residual deviance: 3699.4  on 461  degrees of freedom
## AIC: 4318.5
##
```

```
## Number of Fisher Scoring iterations: 6
```

The results show that all four covariates are associated with the number of sick days at the conventional significance level. Having the raw results, including raw coefficients and their associated precision estimates, can help analysts unravel the overall relationships between the response variable and its covariates. But applied statisticians, especially those in the behavioral, health, and social science disciplines, may want to present such associations with a substantively meaningful way. There are several types of post-estimation analyses we can do, as described in Chapter 2. First, we can focus on the expected rate/count (we use count, frequency, and rate interchangeably henceforth). Because the expected count $\mu = \exp(\mathbf{x}\beta)$, it can be easily obtained that for each unit increase in one generic covariate, x_k, we have $\frac{\mu_{\mathbf{x},x_{k+1}}}{\mu_{\mathbf{x},x_k}} = \frac{\exp(\beta_0+\beta_1 x_1+...\beta_k(x_k+1)+...+\beta_K x_K)}{\exp(\beta_0+\beta_1 x_1+...\beta_k x_k+...+\beta_K x_K)} = \exp(\beta_k)$, which is sometimes called the factor change coefficient. This coefficient can be directly used for the interpretation of β_k that for each unit increase in x_k, ceteris paribus, we would have the expected count to change by a factor of $\exp(\beta_k)$. Depending on whether $\exp(\beta_k)$ is greater or smaller than one, we can then associate it with increase or decrease. Below, we exponentiate the coefficients vector obtained after running the glm function and compute the factor change coefficients. If data analysts are interested in standardized effects, we also illustrate how to use R codes to calculate x-standardized coefficients by multiplying the coefficients with their corresponding sample standard deviations.

```
# count the total number of covariates
varnum = dim(as.matrix(mypoisson$coefficients))[1]
# extract the covariates columns
x <- mypoisson$model[2:length(mypoisson$model)]
# use lapply to compute standard dev for all covariates
xsd <- as.numeric(lapply(x, sd, na.rm=T))
start = 2
# extract coefficients, excluding the intercept
bcoef <- as.matrix(mypoisson$coefficients[start:varnum])
xstand <- bcoef*xsd
# the exponentiated coefficients
exp(bcoef)

##             [,1]
## age    1.0122152
## educ   0.9568724
## impinc 0.8170795
## female 1.5363331

exp(xstand)

##             [,1]
## age    1.1731209
## educ   0.8779468
## impinc 0.8190874
## female 1.2383585
```

To interpret the effect of education, for example, we can state that for each year increase in education, we would expect the number of sick days will decrease by a factor of 0.957, holding all other covariates constant; for the gender effect, we can state that being female as opposed to male increases sick days by a factor of 1.536, net of the effects of other covariates. Because the exponential function is a monotonic increasing function, the confidence interval of $\widehat{\beta}$ can be readily applied to the the factor change coefficient. Here we can use the `confint` function from the base `stats` package to compute the 95% confidence interval for both raw and exponentiated coefficients

```
confint(mypoisson)

##                     2.5 %        97.5 %
## (Intercept)   2.372557745    3.36925752
## age           0.008225144    0.01602292
## educ         -0.062427977   -0.02544954
## impinc       -0.251296942   -0.15120761
## female        0.314272495    0.54596051

exp(confint(mypoisson))

##                    2.5 %        97.5 %
## (Intercept)   10.7247885   29.0569449
## age            1.0082591    1.0161520
## educ           0.9394807    0.9748716
## impinc         0.7777914    0.8596692
## female         1.3692628    1.7262657
```

We can also use the `zelig` function from the `Zelig` package to estimate the same model and calculate the expected rate as well as its simulated confidence interval, given a vector of covariates' values (Imai et al., 2008; Choirat et al., 2018). To estimate the same Poisson regression model, we need to set the model type option of the `zelig` function to be Poisson (`model = ''poisson''`), and we should get the exact same results as we did with the `glm` function.

```
require(Zelig)
z.out <- zelig(physhlth ~ age + educ + impinc + female,
        model = "poisson", data = mydta)
```

As discussed in previous chapters, the `setx` function sets covariates to some observed or hypothetical values to be used by the following `sim` function; the `sim` function then uses model estimates from `z.out` (i.e., estimates) and values of covariates specified in the `setx` function to run simulations and get the expected/predicted rate with its precision estimates (i.e., standard errors and confidence intervals).

```
x.out <- setx(z.out, age=35, female=1, educ=16)
set.seed(47306)
s.out <- sim(z.out, x = x.out)
summary(s.out)

##
##  sim x :
##  -----
## ev
##           mean        sd       50%      2.5%     97.5%
## [1,] 2.691311 0.1175825 2.687324 2.466393 2.921389
## pv
```

```
##        mean        sd 50% 2.5% 97.5%
## [1,] 2.673 1.702048   3    0     7

my.pv <- s.out$get_qi(qi = "pv", xvalue = "x")
pv.ci = quantile(my.pv, probs=c(0.025, 0.975))
pv.ci

##   2.5% 97.5%
##      0     7

# plot(s.out)
```

So in this case, for a 35-year old female with 16 years of education and average logged family income, we would expect her number of sick days to be 2.673 with a 95% confidence of (0, 7). Frequently, analysts are also interested in exploring the effects of some covariates on the count response variable by plotting a series of expected counts given covariates' values. For example, one might want to know how education is associated with the number of sick days by gender. To plot such graphs, we need to supply all covariates' values and the parameter estimates produced by the regression model. We can then calculate the expected counts and plot them.

The R code chunk below illustrates how to plot such a graph. The first `data.frame` function (for `newdataM`) creates a data frame with four variables and 100 hypothetical observations. For these 100 data points, we vary their values in education only, and their age (35), income (sample average), and gender (male) are all set to fixed values. The `predict` function creates an object, `cntPreds`, with three lists, including fit (linear predictor or $\mathbf{x}\beta$), their associated standard errors, and the residual scale (1) for the 100 hypothetical cases. The `within` function adds three new variables, including `daysSick` (expected count), `lo` (lower bound of the expected count), and `hi` (upper bound of the expected count), to `newdataM`. The `plot` function initiates the plotting process by setting up x and y axes, without plotting any data point (`type = "n"`) or marking the two axes (`yaxt = "n"`, `xaxt = "n"`). The `polygon` function fills the confidence region in the graph, and the `lines` function plots and connects all the predictions. Below, we first calculate predictions and their confidence region for males, and then for females.

```
# create data and calculate predictions
newdataM <- data.frame(age = rep(35, each = 100),
    impinc = rep(mean(mydta$impinc),
    each = 100), educ = rep(seq(from = min(mydta$educ),
    to = max(mydta$educ), length.out = 100), 1), female = rep(0,
    each = 100))
cntPreds <- predict(mypoisson, newdataM, type = "link",
    se.fit = TRUE)
newdataM <- cbind(newdataM, cntPreds)
z95q = qnorm(0.975, mean = 0, sd = 1, lower.tail = TRUE)
newdataM <- within(newdataM, {
    daysSick <- exp(fit)
    lo <- exp(fit - z95q * se.fit)
    hi <- exp(fit + z95q * se.fit)
})
# plot predictions
plot(newdataM$educ, newdataM$daysSick, type = "n", yaxt = "n",
```

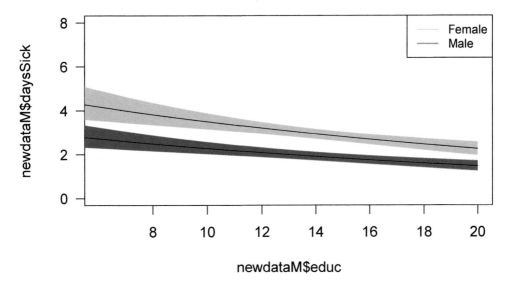

Figure 4.3
Predicted Counts by Gender

```
    xaxt = "n", xlim = c(6, 20), ylim = c(0, 8))
axis(2, at = seq(0, 8, by = 2), las = 2)
axis(1, at = seq(8, 20, by = 2), las = 1)
# add fill
polygon(c(rev(newdataM$educ), newdataM$educ), c(rev(newdataM$lo),
    newdataM$hi), col = "gray40", border = NA)
lines(newdataM$educ, newdataM$daysSick, type = "l")
```

Here we only present the R code chunk for plotting the predictions for males, and one can easily replicate the codes for females by changing "`female = rep(0, each = 100)`" to "`female = rep(1, each = 100)`".

Now we have two expected count prediction lines and their corresponding confidence regions. The light gray area comprises of confidence bands of predictions for females and the dark gray for males. It can be shown from the graph that females are more likely to have higher counts of sick days than those of males across the full range of years of education, given the same covariates' values specified previously. Using crude comparison standards, we can take a step further to argue that the differences at the same levels of covariates are probably statistically significant since none of the confidence bands overlaps, across the whole range of education. Instead of turning to these low-level high-power R utility functions for graphing, we can directly use some customized functions from the `effects` packages to plot predicted counts and their confidence bands (Fox and Weisberg, 2019). Below, we use the `plot` and `predictorEffect` functions from the `effects` package to graph age against the number of sick days using the estimates from `mypoisson`, the estimation object we created previously. The graph shows a general positive association between age and sick days, with a somewhat modest exponential pattern of growth; that is, as age increases, the rate of increase for sick days accelerates.

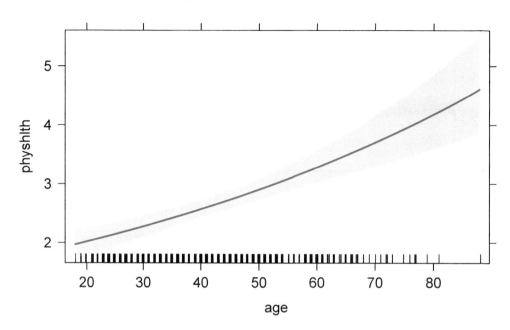

Figure 4.4
The Effects of Age on Predicted Sick Days

```
require(effects)
plot(predictorEffect("age", mypoisson), type="response")
```

Another quantity of interest in the post-estimation analyses of count regression models is the predicted probability of some theoretical count given covariates' values. Below, we use the `predprob` function from the `pscl` package (Zeileis et al., 2008) to help calculate such quantities. We first use the `data.frame` function to create a hypothetical data point with age set to 35, gender to female, education to 16 years, and log of income to its sample mean. We then pass this hypothetical data point onto `predprob`. The resultant matrix, `predcnt`, contains the predicted probabilities for each count from zero to 30. To save space, we request R to print only the predicted probabilities for the first six counts, from zero to five.

```
require(pscl)
newdata <- data.frame(age=35, female=1, educ=16, impinc=mean(mydta$impinc))
print(predcnt <- predprob(mypoisson, newdata)[1:6])
```

```
## [1] 0.06817345 0.18309344 0.24586703 0.22010837 0.14778627 0.07938192
```

After getting such predictions, people might also be interested in knowing their associated precision estimates, or confidence intervals, in this case. Since analytical solutions (e.g., the delta method) to calculating standard errors of predicted probabilities can be cumbersome, we can bootstrap quantities of our interest and construct their empirical confidence intervals. Below, we use the `boot` package (Davison and Hinkley, 1997; Canty and Ripley, 2019) to illustrate how to calculate the predicted probability for zero count given covariates' values—like how it was done in the previous R code chunk box—and its associated precision estimates. One can follow a similar procedure for higher counts up till 30 using the `predprob` function.

The `boot` function from the `boot` package is the workhorse for bootstrapping in this example. Before calling `boot`, one needs to define the function that will return statistics of interest. In the R code chunk below, we first define a new function `bs` with three arguments, including `formula`, `data`, and `index`, corresponding to the regression model formula, the data to be re-sampled from, and the index of observations in the data. The formula and data information will be passed to the `data` and `formula` arguments of the `boot` function used afterwards. In each replicate of bootstrap, the function `bs` re-samples from the given data, estimates the same Poisson regression model, supplies a hypothetical covariate vector, calculates predicted probabilities for counts from zero up till 30, and retains the predicted probability of zero count in this case (the first element in the `predcnt` vector). Then this function is passed onto `boot`, whose arguments specify the formula, the data file name, and the desired replication number.

```
# bootstrap 95% CI for the predicted prob for zero count
# given covariates' values with the boot function
require(boot)
bs <- function(formula, data, index) {
        # allow boot to sample
        d <- data[index,]
        nowfit <- glm(formula, family= "poisson", data=d)
        newdata <- data.frame(age=35, female=1, educ=16,
        impinc=mean(mydta$impinc))
        predcnt = predprob(nowfit, newdata)
        return(predcnt[, 1]) }
# bootstrapping with 1000 replications
set.seed(47306)
results <- boot(data=mydta, statistic=bs, R=1000,
        formula=physhlth ~ age + educ + impinc + female)
# view results results
plot(results)
```

```
# get 95% confidence interval
quantile(results$t, c(0.025, 0.975))
```

```
##      2.5%      97.5%
## 0.02524966 0.15769568
```

In this example, we calculate the predicted probability of a zero count for a 35-year old female with 16 years of education and average log of family income. The results show that her predicted probability of zero count (i.e., not having any sick day) is 0.068 and its bias-corrected and accelerated bootstrapped interval is between 0.022 and 0.151 (results now shown). The histogram and normal Q-Q plots [1] of bootstrapped estimates are graphed in Fig. 4.5, indicative of a slightly right-skewed empirical distribution of the simulated statistic.

---

[1] A Q-Q plot is a scatterplot of the quantiles of some empirical data against the quantiles of another set of data, in this case, a theoretical standard normal distribution.

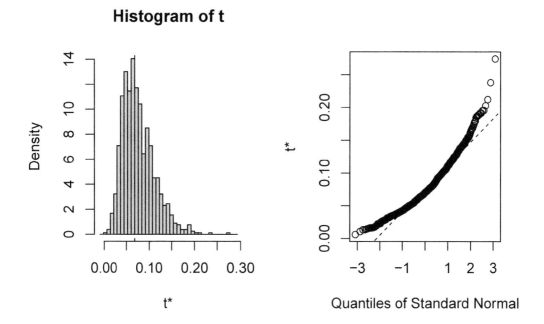

Figure 4.5
Empirical Distribution of Bootstrapped Probability for Zero Count

4.2.4 Contagion, Heterogeneity, and Over-Dispersion

In much of the empirical work, however, quantitative researchers have found it inadequate to use the Poisson distribution and the prototypical Poisson regression model to fit empirical count data. A multitude of studies have shown that the variance of count response variables is often greater than its mean, regardless of whether the conditional or marginal Poisson distribution is used, hence the problem of over-dispersion. In addition, the predictive power of the Poisson distribution for empirical data is somewhat weak, especially for the zero counts. Both Cameron and Trivedi (1998) and Hilbe (2008) provide cogent discussion about possible causes of over-dispersion. First and foremost, postulating that all independent cases share a single parameter, μ, from the same Poisson distribution without any covariate, is in and of itself a very strong assumption, and this concern naturally lends itself to a simple extension that the expected count may vary across cases because of different individual attributes, hence observed heterogeneity. Similarly, when the mean structure is not correctly specified, including the omitted variables problem (i.e., important covariates, such as interaction terms and polynomials, are not included) and unobserved heterogeneity (i.e., unobserved individual attributes), over-dispersion might arise as a result. A second possibility is that data points are correlated or clustered. Not accounting for such implicit data structure in empirical analysis can lead to underestimated standard errors. Another possible source of over-dispersion is the violation of distributional assumptions of the Poisson counting process. For example, one major assumption of the Poisson process is that the event of interest recurs independently across disjoint time intervals; or in other words, the waiting time in the Poisson process is memoryless. This may well be true in theory for the Poisson process, but unfortunately, it is rare in practice. Some events, for example, clinical visits, can recur more frequently once they are above certain threshold count, such as one, probably because the resources required for the initial visit are much greater than recurrences. This is sometimes referred to as contagion, the dependence between successively recurring events. Cameron and Trivedi (1998, p. 106) have luminous discussion about contagion, and further classifies congation into true (positive/negative) and apparent contagion. They also suggest that the count process can have temporal dependence such that events may cluster to transpire during some time spells more so than others.

In general, if over-dispersion occurs with Poisson distributions, mathematically, that means $V(Y) > E(Y) = \mu$. There are several ways to test for over-dispersion. One is to calculate the ratio of residual deviance and its associated degrees of freedom (χ^2/df). If this ratio is greater than one, then there is some evidence of over-dispersion. Hilbe (2008, p. 52) suggests using 1.25 as a threshold for models with a moderate sample size and 1.05 for large sample sizes. Below, we use R to calculate this quantity for the Poisson regression model (`mypoisson`) estimated previously.

```
pr <- residuals(mypoisson,"pearson")
sum(pr^2)/df.residual(mypoisson) # phi
```

```
## [1] 13.20144
```

Our calculation of the dispersion parameter, ϕ, turns out to be 13.201, much greater than one. Thus, it is safe to argue that over-dispersion is a problem for this Poisson regression model. Cameron and Trivedi (1990) propose a regression-based test for checking over-disperson in Poisson regression models. Their idea is simple and straightforward. If the mean-variance equality holds, then we should have $E(Y) = V(Y) = \mu$. We can test this equality as a null against an alternative, $V(Y) = \mu + \alpha \cdot f(\mu)$, and whether α is equal to zero. To implement this test, we use the `dispersiontest` function from the **AER** package (Kleiber and Zeileis, 2008). The first argument in the `dispersiontest` function is to specify the estimation object, `mypoisson` in this case. The `trafo` argument is to indicate whether $f(\mu)$ is a linear (`trafo = 1`) or quadratic (`trafo = 2`) function of μ.

```
require(AER)
# linear specification (in terms of alpha)
dispersiontest(mypoisson, trafo = 1)
```

```
##
##  Overdispersion test
##
## data:  mypoisson
## z = 6.2527, p-value = 2.017e-10
## alternative hypothesis: true alpha is greater than 0
## sample estimates:
##    alpha
## 12.07612
```

```
# quadratic specification (in terms of alpha)
dispersiontest(mypoisson, trafo = 2)
```

```
##
##  Overdispersion test
##
## data:  mypoisson
## z = 6.1744, p-value = 3.32e-10
## alternative hypothesis: true alpha is greater than 0
```

```
## sample estimates:
##    alpha
## 3.833529
```

Both results show that the mean-variance equality can be safely rejected at the conventional significance level. Other alternatives for testing over-dispersion include graphing rootograms [2] and using the likelihood ratio test to compare the Poisson against its negative binomial version of the model, both of which will be covered later.

4.2.5   Quasi-Poisson Regression

One not-so-popular method to account for over-dispersion is to explicitly estimate the dispersion parameter, ϕ, instead of setting it to one in the Poisson regression, thus inflating the standard errors of the coefficient estimates by a factor of $\sqrt{\hat{\phi}}$. In the quasi-Poisson regression model, variance of the count response variable, $V(Y) = \phi\mu$, is assumed to be a linear function of mean, $E(Y) = \mu$.

```
mypoisson.q = update(mypoisson, family=quasipoisson)
out = summary(mypoisson.q)
out$coefficients
```

```
##                 Estimate  Std. Error   t value    Pr(>|t|)
## (Intercept)  2.87624072 0.923725526  3.113740 0.001962326
## age          0.01214117 0.007227075  1.679957 0.093643177
## educ        -0.04408526 0.034281658 -1.285972 0.199098267
## impinc      -0.20201893 0.092768923 -2.177657 0.029937857
## female       0.42939850 0.214690391  2.000083 0.046077505
```

```
out$dispersion
```

```
## [1] 13.20254
```

When we compare the results here with those from their Poisson regression counterpart `mypoisson`, it can be shown that the coefficient estimates still stay the same, but the standard errors are all scaled up by a factor of 3.634, the square root of the estimated dispersion parameter.

4.2.6   Negative Binomial Regression

As illustrated in many empirical studies, the predictive power of Poisson regression for count response variables is often poor. The clear advantage of using a single parameter, even conditional on covariates, also comes with its cost—lack of accuracy in predicting count probabilities, especially zero counts. In Poisson regression models, we assume that $E(Y|\mathbf{x}) = \exp(\mathbf{x}\beta)$. When all major conditions, as discussed previously, are met in the data generation process, the Poisson distribution and regression models can satisfy the equidispersion assumption; that is, mean is equivalent of variance. When this assumption is violated, however, adjustments have to be made. In addition to re-scaling the dispersion

---

[2]Rootograms, or hanging rootogram, were first introduced by John W. Tukey to assess empirical univariate distributions against their corresponding theoretical ones. For count regression models, rootograms are usually used to compare the observed counts with expected ones based on estimated models (Tukey, 1972; Kleiber and Achim, 2016).

parameter, we can turn to one of the most popular methods for correcting over-dispersion—the negative binomial distribution and regression models.

It is documented that Blaise Pascal (1679) probably should be credited with the discovery of the negative binomial distribution, and P. R. Demont (1714) for the first application of the distribution (Dodge, 2008). The negative binomial distribution can be obtained from multiple ways. Boswell and Patil (1970), for example, identified 13 stochastic processes that can generate data following the distribution. Below, we only present two typical data-generating processes. The first and traditional method is to view the negative binomial distribution as generated from a discrete random variable to characterize a series of independent and identically distributed binary (success-failure) random experiments, or Bernoulli trials, with f failures before the sth success. In this case, $y = f$ follows a negative binomial distribution, and its probability mass function (PMF) is calculated as follow,

$$P(y = f) = C(f + s - 1, f) p^s (1 - p)^f \qquad (4.4)$$

where p is the probability of success, $q = 1 - p$ for failure, and $C(f + s - 1, f)$ is $\frac{(f+s-1)!}{f!(s-1)!}$. The mean and variance of this random variable are $\mu = \frac{f(1-p)}{p}$ and $V = \frac{f(1-p)}{p^2}$ respectively. If we use μ and $\alpha = \frac{1}{f}$ to reparameterize, then we can get $V = \mu + \alpha\mu^2$ (Cameron and Trivedi, 1998; Hilbe, 2008).

A second method to derive the negative binomial distribution is through mixing a Poisson with a gamma distribution. Note that in the Poisson regression model, we have the mean structure $\mu = \exp(\mathbf{x}\beta)$. To increase the conditional variance of y, an additional disturbance term, ψ, can be introduced to the mean structure in an additive manner, so $\exp(\mathbf{x}\beta + \varepsilon) = \exp(\mathbf{x}\beta)\exp(\varepsilon) = \mu\psi = \theta$, where $\psi = \exp(\varepsilon)$ following a gamma distribution, and ε is assumed to be independent of $\mathbf{x}$. If the assumption that $E(\psi) = 1$ is conveniently made, then we have $E(\theta) = E(\mu\psi) = \mu E(\psi) = \mu$. Thus by introducing a random error, the expectation of the count response variable stays the same, but the variance increases. So suppose a count random variable Y follows a Poisson distribution, conditional on a random error, ψ, then we have $f(y_i|\theta_i) = \dfrac{\exp(-\theta_i) \times \theta_i^{y_i}}{y_i!}$, $y_i = 0, 1, ...,$ where $\theta_i = \exp(\mathbf{x}\beta + \varepsilon)$ and $\psi_i = \exp(\varepsilon)$. The marginal distribution of y, $h(y_i|\mu_i, \psi_i) = \int f(y_i|\mu_i, \psi_i)g(\psi_i)d\psi_i$ can be obtained by integrating out ψ. For specific choices of $f(\cdot)$ and $g(\cdot)$, the Poisson and gamma densities multiplied together happen to yield an integral with an explicit form. Note that the assumption $E(\psi) = 1$ would imply a one-parameter gamma distribution, $g(\psi_i) = \frac{\delta^\delta}{\Gamma(\delta)}\psi^{\delta-1}\exp(-\psi\delta)$, where $\delta > 0$ is both the shape and scale parameter ($\delta = \phi$) of this gamma distribution and $\Gamma(\delta) = \int_0^\infty t^{\delta-1}\exp(-t)\,dt$. Integrating out ψ, we have

$$h(y_i|\mu_i, \psi_i) = \frac{\Gamma(\delta + y)}{\Gamma(\delta)\Gamma(y + 1)}\left(\frac{\delta}{\delta + \mu}\right)^\delta \left(\frac{\mu}{\mu + \delta}\right)^y \qquad (4.5)$$

If we set $\alpha = \frac{1}{\delta}$, then again we can have $V = \mu + \alpha\mu^2$ (Cameron and Trivedi, 1998; Hilbe, 2008). This is what statistical practitioners usually call NB2; that is, the variance is a second-power function of mean. Although NB2 is a widely used and the most referenced negative binomial distribution and regression model, there is a surprisingly large number of sub-types falling under the name of negative binomial (Hilbe, 2008, p. 78).

There are several functions from multiple packages that are available for estimating negative binomial regression models. The most commonly used one is `glm.nb` from the `MASS` package (Venables and Ripley, 2002) that comes with R default installation.

```
summary(mynbreg <- glm.nb(physhlth~ age + educ + impinc + female, data = mydta))
```

```
##
## Call:
## glm.nb(formula = physhlth ~ age + educ + impinc + female, data = mydta,
##     init.theta = 0.1953406394, link = log)
##
## Deviance Residuals:
##     Min       1Q    Median        3Q       Max
## -1.25672  -1.03007  -0.92408  -0.05042   2.05032
##
## Coefficients:
##               Estimate Std. Error z value Pr(>|z|)
## (Intercept)   3.203252   1.159999   2.761  0.00575 **
## age           0.012692   0.008483   1.496  0.13459
## educ         -0.055291   0.039827  -1.388  0.16505
## impinc       -0.232687   0.122620  -1.898  0.05775 .
## female        0.602673   0.222623   2.707  0.00679 **
## ---
## Signif. codes:  0 '***' 0.001 '**' 0.01 '*' 0.05 '.' 0.1 ' ' 1
##
## (Dispersion parameter for Negative Binomial(0.1953) family taken to be 1)
##
##     Null deviance: 382.45  on 465  degrees of freedom
## Residual deviance: 366.01  on 461  degrees of freedom
## AIC: 1670.8
##
## Number of Fisher Scoring iterations: 1
##
##
##                 Theta:  0.1953
##             Std. Err.:  0.0187
##
##   2 x log-likelihood:  -1658.8250
```

```
(mynbreg$coefficients["age"]) # list a single coefficient
```

```
##        age
## 0.01269188
```

While interpreting the results, we can follow exactly what we did previously with the Poisson regression model. In the results section from the negative binomial regression model, we see that the coefficient for age is 0.013, and we can interpret it as: for each unit (year) increase in age, the expected count of sick days increases by a factor of $\exp(0.013) = 1.013$, while holding all other variables constant. To interpret the effect of the only binary covariate, gender, we can say that being female as opposed to male, ceteris paribus, increases the expected count by a factor of $\exp(0.603) = 1.827$. To get all factor change coefficients, we can use the `exp` base function to carry out computation by exponentiating the raw coefficients vector and listing their associated confidence intervals.

```
exp(mynbreg$coefficients)
```

```
## (Intercept)        age       educ     impinc     female
##  24.6124493  1.0127728  0.9462094  0.7924018  1.8269957
```

```
# A neat way to list estimated coefficients and ci's
nbreg.est <- cbind(Estimates=coef(mynbreg), confint(mynbreg))
```

```
## Waiting for profiling to be done...
```

```
exp(nbreg.est)
```

```
##              Estimates      2.5 %      97.5 %
## (Intercept) 24.6124493 2.5435902 314.182153
## age          1.0127728 0.9967879   1.029346
## educ         0.9462094 0.8774506   1.016118
## impinc       0.7924018 0.6103743   1.004467
## female       1.8269957 1.1749349   2.826269
```

Like with the Poisson regression, one might be interested in calculating the expected count given some covariates' values after running a negative binomial regression model. This can be done with the **predict** function from the base **stats** package again. Below, we create two hypothetical data points, with age, education, and income set to 35, 16, and sample average respectively, while varying the gender status. Then we list the two expected counts along with other covariates.

```
newdata1 <- data.frame(age=35, female=0:1,
                          educ=16, impinc=mean(mydta$impinc))
newdata1$count <- predict(mynbreg, newdata1, type = "response")
newdata1
```

```
##   age female educ    impinc    count
## 1  35      0   16 10.08457 1.516279
## 2  35      1   16 10.08457 2.770235
```

The results show that for a 35 year old female with a roughly college education and average family income, her expected count of sick days is 2.77, and for an otherwise similar male, his expected count is 1.516, with a difference of 1.254 days between the two. Similarly, one might be interested in calculating predicted probabilities for certain counts given a vector of covariates. Below, we use the same **predprob** function used previously for the Poisson model to calculate the predicted probabilities for counts from zero to five, for the same two hypothetical data points from the R code chunk above. It can be shown from the results that males have lower probabilities for low counts (i.e., 0 and 1) and lower probabilities for high counts (3 and above), compared with females. Note that such comparisons have to be contextualized with covariates' values.

```
predcnt <- predprob(mynbreg, newdata1)
rownames(predcnt) <- c("male", "female")
predcnt[,1:6]
```

```
##                0         1          2          3          4          5
## male   0.6544388 0.1132488 0.05996075 0.03887045 0.02750734 0.02044644
## female 0.5878153 0.1072608 0.05988395 0.04093537 0.03054664 0.02394243
```

After calculating all these predictions for substantive interpretation, we can revisit the issue of over-dispersion. In addition to the few methods discussed in the previous two sections, we can use the LR test to compare a negative binomial with a Poisson regression model to see if there is any major sign of over-dispersion. If the LR test shows statistically significant results, then that would suggest the possibility of over-dispersion, and usually the negative binomial is favored over the Poisson regression model. The results below clearly show some sign of over-dispersion, and thus the negative binomial regression model is preferred, as is the case in most empirical applications.

```
2 * (logLik(mynbreg) - logLik(mypoisson))

## 'log Lik.' 2649.72 (df=6)

pchisq(2 * (logLik(mynbreg) - logLik(mypoisson)), df = 1, lower.tail = FALSE)

## 'log Lik.' 0 (df=6)
```

We can also directly use `odTest`—a R function to run LR tests for testing over-dispersion in count regression models—from the `pscl` package to follow through the same calculation and hypothesis-testing procedure. To use it, we need to supply a `glm.nb` estimation object of class `negbin`.

```
odTest(mynbreg)

## Likelihood ratio test of H0: Poisson, as restricted NB model:
## n.b., the distribution of the test-statistic under H0 is non-standard
## e.g., see help(odTest) for details/references
##
## Critical value of test statistic at the alpha= 0.05 level: 2.7055
## Chi-Square Test Statistic =  2649.7202 p-value = < 2.2e-16
```

Another effective method for detecting over-dispersion is to plot observed counts against expected counts for both models and then compare counts fit. Such graphs are called rootograms, which was initially proposed by Tukey (1972) on exploratory data analysis and statistical graphics (Kleiber and Achim, 2016). The general idea is to compare an empirical data distribution with the theoretical one. For plotting such graphs, we can use the `rootogram` function from the `countreg` package. The `rootogram` function for count regression models compares (the square roots of) the observed count, $o_Y = \sum_{i=1}^{N} \mathbf{I}(y_i = Y)$, $Y = 0, 1, 2, ...$, with the model-based expected counts, $e_Y = \sum_{i=1}^{N} f(Y; \widehat{\theta})$, wherein $\mathbf{I}$ is a binary indicator function and $f(Y; \widehat{\theta})$ is the probability mass (for count random variables, and density for continuous variables) function for $y_i = Y$ and parameter estimates $\widehat{\theta}$ from our estimated model (Kleiber and Achim, 2016, p. 272). In simple terms, o_Y is a simple count of $y = Y$ in the empirical data, and e_Y is the sum of the model-based predicted probabilities for $y = Y$ across all individual cases. These statistics can be easily transformed into proportions by dividing by the estimation sample size. Note that the `rootogram` function provides three options for graphing, including standing, hanging, and suspended. The standing option graphs the square roots of observed counts against the square roots of their corresponding expected counts. The hanging and suspended options provide additional ways to compare the two sets of counts using graphing techniques that provide assessment information almost identical to that of the standing rootograms.

```
require(countreg)
par(mfrow=c(1,2))
rootogram(mypoisson, style="standing")
rootogram(mynbreg, style="standing")
```

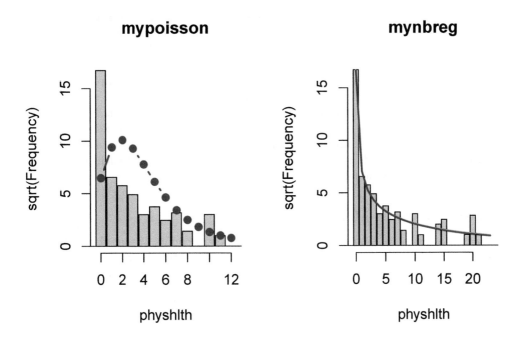

Figure 4.6
Rootograms of Negative Binomial vs. Poisson Regression

It can be shown from this combined graph Fig. 4.6 that the negative binomial regression model fits the empirical data much better than does the Poisson regression model. The expected counts (curved lines) calculated from the negative binomial model are closer to the observed counts, (vertical bars), and for some of the counts, such as zero–three and five, the observed and expected counts match almost exactly. The theoretical distribution derived from the Poisson regression model, however, fares much worse, and the most obvious problem is the serious underestimate of the zero count.

4.3 Zero-Modified Count Regression

When there is a large number of zeros in the count response variable, further adjustments are needed. Hurdle and zero-inflated count regressions are the two most frequently used models to address the issue of having excess zeros. Usually, there are two components for both hurdle and zero-inflated models. In the hurdle model, the data are presumed to be generated through some seemingly (quasi-) sequential process; the first step is to determine whether the response belongs to a zero or non-zero (positive) group, and the second is to predict a positive count among the non-zero group. This data-generation process can occur frequently in consumer behavior. For example, a consumer first has to decide whether to purchase a product or not. Once the consumer decides to go with consumption, then the quantity has to be determined. The zero-inflated count regression model corresponds to a somewhat different data-generation process, in which two distributions—usually one binomial and the other a count distribution (e.g., Poisson or negative binomial)—are mixed to generate observed data. In empirical analysis, we usually have uncertainty (i.e., probability) about whether one belongs to one group or the other. In the consumer behavior example described above, that would imply that the zero count (i.e., no purchase) might come from either the no-purchase group or the possible-purchase group with some uncertainty.

A lesser-discussed but popular scenario is when the zero count is completely missing by design; for example, a study explores the association between individual characteristics and the number of products purchased among those who have consumed. The minimum number of products purchased is one among this group of people, and zero counts are simply truncated by design or under structural constraints. In this case, some common normalization procedure needs to be applied in modeling, and we call such models zero-truncated count regression models, which will be discussed next.

4.3.1   Zero-Truncated Models

Zero-truncated count regression models are used when zero counts are un-observable for a count response variable due to sampling design or other constraints. So the sample data only have non-zero positive counts for the response variable. There are many empirical applications, especially in healthcare utilization; for example, the number of days patients stay in the hospital (i.e., inpatients) that excludes any information about those who receive medical treatment without staying in hospital or healthy individuals. More formally, for a response variable that follows a counting process with zeros structurally excluded, such as zero-truncated Poisson or negative binomial distribution, we have its probability mass function $f(y, \theta | y > 0) = \frac{f(y,\theta)}{1-f(y=0,\theta)}$, where $1 - f(y = 0, \theta)$ is the normalizing factor. For example, for a zero-truncated Poisson random variable, the probability for any non-zero positive count needs to be normalized by $1 - e^{-\mu}$, and thus we have $P(Y = y | y > 0) = e^{-\mu} \frac{\mu^y}{y!(1-e^{-\mu})}$. A similar normalization procedure can be applied to negative binomial distribution/regression models with zero truncation. The `zerotrunc` function in the `countreg` package can be used to estimate zero-truncated count regression models. The following few lines of R codes estimate a zero-truncated negative binomial regression model with the same set of variables used previously, except that now we exclude all zero counts from the count response variable.

```
# delete zero counts for physical health
ztdta <- mydta[ which(mydta$physhlth > 0), ]
negbin.zerotrunc <- zerotrunc(physhlth~ age + educ + impinc + female,
                                    data = ztdta, dist=c("negbin"))
summary(negbin.zerotrunc)$coefficients
```

```
##                 Estimate  Std. Error   z value     Pr(>|z|)
## (Intercept)   3.2356537  1.371468882  2.359261  0.018311348
## age           0.0166426  0.008992205  1.850781  0.064201076
## educ         -0.0479155  0.041722673 -1.148428  0.250791810
## impinc       -0.2227706  0.140088668 -1.590211  0.111787154
## female        0.2859653  0.257491754  1.110580  0.266749038
## Log(theta)   -1.3238243  0.434853486 -3.044300  0.002332228
```

When it comes to interpretation, there is almost no difference between zero-truncated and other count regression models without truncation. For example, for each year increase in age, the expected count will grow by a factor of $\exp(0.017) = 1.017$, net of the effects of all other variables. We can follow similar procedures for other types of post-estimation analysis, including graphing predictions or calculating predicted probabilities. If we assume or find out that over-dispersion is not a serious problem and the count response variable follows a conditional truncated Poisson distribution, then we can replace "`negbin`" with "`Poisson`" to estimate a zero-truncated Poisson regression model instead.

4.3.2 Hurdle Models

The hurdle model (or sometimes called zero-altered model) arises from the consideration that the count data can be generated through a process, in which the non-zero counts emerge only after crossing the zero "hurdle." The hurdle model thus has two components. The first is a hurdle component that uses either binomial or a censored (right censored at $y = 1$) count distribution to predict zero vs. non-zero positives, and a second component using truncated (left truncated at $y = 1$) Poisson or negative binomial distribution to model positive counts. Rational decision-making, including those in consumer behavior, can be viewed as following such a data-generating process; that is, individuals first decide whether they would consume or not, and once opting for consumption, they then decide the amount of consumption, depending on how it is measured. So for hurdle models, we have the following probability mass function (Zeileis et al., 2008),

$$f_{\text{hurdle}} = \begin{cases} f_{\text{zero}}(0; \mathbf{z}, \gamma) & y = 0 \\ (1 - f_{\text{zero}}) \cdot f_{\text{count}}(y; \mathbf{x}, \beta)/(1 - f_{\text{count}}(0; \mathbf{x}, \beta)) & y > 0 \end{cases} \quad (4.6)$$

where $\mathbf{z}$ and γ are covariates and parameters for the zero hurdle part (usually a binary regression model), and $\mathbf{x}$ and β are covariates and parameters in the non-zero part of the hurdle model. In many empirical applications, $\mathbf{z}$ and $\mathbf{x}$ are exactly the same, or they overlap for the majority of covariates. While the probability mass function of the zero hurdle part is written for having zeroes, most statistical software applications explicitly model the probability for having positive counts instead. To estimate the hurdle model, we can use the `hurdle` function again from the `countreg` package. The syntactic structure of `formula` in the hurdle function is exactly the same as that of most other estimation functions in R, and the `dist` option is to specify the distribution for the non-zero count part, in this case `poisson`. We can also specify the distribution for the zero hurdle part, which is set to binomial or logit link by default (i.e., `zero.dist="binomial"`).

```
poisson.hurdle <- hurdle(physhlth~ age + educ + impinc + female, data = mydta,
                         dist=c("poisson"), zero.dist="binomial")
summary(poisson.hurdle)$coefficients
```

```
## $count
##               Estimate   Std. Error    z value      Pr(>|z|)
## (Intercept)  3.23086945 0.245549761  13.157697  1.536988e-39
## age          0.01558246 0.002126190   7.328819  2.321902e-13
## educ        -0.03007969 0.008995291  -3.343937  8.259847e-04
## impinc      -0.15957898 0.024358304  -6.551318  5.703164e-11
## female       0.07714599 0.059166890   1.303871  1.922776e-01
##
## $zero
##               Estimate    Std. Error    z value     Pr(>|z|)
## (Intercept)  0.195598953 1.019314171  0.1918927  0.847826252
## age         -0.002874477 0.007492045 -0.3836705  0.701222686
## educ        -0.026949901 0.035079213 -0.7682584  0.442333676
## impinc      -0.045777562 0.107863956 -0.4244009  0.671273453
## female       0.639296946 0.196834244  3.2478949  0.001162622
```

```
summary(poisson.hurdle)$coefficients$count[2, 1]
```

```
## [1] 0.01558246
```

```
summary(poisson.hurdle)$coefficients$zero[2, 1]
```

```
## [1] -0.002874477
```

```
# newdata <- data.frame(age=35, female=1, educ=16, impinc=mean(mydta$impinc))
# type: count/prob/response/zero
print(cntprob <- predict(poisson.hurdle, newdata, type = "prob")[1:6])
```

```
## [1] 0.539556252 0.007916941 0.023073812 0.044832201 0.065331411 0.076162992
```

Interpretation of the results is straightforward. For the zero hurdle part, we can employ all the post-estimation analytic methods used previously for binary regression to interpret the odds of having non-zero counts vs. zero and the probabilities of both; for the truncated count component, one can use the few methods illustrated in the previous few sections. For example, for each year decrease in age, we would predict that the expected count is going to increase by a factor of $\exp(0.016)$, and the odds of having non-zero as opposed to zero counts is expected to decrease by a factor of $\exp(-0.003)$, holding all other variables constant. We can also use the `predict` function from the `countreg` package to calculate predicted probabilities and expected counts given a set of covariates' values, like how we did previously for the Poisson and negative binomial regression models. Towards the end of the R code chunk above, we use the model estimates to calculate the predicted probabilities of having zero to five counts for a 35-year old female with roughly college education and sample average income.

Whether the hurdle model is a true stand-alone model or a simple combined one with two separate components is arguable. Below, we re-create the zero hurdle part by resetting the non-zero positives of `physhlth` to one, and use a logit model to predict the odds of non-zero vs. zero.

```
mydta$physhlth1 = as.numeric(mydta$physhlth > 0)
hurdlePart <- glm(formula = physhlth1 ~ age + educ + impinc + female,
                  data    = mydta,
                  family  = binomial(link = "logit"))
summary(hurdlePart)$coefficients
```

```
##                  Estimate  Std. Error    z value     Pr(>|z|)
## (Intercept)   0.195598953 1.019306727  0.1918941  0.847825154
## age          -0.002874476 0.007490441 -0.3837526  0.701161824
## educ         -0.026949901 0.035078638 -0.7682710  0.442326187
## impinc       -0.045777562 0.107862508 -0.4244066  0.671269300
## female        0.639296946 0.196834284  3.2478943  0.001162625
```

One can see that the results are almost identical to those from the zero hurdle part of the hurdle model previously estimated.

4.3.3 Zero-Inflated Models

Unlike hurdle models, whereby there is only a single source of zeros, zero-inflated models presume that there are two sources of zeros through mixing different distributions. For count distributions and regression models, it is postulated that the excess zeros are resulted from mixing usually a binomial distribution and some probability distribution for a count random variable (e.g, Poisson and negative binomial). In essence, zero-inflated models are sub-types of finite mixture models, whereby populations with different parameters or distributions are mixed together to produce observed data, thus increasing the unobserved heterogeneity for statistical modeling. For example, for the zero count of certain product purchased, one source of zeros could come from those who never like the product at all and thus have never made any purchase, and the other type is from those who are interested in the product but decide not to purchase it simply by chance or due to some unobserved individual attributes. In this case, we have two populations contributing to zeros; one is the never-buyers and the other is possible-buyers; usually we do not have information about whether a consumer belongs to one group or the other, but we can use other information available in the data to predict with uncertainty. So the probability mass function for a zero-inflated count regression is usually formulated as:

$$f_{\text{zeroinfl}} = \pi \cdot \mathbf{I}_{\{0\}}(y) + (1 - \pi) \cdot f_{\text{count}}(y; x, \beta) \tag{4.7}$$

where $\pi = f_{\text{zero}}(0; z, \gamma)$ and $\mathbf{I}_{\{0\}}(y)$ is a point mass at zero (Zeileis et al., 2008). To estimate zero-inflated count regression models, we can use the `zeroinfl` function from the `countreg` package.

```
poisson.zeroinfl <- zeroinfl(physhlth ~ age + educ + impinc + female,
    data = mydta,
    dist = c("poisson"))
summary(poisson.zeroinfl)$coefficients
```

```
## $count
##               Estimate  Std. Error    z value     Pr(>|z|)
## (Intercept)  3.23039967 0.245502010 13.158343 1.523913e-39
## age          0.01559264 0.002126653  7.332009 2.267287e-13
## educ        -0.03008637 0.008990321 -3.346529 8.183022e-04
## impinc      -0.15959953 0.024349196 -6.554612 5.578681e-11
## female       0.07751379 0.059180600  1.309784 1.902690e-01
##
## $zero
##                Estimate  Std. Error    z value    Pr(>|z|)
## (Intercept) -0.177861771 1.020938227 -0.1742140 0.861697253
## age          0.003152471 0.007508315  0.4198640 0.674584793
## educ         0.026284744 0.035129806  0.7482177 0.454328844
## impinc       0.043245612 0.108028959  0.4003150 0.688924525
## female      -0.638274942 0.197189942 -3.2368534 0.001208555
```

If we opt to use the conditional negative binomial instead of the Poisson distribution, we can replace "poisson" with "negbin" in the `dist` option of the `zeroinfl` function. We can also conduct a series of post-estimation analyses, such as calculating the expected count or probability for some count given a vector of covariates' values. Below, we first estimate a zero-inflated negative binomial regression model. We then use the `data.frame` function to construct the same hypothetical individual that has been used consistently in this chapter, a 35-year-old female with 16 years of education and sample average income. We then pass the hypothetical data point onto the `predict` function to calculate the predicted probabilities for having counts zero to five as well as its expected count.

```
negbin.zeroinfl <- zeroinfl(physhlth ~ age + educ + impinc + female,
    data = mydta,
    dist = c("negbin"))
summary(negbin.zeroinfl)$coefficients
```

```
## $count
##                 Estimate Std. Error   z value     Pr(>|z|)
## (Intercept)   2.82173125 1.38754538  2.033614 0.041990563
## age           0.01877299 0.01026112  1.829526 0.067320775
## educ         -0.05249402 0.04276777 -1.227420 0.219664684
## impinc       -0.18820468 0.13058356 -1.441259 0.149511632
## female        0.35229626 0.33032853  1.066503 0.286196465
## Log(theta)   -1.32267715 0.43840152 -3.017045 0.002552518
##
## $zero
##                 Estimate Std. Error    z value   Pr(>|z|)
## (Intercept)  -3.10332860 7.24186877 -0.42852594 0.6682683
## age           0.03767094 0.06807547  0.55337019 0.5800099
## educ          0.00765428 0.11258180  0.06798861 0.9457947
## impinc        0.05043229 0.44204113  0.11408958 0.9091668
## female       -1.38880327 1.15149208 -1.20609016 0.2277827
```

```
# count probs can go up to the observed max
predprob = predict(negbin.zeroinfl, newdata, type = c("prob"), at = 0:5)
print(predprob, digits = 3)
```

```
##        0     1      2      3      4      5
## 1  0.549 0.116 0.0677 0.0469 0.0352 0.0276
```

```
predict(negbin.zeroinfl, newdata, type = c("count"))
```

```
##        1
## 2.983849
```

The results show that the predicted probabilities for having zero to five counts are 0.549, 0.116, 0.068, 0.047, 0.035, and 0.028 respectively, and the expected count is 2.984. One can use similar methods showcased in previous chapters to conduct more elaborate post-estimation analyses, such as computing marginal effects, group differences in predictions or effects, and their associated precision estimates using either analytic, closed-form methods or simulation (e.g., bootstrap).

4.4   Bayesian Estimation of Count Regression

4.4.1   Bayesian Estimation of Negative Binomial Regression

Bayesian estimation of count regression models is quite similar to that of multi-category response regression models in its workflow. As usual, one needs to supply the likelihood function and priors for parameters. What follows is a working example of estimating a Bayesian negative binomial regression model (NB2) using weakly informative priors. We retain the same five variables used in this chapter throughout and clean the data with pairwise deletion. Once we have a cleaned data frame without any missing data, we can begin the Stan model code chunk. Below, in between the `modelString` and `writeLines` statements are the Stan codes to be saved and called by the `stan` function from the `rstan` package (Stan Development Team, 2020) later in the process.

```
#rm(list=ls(all=TRUE))
library(rstan)

modelString = "
```

As discussed in previous chapters, Stan model codes usually follow a similar syntactic structure with three basic blocks, including `data`, `parameters`, and `model` blocks. The `data` block declares data variables and scalars along with their boundary information, number types, and dimensions if they are matrices. Below we declare K to be the number of covariates, N for the number of cases, y for the count response variable, and X for a row vector with N rows and K columns. Towards the end, we also declare a row vector, `xvec`, with four elements corresponding to the four covariates, and we will use this row vector later to generate a predicted count.

```
data {
int K;
int N;
int<lower=0> y[N];
row_vector[K] X[N];
row_vector[4] xvec;
}
```

Next is the `parameters` block, which declares parameters to be estimated and sets their boundary information. For the negative binomial regression model, we declare two scalar parameters, including `phi`—the shape and scale parameter of the negative binomial distribution and `alpha`—the intercept parameter for the unit column, and a vector `beta` containing slope parameters. Since we have K covariates, the dimension of `beta` is set to K.

```
parameters {
real alpha;
vector[K] beta;
real<lower=0> phi;
}
```

Next follows the `model` block, which sets the priors for `phi`, `alpha`, and `beta`, and specifies the likelihood. Since `phi` needs to be a positive number, we use the right half of the symmetric Cauchy distribution, or the half-Cauchy distribution, following the advice by Andrew Gelman.[3] All parameters are assigned with weakly informative priors. We use Stan's `neg_binomial_2_lpmf` distributional function to set up the NB2 regression model. This function includes two arguments for the mean/location parameter, `eta`, and the shape and scale parameter, `phi`, respectively. Since `phi` is a scalar parameter, we can directly add it in the function. We construct `eta` as an exponentiated function of the product of X and its corresponding parameter vector, `beta`. `alpha` is included as the parameter for the unit column of the linear predictor.

---

[3]Pleast see the details at https://github.com/stan-dev/stan/wiki/Prior-Choice-Recommendations.

```
model {
real eta;
phi ~ cauchy(0, 5);
alpha ~ normal(0, 10);
beta ~ normal(0, 10);
for (n in 1:N) {
// exponentiate xb for _lpmf
eta = exp(alpha + X[n]*beta);
// y[n] ~ neg_binomial_2_log(eta, phi);
y[n] ~ neg_binomial_2_lpmf(eta, phi);
}
}
```

Below the Stan `model` block, we can create an additional code chunk, the `generated quantities` block, which is used to calculate quantities of interest based on the preceding Bayesian MCMC simulation. In the block, we request to calculate the expected count, given covariates' values provided in the `xvec` vector. We can also calculate the predicted probability for some count, given covariates' values.

```
generated quantities {
real<lower=0> mu;
real<lower=0> yPred;
mu = exp(alpha + xvec*beta);
// yPred = exp(neg_binomial_2_lpmf(2 | mu, phi));
}
```

Once the Stan model is set up, we can close it out and use the `writeLines` function to write the Stan model codes into a text file.

```
" # ending quotation mark
writeLines(modelString,con="model.txt")
```

As discussed in previous chapters, the next major step again is to link empirical data to the data matrices and scalar names declared in the data block of Stan codes. The first few lines below extract data and scalar information from the empirical data. The `list` function towards the end creates several lists of data elements to be linked to the variable and scalar names used in the Stan model setup step. There are several equality signs within the `list` function. On the left of the equal signs are the names used in the Stan code blocks, and on the right are the data objects created from the empirical data right above the `list` function. It would be a good practice to use the same names throughout, but below we slightly alter naming convention so as to illustrate and help distinguish the names used in the Stan code blocks and those directly created from the empirical data. Note that we create a row vector, `xvec`, with four elements, 35, 16, 13, and 1, corresponding to values for age, education, income, and gender respectively, as how they are arranged in the `X = dataMat[,-1]` matrix.

```
dataMat = as.matrix(mydta)
N = NROW(dataMat)
X = dataMat[,-1]
predictorNames = colnames(dataMat)[-1]
nPredictors = NCOL(X)
```

```
y = as.matrix(dataMat[,1])
predictedName = colnames(dataMat)[1]
nYlevels = max(y)
xvec = c(35, 16, 13, 1)
# THE DATA.
dataList = list(
        K = nPredictors ,
        N = N ,
        X = X ,
        y = as.vector( y ) ,
        xvec = xvec )
```

Once we have the Stan model set up and empirical data linked to the matrices and scalars declared in the Stan model block, the next and last step is to run the MCMC simulation using the **stan** function. The first few lines in the following R code chunk set up options for the **stan** function, and the **summary** and **print** functions present summary results of the MCMC simulation.

```
parameters = c("alpha", "beta", "phi", "mu", "yPred")
# parameters to be collected.
adaptSteps = 1000   # Number of steps to 'tune' the samplers.
burnInSteps = 2000   # Number of steps to 'burn-in' the samplers.
nChains = 3   # Number of chains to run.
numSavedSteps = 10000   # Total number of steps in chains to save.
thinSteps = 1   # Number of steps to 'thin' (1=keep every step).
nPerChain = ceiling((numSavedSteps * thinSteps)/nChains)   # Steps per chain.
time.used <- proc.time()
mcmcSamples <- stan(model_code = modelString, data = dataList,
    seed = 47306, pars = parameters, chains = nChains, iter = nPerChain,
    warmup = burnInSteps)   # init=initsChains
summary(mcmcSamples)
print(mcmcSamples, digit = 3)
# mcmcChain = as.matrix(mcmcSamples) Stop the clock
proc.time() - time.used
```

In addition to calculating the expected count and predicted probabilities for counts of our interest, we can take a step further to make a Bayesian assessment of some null values, such as the expected count, $\mu_0 = 2.5$. We invoke the **BEST** package to plot highest density intervals (HDIs) along with region of practical equivalence (ROPE) while setting the ROPE to be from 2.4 to 2.6 with a radius of 0.1 just for illustrative purposes (Kruschke, 2013, 2015). It can be shown from the graph that only about 18 percent of the highest density interval is within the chosen ROPE interval, and thus we have to withhold from providing a conclusive statement about whether the parameter is not different from 2.5.

```
require(BEST)
plotPost(mcmcChain[, "mu"],
            ROPE=c(2.4, 2.6),
            xlab="Hypothetical Count",
            title="this is good" ) # col="grey"
```

We can also use the customized **stan_glm.nb** function from the **rstanarm** package (Goodrich et al., 2018) to estimate the same model.

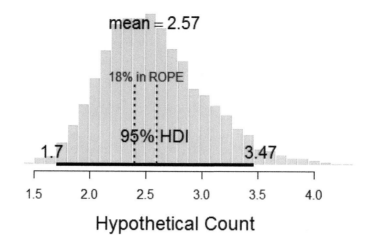

Figure 4.7
HDI and ROPE Plot for Predicted Count

```
require(rstanarm)
stan.nb <- stan_glm.nb(physhlth ~ age + educ + impinc + female,
                   data = mydta, seed=47306)
```

After estimation, one can use the **summary** function to display the results.

```
summary(stan.nb, probs = c(0.025, 0.5, 0.975), digits=3)
```

```
Model Info:
function:     stan_glm.nb
family:       neg_binomial_2 [log]
formula:      physhlth ~ age + educ + impinc + female   algorithm:
sampling  sample:      4000 (posterior sample size)
priors:       see help('prior_summary')
observations: 466
predictors:   5
```

```
Estimates:
                      mean    sd      2.5%    50%     97.5%
(Intercept)           3.328  1.235   1.072   3.279   5.939
age                   0.013  0.008  -0.003   0.013   0.029
educ                 -0.056  0.038  -0.134  -0.055   0.017
impinc               -0.241  0.126  -0.502  -0.237  -0.005
female                0.598  0.230   0.122   0.603   1.050
reciprocal_dispersion 0.193  0.019   0.159   0.192   0.232
```

```
Fit Diagnostics:
          mean    sd    2.5%    50%   97.5%
mean_PPD 3.132  0.605 2.169  3.060 4.536
```

The `mean_ppd` is the sample average posterior predictive distribution of the outcome variable (for details see `help('summary.stanreg')`).

```
MCMC diagnostics
                      mcse  Rhat  n_eff
(Intercept)           0.019 1.000 4127
age                   0.000 1.000 4021
educ                  0.001 1.001 4136
impinc                0.002 1.001 3380
female                0.004 1.000 3879
reciprocal_dispersion 0.000 1.000 4496
mean_PPD              0.010 1.000 3796
log-posterior         0.043 1.000 1688
```

For each parameter, mcse is Monte Carlo standard error, n_eff is a crude measure of effective sample size, and Rhat is the potential scale reduction factor on split chains (at convergence Rhat=1).

4.4.2 Bayesian Estimation of Zero-Inflated Poisson Regression

Next we illustrate how to estimate a zero-inflated Poisson regression model using `rstan`. Since most of the R codes before and after the Stan code blocks are quite similar, below we only present the Stan model setup in this section. As usual, the first Stan code block, the `data` block, declares data matrices and scalars (e.g., K for the number of covariates or the column number, and N for the total number of cases or the row number). The `parameters` block declares the scalar and vector parameters to be estimated. `thetaCons`, `thetaBeta`, `lambdaCons` and `lambdaBeta` correspond to the intercept and the slope vector of the binomial part, and the intercept and the slope vector of the Poisson part of the zero-inflated regression model, respectively.

```
data {
int<lower=0> N;
int<lower=0> K;
matrix[N, K] X;
int<lower=0> y[N];
}

parameters {
vector[K] thetaBeta ;
vector[K] lambdaBeta;
real thetaCons;
real lambdaCons;
}
```

The second Stan code block is the `transformed parameters` block, which includes quantities that are functions of parameters declared in the parameters block, for the ease of MCMC simulation.

```
transformed parameters {
vector[N] theta;
vector[N] lambda;
theta = inv_logit(thetaCons + X * thetaBeta);
lambda =  exp(lambdaCons + X * lambdaBeta);
}
```

Because zero-inflated models are finite mixture models in nature and their likelihood functions usually involve mixing distributions from different families or with different parameters, there are usually no customized Stan distributional functions for these likelihoods. For such models, one needs to use some low-level Stan utilities, including functions and key words, to program. First, the increment log density key word (not a variable), `target`, is required to be used for summing across individual log likelihoods. The `+=` notation right after `target` follows the usual programming language convention and explicitly requests Stan to add up log likelihoods to `target` incrementally. For example, `target += pmf` is equivalent of `target = target + pmf` semantically, although the latter is not a legitimate statement in Stan, syntactically. The `log_sum_exp` function is a Stan function to define some mixture of distributional functions on the log scale. For example, `log_sum_exp(f, g) = log(exp(f) + exp(g))`. While programming the mixture component for modeling zero counts, we follow the common practice to reverse the one-zero coding such that one denotes zero count and zero for non-zero counts for the binary response variable in the binomial part.

```
model {
thetaBeta ~ normal(0,10) ;
lambdaBeta ~ normal(0,10);
thetaCons ~ normal(0,10) ;
lambdaCons ~ normal(0,10);
for (n in 1:N) {
if (y[n] == 0)
target += log_sum_exp(bernoulli_lpmf(1 | theta[n]),
bernoulli_lpmf(0 | theta[n])
+ poisson_lpmf(y[n] | lambda[n]));
else
target += bernoulli_lpmf(0 | theta[n])
+ poisson_lpmf(y[n] | lambda[n]);
}
}
```

Once the Stan model block is completed, we can link the empirical data to the Stan model and run the MCMC simulation using the `stan` function from the `rstan` package as usual.

5

Survival Regression

> From tenderness arises survival, and
> within demise lurks rigidity.
>
> in Chapter 76 of *Dao De Jing* by Laozi

5.1 Introduction

There are several names used for survival analysis across different disciplines, including duration or transition models in econometrics, event history analysis in sociology, survival analysis in biostatistics, quality control and reliability analysis in manufacturing and technology (Greene, 2007; Hosmer et al., 2008; Allison, 2014). All these refer to the same ensemble of statistical techniques for analyzing time-to-event, or most often, survival data, for example, the duration of a disease remission, age to death or divorce, or time to a failure of some machinery. The response variables in such analyses are usually measured as time taken to some milestone event, and they are sometimes called event or failure time variables, which we will use throughout the rest of the chapter.

Conceptually, survival analysis can be viewed as an inverse modeling technique of count regression models. For count or event-count response variables, we mainly look at the total number of counts a similar (in nature) or same event occurs during a specified time frame, for example, the number of clinical visits, alcohol consumption, or publications one has during a year. Event time response variables usually measure how much time it takes to change from one status to another, or the amount of time taken for the milestone event to occur, for example, time to death, to first marriage, or to first sexual intercourse. The transition from one status to another or the occurrence of the event needs to be short and crisp, relative to the duration before and after the change.

The nomenclature of survival analysis does not arise without a good reason, and it is about survival and demise indeed. The origin of survival analysis that has a clear record could date back to England in the 16th century, when this island state and the whole European continent were still being struck intermittently by the bubonic plague, and the government needed to tally death counts so as to monitor its prevalence and take measures to mitigate or even prevent it from spreading. In this sense, statistics was indeed born for politics. A warning system was established in London in the 1530s and then was expanded to the rest of England within the Church (Hald, 1990).

Counts of mortality for the parishes of London were released on a weekly basis since 1604. It was not until the mid of the 17th century did scholars take a first serious analytic look at such cumulative data, with the publication of the 85-page thin volume of *Observations Made upon the Bills of Mortality* by John Graunt (1662). While attempting to estimate the size of the pool of eligible men to serve in combat, Graunt in this book derived a life table (Table 5.1) showing the distribution of a hypothetical cohort of 100 newborn infants

DOI: 10.1201/9780429056468-5

Table 5.1
Graunt's Life Table

Viz. of 100 there dies within		The fourth	6
the first six years	36	The next	4
The next ten years, or Decad	24	The next	3
The second Decad	15	The next	2
The third Decad	09	The next	1

as time elapses over decades starting with ages zero and six (Graunt, 1662; Hald, 1990). Based on these numbers, he continued to fill out the frequency distribution of the said 100 that remain alive at each decade in Table 5.2.

In these two quaint tables, the two pairs of values that Graunt had at hand are (6, 64) and (76, 1) with the first number referring to age and the second for number of survivors. It is still not clear as to how Graunt came up with all the numbers in between. Although he mysteriously "sought six mean proportional numbers" to fill in the gap in between, later scholars surmised that Graunt probably used some geometric series for the living to fill in all the blanks in between (Hald, 1990).

We can use simple base functions and those from the `survival` package (Therneau and Grambsch, 2000) in R to reconstruct and plot the data as illustrated in Fig. 5.1. Graunt's life table, constructed almost four centuries ago, might look primitive and antique to some today. The statistical techniques demonstrated in his analyses nonetheless present probably the first rigorous systematic investigation of the association between age and death rates.

5.1.1   Censoring and Truncation

Without censoring or truncation, survival analysis might have been simply listed as a regular section under the general framework of categorical and limited response data analysis. Censoring and truncation are often considered as nuisance insomuch as they pose additional challenges in survival analysis that other major types of data analysis usually do not have; they are also very "popular" since data with missing information, or missing data, have to be dealt with in almost all kinds of data analysis, given that censoring and truncation could be viewed as special types of missing data issues. Like in many other types of statistical modeling, censoring in survival analysis refers to cases (or rows) containing only partial information. For censoring, usually some information is collected, but not in its complete form. For example, a cancer patient enters a study and then survives after the study terminates. Thus, we cannot retrieve any additional information about what would happen to this patient after the study. Truncation is an issue usually related to sample bias or

Table 5.2
Graunt's Life Table Continued

At Sixteen years end	40	At Fifty six	6
At Twenty six	25	At Sixty six	3
At Thirty six	16	At Seventy six	1
At Forty six	10	At Eighty	0

Note: This table is reproduced from p. 62 of Graunt's 1662 *Bill of Mortality*

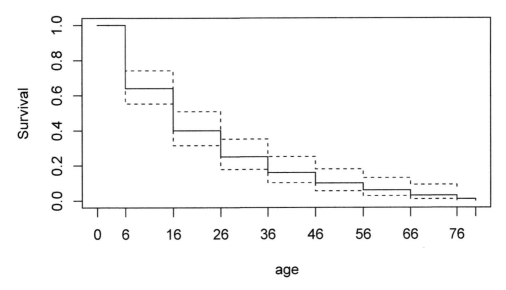

Figure 5.1
Plotted Curve Based on Graunt 1662

selection. When truncation occurs, we do not have any information about some cases that are supposed to be in our sample, for example, a longitudinal study that explores whether certain medicare coverage increases a patient's life expectancy or time to death. Although data analysts usually intend to make inferences about the general patient population, those who die before 65 are impossible to be in the study sample, because by law patients have to reach that cut-off age to be able to receive the benefits of medicare. Thus, we run into the problem of truncation.

There are different kinds of censoring and truncation, including left, right, interval censoring and truncation, and even more complicated types. There are also sub-categories within almost each of the major types of censoring and truncation. The most common form of censoring (as well as truncation) is probably right-censoring, followed by left truncation and interval censoring (Greene, 2007; Hosmer et al., 2008). In the text, we primarily consider survival regression models with right censoring, with brief discussion of adjustments made to examine survival data with other types of missing information, wherever appropriate. For right censoring, observation usually starts at a known time, and cases survive after the termination of a study without changing status (e.g., from life to death) within the observed time frame, and it is also called end-of-study censoring. Sometimes, observations are lost to followup before the milestone event occurs, for example, random dropouts due to migration. This particular type is called loss-to-followup censoring. Fig. 5.2 gives us the graphical presentation of a hypothetical longitudinal study with right censoring. Cases x_1, x_4, and c_{r3} all enter the study at the inception, and both x_1 and x_4 have the milestone event occurring in the middle of the observation, whereas c_{r3} is lost due to study termination, and we are not sure whether and how long it takes for c_{r3} to have the event occur after all. As for c_{r2}, it enters in the middle of the study and disappears due to reasons other than the milestone event occurring or study termination. Both c_{r2} and c_{r3} belong to right-censoring, and our text primarily focuses on cases such as c_{r3}.

Right Censoring Examples

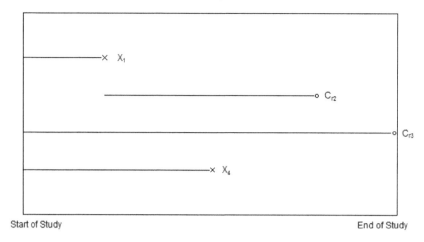

Figure 5.2
Examples of Right-Censoring

5.2 Basic Concepts

The complexity with survival data analysis does not only lie with the convoluted longitudinal data structure with censoring, but also some level of the obscurity with the response variable and accordingly interpretation of results. In survival analysis, sometimes it can be confusing for beginners about what the response variable really is. In survival data, event time (time to event) is calculated based on entry and exit time, and status is about whether the milestone event occurs or not. Both are usually recorded and required for survival analysis. So a natural candidate for analysis is simply time—the time having elapsed for the milestone event to occur. Of course, it is possible that the event does not occur by the time when the study terminates, or at all. A second, probably more popular quantity of interest is how likely the milestone event would occur during the instantaneous moment right around some amount of time elapsed. This is usually called hazard, hazard rate, or hazard function in survival analysis. Time and hazard are the two most commonly used response quantities of interest to be predicted. Based on these two quantities, one can derive other statistics of interest for analysis and interpretation. Unless noted otherwise, we will use survival (e.g., time to death) as an example throughout the rest of this chapter to simplify our discussion.

5.2.1 Time and Survival Function

Given some amount of survival data, which usually contain event time elapsed from a defined time point until the observation terminates with or without the milestone event occurring, one might be interested in several quantities other than the event time, especially given the stochastic nature of the whole process. For example, one might be interested in the probability of surviving past a certain time point, or the probability that the event could

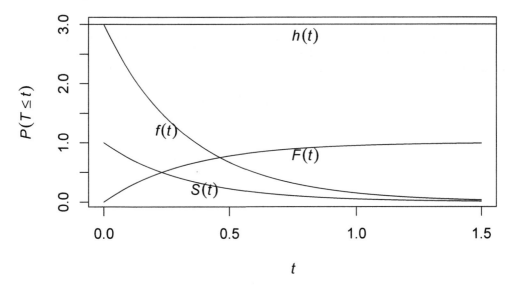

Figure 5.3
An Exponentially Distributed Event Time Variable

occur at a certain time point. Survival analysis has its nomenclature reserved for such quantities; so if we define T as a random variable denoting the time elapsed from a defined entry time point (e.g., the starting time point of observation), then for this random variable, we could be interested in the probability of survival up to a specific time point t, or, what is usually defined as the cumulative density function of this event time variable T,

$$F(t) = P[T < t]$$
$$= \int_0^t f(z)dz \tag{5.1}$$

Survival function, on the other hand, corresponds to the probability of surviving through and after time t. So the survival function is the complement of $F(t)$ defined above,

$$S(t) = P[T \geq t]$$
$$= 1 - F(t) \tag{5.2}$$

Note that Fig. 5.3 graphs the probability density, cumulative probability density, and corresponding hazard and survival functions of a hypothetical random event time variable that follows an exponential distribution with its parameter set to 3.

It can be shown from the figure that the hazard function, $h(t)$, derived based on our distributional assumption, is also a constant of 3. The probability density function, $f(t)$, decreases from 3 at $t_0 = 0$ onward, though in a decelerated manner. Since $f(t)$ determines the functional form of the cumulative density, $F(t)$, and the survival function, $S(t)$, it is not surprising to see that $F(t)$ also increases at a diminishing rate, whereas $S(t)$ decreases as a mirroring curve of $F(t)$, symmetric about the horizontal line $y = 0.5$.

5.2.2 Hazard Function

In survival analysis, hazard function can be approximately understood as the instantaneous probability of the milestone event occurring as a function of the event time variable, T, given surviving up to time point t without any status change. Given time t, we can then calculate the hazard or hazard rate. Hazard is probably the most targeted quantity in survival analysis and is also known as the failure rate in quality control, hazard rate in most disciplines, or force of mortality in demography (Hosmer et al., 2008; Cleves et al., 2016). Hazard is usually viewed as more informative than other quantities related to an event time variable, and that is why most parametric and semi-parametric regression models use hazard rates as the quantity of interest. Formally, hazard as a function of time variable, T, is defined as

$$
\begin{aligned}
h(t) &= \lim_{\Delta t \to 0} \frac{P[t \leq T < t + \Delta t | T \geq t]}{\Delta t} \\
&= \lim_{\Delta t \to 0} \frac{F(t + \Delta t) - F(t)}{\Delta t \, S(t)} \\
&= \frac{f(t)}{S(t)}
\end{aligned}
\tag{5.3}
$$

Since this quantity is essential in survival analysis, it is worthwhile to study its definition and related quantities thoroughly. The first line of the above equation mathematically re-states the definition of hazard discussed previously; it is the limit of a ratio of the probability that the milestone event is going to occur during the instantaneous moment $t \leq T < t + \Delta t$, given that one has survived up till time point t ($T \geq t$) while passing into the next very small amount of time interval, denoted by $t + \Delta t$, over that usually minuscule time interval, Δt. As Δt approaches zero, $\Delta t \to 0$, $t + \Delta t$ gets closer to t asymptotically. So $t \leq T < t + \Delta t$ is the next instantaneous moment right after t. Note that T denotes the event time variable, viz., the time it takes for the milestone event to occur. That means a change of status occurs in between time, $t \leq T < t + \Delta t$. The second line of Eq. 5.3 turns the conditional probability, $P[t \leq T < t + \Delta t | T \geq t]$, into a ratio of the difference in two cumulative density functions (usually equivalent of probabilities), $F(t + \Delta t) - F(t)$, normalized by the survival function, $S(t)$, based on the condition $T \geq t$; thereby, $P[t \leq T < t + \Delta t | T \geq t] = \frac{F(t+\Delta t)-F(t)}{S(t)}$. Based on the relationship between a cumulative density and its probability density functions, $F'(x) = f(x) = \lim_{\Delta t \to 0} \frac{F(t + \Delta t) - F(t)}{\Delta t}$, finally we can have $h(t) = \frac{f(t)}{S(t)}$. Here, $f(t)$ can be roughly understood as the proportion of cases having survived to *and* having their status changed at (right around) time t, whereas $S(t)$, as defined previously, is the proportion of cases surviving past t. So $h(t)$ could be viewed as the proportion of cases with their status changed at time t, of all those that have survived up to time t, exactly a measure of rate. One can refer back to Fig. 5.3 to study the curves corresponding to these functions for an exponential random variable to have more intuitive understanding of the relationships among the few functions.

5.3   Descriptive Survival Analysis

5.3.1   The Kaplan-Meier Estimator

Before we investigate survival regression analysis, it would be useful to review descriptive methods for survival data. Like in any type of data analysis, with survival data, it is highly recommended that one begins with descriptive methods. The most commonly used descriptive statistic for survival data is the Kaplan-Meier estimator, which is also called the product limit estimator. The Kaplan-Meier estimator is a non-parametric statistic to calculate the

Table 5.3

Selected 10 Cases from the 2000 GSS-NDI Database

Subject	Time	Death 2008
1	34	1
2	47	0
3	75	1
4	75	1
5	75	1
6	78	0
7	79	1
8	86	1
9	91	0
10	94	1

survival function, since no parametric functional form is assumed. The Kaplan-Meier estimator begins with a sample of subjects, be they human beings, animals, or machinery under study, starting from time point zero, or the entry time, and follows through this sample as cases reach the milestone event or disappear due to either follow-up or study termination loss, with a step-wise procedure as the time unit progresses. Below, we use 10 selected cases from the General Social Survey data linked to the National Death Index (henceforth GSS-NDI) to illustrate how to calculate the empirical survival function and accordingly the Kaplan-Meier estimator (Muennig et al., 2011; Smith et al., 2019). The data are given in Table 5.3.

Column 1 lists renumbered subject IDs, column 2 their corresponding death age, and column 3 for censoring status, with one denoting death and zero for censored cases (i.e., still alive in 2008 in this subset). These ten cases were already pre-processed and are ordered by time (or death age). The Kaplan-Meier estimator follows through a sample of observations, constructs conditional probability for each time interval record available, and then computes the empirical survival function. For each empirically given time interval, albeit how irregular it may be, one computes the proportion of the subset that survives to the end of the time interval, out of the total that are at risk at the beginning of this particular time interval. To carry through and illustrate such computation, we aggregate data from Table 5.3 and turn it into the following computational table:

It can be shown in Table 5.4 that the columns record the age at which observations died, the number of observations at risk at the beginning of the time interval, the number that

Table 5.4

The Kaplan-Mier Estimation: An Example Using the 10-Case GSS-NDI Data

Time	Num at Risk	Num Died	Num Censored	Cond. Prob.	Survival Function
34	10	1	0	0.9	0.9
47	9	0	1	1	0.9
75	8	3	0	0.625	0.5625
78	5	0	1	1	0.5625
79	4	1	0	0.75	0.4219
86	3	1	0	0.667	0.2813
91	2	0	1	1	0.2813
94	1	1	0	0	0

died during this interval, the number censored (i.e., cases living through this interval and we do not know whether they died after 2008), the conditional probability of survival for each time interval (i.e., the probability of survival during the interval given one had survived to the beginning of this time interval), and the empirical survival function estimated through computing conditional probabilities sequentially for all time intervals.

For example, in the first row, during the age interval from birth up to, but not including, the 35th birthday (written as $I_0 = \{t : 0 \leq t < 35\} = [0, 35)$), we have ten observations, and one of them died between the age of 34 and 35. So as we calculate, the conditional probability of survival is $1 - 1/10 = 0.9$. For the second row or time interval ($I_1 = \{t : 35 \leq t < 48\} = [35, 48)$), we begin with nine observations. None of the nine individuals died, but we lost one case to censoring. This individual was still alive while being checked through the National Death Index in 2008, when he/she was 47 years of age. Although the information is incomplete as to when this person died, we do know though that this individual lived past the second time interval. During the second time interval, the conditional probability for survival is 1 since all nine cases survived through this interval. When we move down to the third row or time interval ($I_2 = \{t : 48 \leq t < 76\} = [48, 76)$), we see that three observations died during this time interval. Note that we lost one case to death during the first time interval, and one case to censoring during the second time interval, so we are left with eight cases at the beginning of the third time interval. Since three died during this interval, we have five surviving through. The conditional probability of survival given one had survived until the beginning of the third time interval is $1 - 3/8 = 0.625$. So far, all we have dealt with are conditional probabilities; that is, the probability of surviving through a particular time interval, conditioned on being alive at the beginning of that particular time interval. Then what about the probability of surviving from $t = 0$, age zero, to the end of the third time interval, $t = 75$? Here we need to apply one of the basic properties of probability, the multiplication rule for conditional probabilities. The probability of surviving through the first, second and third, three consecutive time intervals, $P(I_0\&I_1\&I_2)$, is equivalent of $P(I_0)P(I_1|I_0)P(I_2|I_1) = 0.9 \times 1 \times 0.625 = 0.5625$, the last number in row 3. Note that all numbers in the last column (column 6) can be computed as a *product* of the number in their corresponding row under the second to the last column (i.e., column 5) and the number immediately above (i.e., the previous row) in the last column. As such, we can get all the marginal probabilities, or the empirical survival function. This example gives us some sense about how generally the Kaplan-Meier estimator of the survival function works. Mathematically, the survival function is defined as

$$\widehat{S}(t) = \prod_{k|t_k \leq t} \frac{n_k - d_k}{n_k} \tag{5.4}$$

where n_k is the number of cases at risk at time t_k, d_k is the number of deaths at time t_k, and $\frac{n_k - d_k}{n_k}$ is the conditional probability of survival at time t_k. Then $\widehat{S}(t)$ is a product of a series of such conditional probabilities for each t_k elapsed successively by time t. And as described previously, $S(t_k) = S(t_{k-1})\left(1 - \frac{d_k}{n_k}\right)$. For details about calculating standard errors of $\widehat{S}(t)$, please refer to Hosmer et al. (2008, pp. 27-44).

There are several R packages that are often used for survival analysis, including `survival`, `flexsurv`, and `rms` (Harrell, 2015; Jackson, 2016). Each of the packages has its own advantages and limitations, including types of models estimated and the techniques used for post-estimation analyses. We can use functions from the `survival` package to calculate the Kaplan-Meier estimator of the ten data points discussed previously and get estimates very close to those that we hand-calculated (Therneau and Grambsch, 2000; Therneau, 2021). We first create two vectors, `time` and `death`, to contain the values for age and censoring status, with one denoting death and zero for censoring. Both `Surv` and `survfit` are functions from the `survival` package. `Surv` is used to create a survival object, and it is also a function that other R survival analysis packages usually inherit and build upon, for example, `flexsurv` (Jackson, 2016), `rms` (Harrell Jr, 2021), and `spBayesSurv` (Zhou et al., 2020). For right-censored data, the first argument in the `Surv` function is usually the follow-up time variable and the second is the status or what the default calls the event variable, with one denoting event time and zero censored. Of course, there will be some syntactic differences if other types of censoring or two time points are involved (e.g., entry and exit/censoring time). Since no covariates are used in this case, one can specify a 1 to the right of the tilde sign for an intercept only model. The `survfit` function estimates the non-parametric survival probability for each empirically observed time interval and its corresponding standard error as well as confidence interval. We can then create an `survfit` object, `km.fit` in this case, which stores all the estimated statistics for the `plot` function to draw a Kaplan-Meier plot.

```
require(survival)
time = c(34, 47, 75, 75, 75, 78, 79, 86, 91, 94)
death = c(1, 0, 1, 1, 1, 0, 1, 1, 0, 1)
km.fit <- survfit(Surv(time, death) ~ 1)
summary(km.fit)

## Call: survfit(formula = Surv(time, death) ~ 1)
##
##   time n.risk n.event survival std.err lower 95% CI upper 95% CI
##     34     10       1    0.900  0.0949       0.7320        1.000
##     75      8       3    0.562  0.1651       0.3165        1.000
##     79      4       1    0.422  0.1737       0.1883        0.945
##     86      3       1    0.281  0.1631       0.0903        0.876
##     94      1       1    0.000     NaN           NA           NA

plot(km.fit, main="Kaplan-Meier Plot")
```

In Fig. 5.4, the solid line is the estimated survival function and the two dotted lines sketch the confidence bands of the survival function. It is not surprising to see that as one ages, the probability of survival decreases.

5.3.2 The Log-Rank Test

In addition to plotting single descriptive survival curves, sometimes we may want to compare survival curves across two and sometimes more groups to see if they are statistically different, and hence the log-rank test (Clark et al., 2003; Hosmer et al., 2008; Cleves et al., 2016). The whole idea of conducting the log-rank test is quite similar to computing the chi-squared statistic for testing the association between two categorical variables. To illustrate, we focus on one time interval with two groups in the following table.

Table 5.5
A 2x2 Survival Table in Time Interval t_j

Status	Group 1	Group 2	row marg.
Died at t_j	o_{d1}	o_{d2}	r_d
Survived at t_j	o_{s1}	o_{s2}	r_s
col. marg.	c_1	c_2	Total $= n_j$

Kaplan–Meier Plot

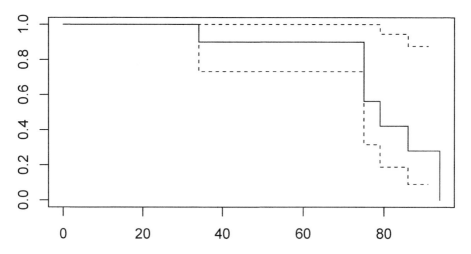

Figure 5.4
The Kaplan-Meier Plot

where $n_j = o_{d1} + o_{d2} + o_{s1} + o_{s2}$ is the total count of the risk set (cases at risk), $r_d = o_{d1} + o_{d2}$ and $r_s = o_{s1} + o_{s2}$ are row marginals, $c_1 = o_{d1} + o_{s1}$ and $c_2 = o_{d2} + o_{s2}$ are column marginals for the time interval t_j. For the null hypothesis to hold that there is no difference in the survival function between the two groups, we would expect that the column percentages for both groups should be equal to their corresponding row marginals divided by total n_j. For example, if the null hypothesis is valid, then for the expected cell count of those in Group 1 who died, e_{d1}, we should have $e_{d1}/c_1 = r_d/n_j$, and thus $e_{d1} = r_d{\cdot}c_1/n_j$. This is for a particular time interval t_j. Then for every time interval with G groups under examination, the test statistic first calculates the difference between the observed failure count and the expected count under the null (group membership is independent of death/failure rates), given the total number at risk at the beginning of the time interval t_j, and then sums such squared difference adjusted by its corresponding variance for all groups, thus leading to a χ^2 statistic with $G - 1$ degrees of freedom for G groups (Mantel, 1966; Clark et al., 2003),

$$\chi^2 = \sum_{g=1}^{G} \frac{(o_g - e_g)^2}{e_g} \tag{5.5}$$

Note that to have an omnibus test of all time intervals combined to test the statistical difference in *two* survival curves, one may have several versions of the testing procedure, proposed, revised, and improved by Mantel (1966), Peto and Peto (1972), and Hosmer et al. (2008). The most predominant form of this test is to calculate a χ^2 statistic with one degree of freedom, $M = \frac{\left[\sum_{j=1}^{J} w_j (o_{d1} - e_{d1})\right]^2}{\sum_{j=1}^{J} w_j^2 v_{d1}}$, where o_{d1} and e_{d1} are calculated as described previously, and v_{d1}, the estimator of the variance of o_{d1}, is defined as $\frac{c_1 c_2 r_d r_s}{n_j^2 (n_j - 1)}$ using hypergeometric probability distribution, and w_j is the weight used for adjusting for the difference between observed and expected counts in the numerator and the variance in denominator, depending on the selected test; in most cases, the weights are set to one (Hosmer et al., 2008, p. 47).

Another version of the test constructs the test statistic as $\sum_g \frac{(O_g - E_g)^2}{E_g}$, where $O_g = \sum_{j \in g} d_j$ and $E_g = \sum_{j \in g} e_j$, and it is premised to follow a χ^2_{G-1} distribution (Peto and Peto, 1972, pp. 192-193).

Below, the 2008 GSS-NDI data are used again to illustrate. Since samples from later years have many censored cases, in this example we only select birth cohorts from 1930 to 1939. The response variable is survival time from birth, and the predictors chosen include education (educ) in years, income (impinc) on a natural logarithm scale, and gender (female = 1). Note that our selection of these variables serves illustrative and presentational purposes mostly, not to achieve high explanatory power. Below, the first few lines of R codes read in, subset, and summarize the data for analysis. We can then use the survdiff function from the survival package to conduct the log-rank test. Again, the Surv function first sets up the event time response variable along with its censoring status information to the left of the tilde sign, and the group variable, female, on the right. For the log-rank test, we use the gender variable to test whether the survival curves between males and females are statistically different. The few other predictors will be used in the survival regression models to be discussed later. To see all the details of the survival table, we can use the summary function that is silenced below with the "#" sign in front of it.

```
mygss <- read.dta("data/gssndi7802rec.dta", convert.factor=F)
mydta <- subset(mygss, cohort > 1930 & cohort <= 1940,
        select=c(year, cohort, time, agedeath, death, age,
        educ, impinc, female, race, black, white, other))
# summary(mydta)
log.rank <- survdiff(Surv(time, death) ~ female, data=mydta)
log.rank

## Call:
## survdiff(formula = Surv(time, death) ~ female, data = mydta)
##
##
##             N Observed Expected (O-E)^2/E (O-E)^2/V
## female=0 1686      641      549      15.4      27.5
## female=1 2128      645      737      11.5      27.5
##
##   Chisq= 27.5  on 1 degrees of freedom, p= 2e-07

# summary(genderSurv)
```

The results show that for one degree of freedom (a total of two groups), we have a $\chi^2 = 27.538$, and thus we can reject the null that there is no difference in empirical survival functions between males and females. We can further plot the two survival curves using the survfit and plot functions from the survival package.

```
kmSexFit <- survfit(Surv(time, death) ~ female, data=mydta)
plot(kmSexFit, col=c("grey80", "grey50"), xlim=c(50, 80), ylim=c(0.4, 1))
legend("topright", legend=c("Male", "Female"),
        col=c("grey80", "grey50"), lty=1, cex=0.8)

# autoplot(genderSurv, xlim=c(50, 80))
```

It can be shown in Fig. 5.5 that the survival curve for females is consistently above that of males from 50 to 80 years of age, thus graphically corroborating our findings from the log-rank test.

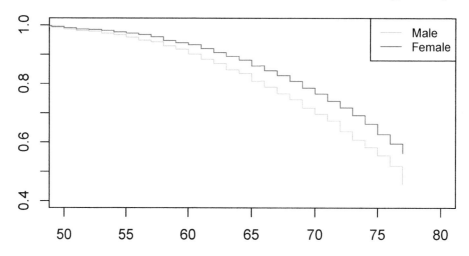

Figure 5.5
Kaplan-Meier Plot by Gender

5.4   Accelerated Failure Time Model

After one conducts descriptive analysis of survival data, the next step is to model the event time response variable within a regression framework. As discussed previously, there are usually two quantities of interest in survival models—time and hazard, and these two quantities lead to different parameterizations of survival regression models. We start with the most obvious one, event time, T, as the response variable to be modeled directly. Because time has to be a positive number, taking the natural logarithm of the event time response variable might be the natural choice in a structural model so that the natural log of time matches the domain on the right-hand side—any number on the real number line—of the structural model, for instance, $\ln(t) = \mathbf{x}\beta + \varepsilon$; equivalently, one can have $t = \exp(\mathbf{x}\beta + \varepsilon)$ to satisfy the range requirement for time. If we assume the error term, ε, follows a normal distribution, then T follows a log-normal distribution. Without censoring, we can directly use classical linear regression to model event time. This actually can be done easily in most popular statistical environments, such as R or Stata. Since event time has its unique features, assuming a normal distribution for the error can lead to information loss and biased estimates. Although the log-normal distribution is quite convenient, it usually does not accurately describe most (empirical) distributions of event time response variables. In the empirical modeling process, ε is often understood and re-written as a quantity also on a log scale, $\varepsilon = \ln(\epsilon)$, so the structural model becomes

$$\ln(t) = \mathbf{x}\beta + \ln(\epsilon) \tag{5.6}$$

This model is called the accelerated failure time (AFT) model. By exponentiating both sides of the equation above, we have $t = \epsilon \cdot \exp(\mathbf{x}\beta)$. Note that the effects of covariates on time are on a multiplicative scale, thus accelerated failure time. Suppose $\exp(\mathbf{x}\beta) = \exp(\beta_0 + \beta_1 x_1 + \beta_2 x_2 + \ldots + \beta_k x_k + \ldots + \beta_K x_K)$, and if we increase the value of a generic variable x_k by one unit to $x_k + 1$, while holding all other variables unchanged, and take a ratio of these two time-point values, then we get $\frac{t_{\mathbf{x}\beta, x_k+1}}{t_{\mathbf{x}\beta, x_k}} = \exp(\beta_k)$. This gives us the exact interpretation of the effect of any covariate on time; that is, for each unit increase in x_k, we would expect time to increase by a factor of $\exp(\beta_k)$, holding all other variables constant.

This interpretation applies to both exponential and Weibull AFT models to be discussed in the next two subsections.

Combining $\exp(\mathbf{x}\beta)$ with t to solve for ϵ, we have $\epsilon = t \cdot \exp(-\mathbf{x}\beta)$. Here $\exp(-\mathbf{x}\beta)$ is called the acceleration parameter. If $\exp(-\mathbf{x}\beta) = 1$, we have a model with time passing at a constant or normal rate. If $\exp(-\mathbf{x}\beta) > 1$, then time elapses at an accelerated rate. If however $\exp(-\mathbf{x}\beta) < 1$, then we have time running at a decelerated rate (Cleves et al., 2016). Because the event time variable is on a natural log scale, this model is also called a log-time model. Depending on the distributional assumption made about ϵ and thereby $\ln(\epsilon)$, we can have various parametric AFT survival regression models. For AFT survival regression models to be discussed below, we will focus on R implementation, interpretation of coefficients, and graphing survival curves since most empirical research uses hazard instead of time as the quantity of interest. As much of the idea behind distributional assumption, estimation, post-estimation analysis, and the transformation from the response quantity of interest to survival probabilities is somewhat or even very similar across parametric survival regression models, we will defer our discussion about such issues till when we cover parametric proportional hazard regression models.

5.4.1 Exponential AFT Regression

Based on the previous formulation, if we assume ϵ follows an exponential distribution, which has a single parameter λ, then we have an exponential AFT survival regression model. Using the same 2008 GSS-NDI data, below we illustrate how to estimate an exponential AFT model using the `survreg` function from the `survival` package. Similar functions can also be found in the `flexsurv`, `rms`, and other packages for survival analysis. With `suvreg`, the default type is an AFT model. The first argument in the `survreg` function is to specify the `formula`. We start with the concatenated function, `Surv`, which has two arguments, the first (`time`) of which tells the event time, and the second (`death`) for censoring status. With a variety of set-ups for censoring types, the second variable for censoring time sets alive to zero (censored) and dead to one. The tilde sign, "`~`", separates the event time response variable from a list of covariates, followed by the data file loaded in the memory, `mydta`, and the distribution type for the error term, which is exponential in this case (`dist = "exponential"`).

```
aftexp <- survreg(Surv(time, death) ~ educ + impinc+ female,
        data=mydta, dist = "exponential")
summary(aftexp)

##
## Call:
## survreg(formula = Surv(time, death) ~ educ + impinc + female,
##     data = mydta, dist = "exponential")
##                  Value Std. Error     z        p
## (Intercept) 3.60531    0.28380 12.70 < 2e-16
## educ        0.02183    0.00942  2.32     0.02
## impinc      0.13064    0.03072  4.25 2.1e-05
## female      0.27933    0.05636  4.96 7.2e-07
##
## Scale fixed at 1
##
## Exponential distribution
## Loglik(model)= -8093.9   Loglik(intercept only)= -8121.7
##  Chisq= 55.61 on 3 degrees of freedom, p= 5.1e-12
```

```
## Number of Newton-Raphson Iterations: 4
## n=3801 (13 observations deleted due to missingness)

print(expaftexp <- exp(aftexp$coef), digit=3)

## (Intercept)        educ       impinc       female
##       36.79        1.02         1.14         1.32
```

There are several parts in the results section. Right below the recalled formula, there are five columns, corresponding to variable names, raw coefficients, standard errors, z statistics, and their associated p-values. Following the convention in statistics, any estimate with a p-value equal to or smaller than .05 is considered statistically significant. In this case, all parameter estimates are statistically significant; that is, education, income, and gender all have statistically significant associations with the event time response variable. To provide substantively meaningful interpretation, we can exponentiate raw coefficients and discuss the effects of predictors on time directly. For example, after exponentiating the coefficient for education, $\exp(0.022)$, we get 1.022. To interpret the effect of education on the time scale, one can phrase that for each year increase in education, we would expect time to death rises by a factor of 1.022, holding all other variables constant. For dummy variables, in this case `female` (female =1; otherwise = 0), one can state that being female as opposed to male, we would expect the time to death to increase by a factor of $\exp(0.279) = 1.32$, net of the effects of all other covariates. Here we have all "increases" as we have positive raw coefficients. When we have negative raw coefficients or exponentiated coefficients between zero and one, then we have "decrease" by a factor of exponentiated coefficients.

Like for other MLE-based models, we can use some classical hypothesis testing procedures, such as the LR test, or scalar fit measures to compare models. To conduct an LR test, we can use the `anova` function from the base `stats` package by simply listing estimation object names.

```
aftexp0 <- update(aftexp, .~.-impinc)
summary(aftexp0)

##
## Call:
## survreg(formula = Surv(time, death) ~ educ + female, data = mydta,
##     dist = "exponential")
##                Value Std. Error     z        p
## (Intercept) 4.72408    0.11172 42.29 < 2e-16
## educ        0.03843    0.00854  4.50 6.8e-06
## female      0.24733    0.05585  4.43 9.5e-06
##
## Scale fixed at 1
##
## Exponential distribution
## Loglik(model)= -8102.6   Loglik(intercept only)= -8121.7
##  Chisq= 38.26 on 2 degrees of freedom, p= 4.9e-09
## Number of Newton-Raphson Iterations: 4
## n=3801 (13 observations deleted due to missingness)

test = anova(aftexp0, aftexp)
test
```

```
##                   Terms Resid. Df    -2*LL Test Df Deviance     Pr(>Chi)
## 1         educ + female    3798 16205.21      NA      NA          NA
## 2 educ + impinc + female    3797 16187.86    =  1 17.34598 3.115559e-05
```

For 1 degree of freedom and a χ^2 of 17.346, the null hypothesis that the parameter for income is zero can be rejected at the conventional .05 level. We can also use information theory-based scalar measures, such as AIC or BIC, to compare the two models. The `AIC` and `BIC` functions from the R `base` package `stats` can compute the AIC and BIC statistic respectively. Recall that a smaller AIC or BIC statistic corresponds to a better-fitting model. Below, both AIC and BIC prefer the model with income.

```
AIC(aftexp0, aftexp)
```

```
##         df    AIC
## aftexp0  3 16211.21
## aftexp   4 16195.86
```

```
BIC(aftexp0, aftexp)
```

```
##         df    BIC
## aftexp0  3 16229.94
## aftexp   4 16220.83
```

5.4.2   Weibull AFT Regression

If instead we assume ϵ follows a Weibull distribution, which has a shape parameter η and a scale parameter σ, then we have a Weibull AFT survival regression model. With the same GSS-NDI data used in the preceding section, below we illustrate how to estimate a Weibull AFT model using the same `survreg` function. With everything else staying the same as they are in the exponential AFT model, we only change the `dist` option from `exponential` to `weibull`.

```
aftweib <- survreg(Surv(time, death) ~ educ + impinc+ female,
        data=mydta, dist = "weibull")
expaftweib <- exp(aftweib$coef)
```

Interpretation-wise, there is little difference between the Weibull AFT and exponential AFT regression models. So for example, for education, we can say that holding all other variables constant, for each year increase, we would expect the time to death to increase by a factor of $\exp(0.001) = 1.001$ (output omitted). We can use AIC, for example, to compare across models with different parameterization, and the results show that the AFT Weibull regression outperforms the AFT exponential regression.

```
AIC(aftexp, aftweib)
```

```
##         df    AIC
## aftexp   4 16195.86
## aftweib  5 12560.08
```

We can also use survival curves to examine and compare models. To more easily graph survival curves from AFT models, we use the `flexsurvreg` function from the `flexsurv`

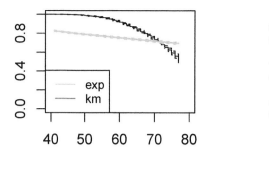 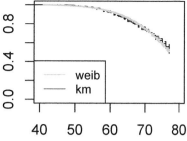

exponential AFT   weibull AFT

Figure 5.6
Survival Curves

package to run the model first and then plot. The syntactic structure of `flexsurvreg` is very similar to that of `survreg`. After fitting multiple models, we can use the `plot` function from the same package to graph survival curves. To plot, we first specify the estimation results object name, followed by other arguments for various graphing as well as aesthetics parameters. The `plot` function usually draws a survival curve using mean values for all independent variables. To more easily compare survival graphs, we reconfigure the graph parameter, `mfrow`, which follows the `c(nr, nc)` format; the first number in the vector determines the number of rows (1 in this case), and the second number sets the number of columns (2 in this case). Below, we plot the survival curves based on results from the exponential and Weibull AFT regression models.

```
library(flexsurv)
exp.fit <- flexsurvreg(Surv(time, death) ~ educ + impinc + female,
        data=mydta, dist = "exponential")

weib.fit <- flexsurvreg(Surv(time, death) ~ educ + impinc + female,
        data=mydta,dist = "weibull")

par(mfrow=c(1,2))
plot(exp.fit, xlim=c(40, 80), col=c("grey80", "grey50"), sub="exponential AFT")
        legend("bottomleft", legend=c("exp", "km"),
        col=c("grey80", "grey50"), lty=1, cex=0.8)
plot(weib.fit, xlim=c(40, 80), col=c("grey80", "grey50"), sub="weibull AFT")
        legend("bottomleft", legend=c("weib", "km"),
        col=c("grey80", "grey50"), lty=1, cex=0.8)
```

From the two collated graphs in Fig. 5.6, it can be shown that the fitted survival curve in light gray based on the Weibull AFT model fits the observed non-parametric Kaplan-Meier plot much better than the one produced by the exponential AFT model. Sometimes we want to examine survival curves with covariates set to other values than the middle of the measures of central tendencies. We can draw such curves with an `flexsurvreg` object using the `plot` function. For example, we could set gender to be female, education to be 16, income to its sample mean, and then plot its corresponding survival curve,

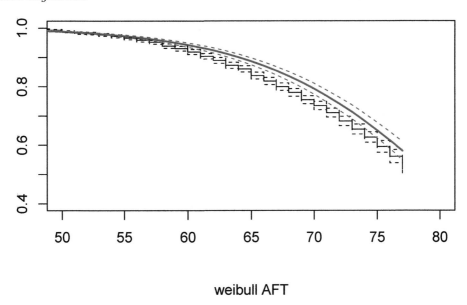

Figure 5.7
Weibull AFT Survival Curve for A Hypothetical Individual

```
covs <- data.frame(educ = 16, impinc=mean(mydta$impinc), female = 1)
plot(weib.fit, newdata=covs, xlim=c(50, 80), ylim=c(0.4, 1), sub="weibull AFT")
```

In addition to interpreting the effects of covariates on the event time variable, sometimes we may be interested in predicting this response variable. There are several packages available for this purpose, and the `Zelig` package probably meets the needs best. In general, the `zelig` function from the `Zelig` package is quite similar to the `survreg` function from the `survival` package except that the option that determines model types changes from `dist` in `survreg` to `model` in `zelig`. The concatenated function `Surv(time, death)` sets the event time variable with time and censoring status supplied. The results are stored in the object `z.out`.

```
library(Zelig)
z.out <- zelig(Surv(time, death) ~  educ + impinc + female,
               data=mydta,model = "weibull")
```

Since almost all survival regression models are non-linear in nature, one has to supply both values of the covariates and parameter estimates to calculate the predicted event time. The `setx` function from the `Zelig` package serves this function. The first argument of the `setx` function is to specify a `zelig` output object that stores model-specific estimation results, such as the parameter estimates, and then a list of values for covariates. By default, the values of the covariates that are used in estimation but unspecified in the `setx` function are set to their corresponding sample averages. In this case, we want to compare the survival time of males with that of females.

```
man <- setx(z.out, female = 0)
wom <- setx(z.out, female = 1)
set.seed(47306)
```

```
s.out <- sim(z.out, x = man, x1 = wom)
summary(s.out)

##
##  sim x :
##  -----
## ev
##      mean        sd      50%      2.5%     97.5%
## 1 74.59945 0.3483371 74.60242 73.93256 75.27524
## pv
##        mean        sd      50%      2.5%     97.5%
## [1,] 74.70144 10.37222 75.68379 50.23101 91.56357
##
##  sim x1 :
##  -----
## ev
##      mean        sd      50%      2.5%     97.5%
## 1 77.44695 0.360064 77.44383 76.75856 78.15813
## pv
##        mean        sd      50%      2.5%     97.5%
## [1,] 77.95077 10.31788 79.14406 54.99495 94.75468
## fd
##      mean        sd      50%      2.5%     97.5%
## 1 2.847498 0.4761466 2.857504 1.891036 3.791161
```

The results show that the expected survival time for male is 74.599 and for female is 77.447, so females with the given characteristics on average live 2.847 years longer than males with otherwise similar characteritics.

5.5   Parametric Proportional Hazard Regression

As discussed previously, the two quantities of interest in survival analysis are time and hazard. Some popular parametric models based on time and hazard actually can be translated into one another, such as exponential and Weibull regression models. Such equivalence only occurs for the Weibull regression models, if exponential regression is viewed as a special case of Weibull where the shape parameter, η, is set to be one. In practice, survival regressions directly built on on hazard function appear to be more popular than those based on time.

Once we make an assumption about the functional form of the hazard function, $h(t)$, we can go from $h(t)$ to $f(t)$, the probability density function, as shown in section 5.2.2. Conceptually, $h(t)$ is a rate, so some form of the exponential function becomes a natural candidate for $h(t)$, such as the following widely used formulation,

$$h(t|x_i) = h_0(t) \exp(\mathbf{x}\beta) \qquad (5.7)$$

in which $h_0(t)$ is called the baseline hazard. Models built on this functional form for the hazard function are called proportional hazard (henceforth PH) survival models, because the hazards of two different $\mathbf{x}$ vectors, for example $\mathbf{x}_m$ and $\mathbf{x}_n$, are proportional to each other by a factor of $h(t|\mathbf{x}_m)/h(t|\mathbf{x}_n) = \exp[(\mathbf{x}_m - \mathbf{x}_n)\beta]$. It can also be shown from the mathematical formulation that the hazard at a certain time point is a product of the baseline hazard, which is a function of time, and a positive number calculated from covariates; so the hazard is a result of the nonlinear interaction between time and the exponential function of a linear combination of the covariates.

Depending on the assumption we make about $h_0(t)$, we can have different PH regression models. Also because $h(t) = \frac{f(t)}{S(t)}$, we can construct the likelihood function as follows,

$$
\begin{aligned}
L(\beta|t, \mathbf{x}) &= f(t_1|\beta, \mathbf{x}_1)f(t_2|\beta, \mathbf{x}_2)...f(t_N|\beta, \mathbf{x}_N) \\
&= S(t_1|\beta, \mathbf{x}_1)h(t_1|\beta, \mathbf{x}_1)S(t_2|\beta, \mathbf{x}_2)h(t_2|\beta, \mathbf{x}_2)...S(t_N|\beta, \mathbf{x}_N)h(t_N|\beta, \mathbf{x}_N)
\end{aligned}
\tag{5.8}
$$

For censored cases, we simply retain their corresponding survival functions and drop the hazard functions. Then we can use MLE to estimate the parameter vector, β, and other related quantities or statistics.

5.5.1 Exponential PH Regression

The exponential PH regression is probably the simplest form of survival regression models and often used as the naive baseline for exploratory or illustrative purposes. As the name of the model suggests, the exponential PH regression assumes that the event time variable, T, follows an exponential distribution, $T \sim \exp(\lambda)$, with a single parameter $\lambda > 0$. As such, its probability density function $f(t)$ and cumulative density function $F(t)$ are $\lambda e^{-\lambda t}$ and $1 - e^{-\lambda t}$ respectively. Since $S(t) = 1 - F(t)$, we can get $h(t) = f(t)/S(t) = \lambda$. As it is further assumed that variation in the hazard is determined by individual heterogeneity (characteristics) in the covariates, we set $h_0(t) = c$ to be a constant such that

$$
\begin{aligned}
h(t|\mathbf{x}) &= h_0(t)\exp(\mathbf{x}\beta) \\
&= \exp(\beta_0)\exp(\mathbf{x}\beta) \\
&= \exp(\beta_0 + \mathbf{x}\beta)
\end{aligned}
\tag{5.9}
$$

With the relationship established in Eq. 5.3, we have $S(t|\mathbf{x}) = \exp(-\exp(\beta_0 + \mathbf{x}\beta)t)$ and other related quantities. Researchers often focus on the hazard rate given its ease of interpretation. Relying on the exponential functional form, we can interpret the effects of β on the hazard rate on a multiplicative scale. To make things simpler, let us focus on a generic variable x_k in $\mathbf{x}$ for a series of predictors indexed from 1 to K. To fill in all the details, the equation above becomes $h(t|\mathbf{x}, x_k) = \exp(\beta_0 + \beta_1 x_1 + \beta_2 x_2 + ... + \beta_k x_k + ... + \beta_K x_K)$. For a single unit increase in x_k, say from x_k to $x_k + 1$ for any reasonable value of x_k, we have $h(t|\mathbf{x}, x_k + 1) = \exp(\beta_0 + \beta_1 x_1 + \beta_2 x_2 + ... + \beta_k(x_k + 1)... + \beta_K x_K)$. If we take a ratio of these two hazards, then $\frac{h(t|\mathbf{x}, x_k+1)}{h(t|\mathbf{x}, x_k)} = \exp(\beta_k)$. To interpret, we can say for each unit increase in x_k, one would expect the hazard to change by a factor of $\exp(\beta_k)$, while holding all other variables constant. This quantity is also called hazard ratio or hazard ratio coefficient.

Note that with the `survival` package in R, the default and only model option for the exponential distribution/regression is AFT, and the direct estimation of its PH counterpart is not available. Fortunately, the PH and AFT versions of the exponential regression model are mathematically equivalent. To be exact, the parameters we get from a PH exponential regression model are the same in magnitude with opposite signs, $\beta_{PH} = -\beta_{AFT}$.

```
phexpcoef <- -(aftexp$coef)
exphr <- exp(phexpcoef)
phexpcoef

## (Intercept)        educ       impinc      female
## -3.60531028 -0.02183322 -0.13063719 -0.27933133
```

```
exphr
```

```
## (Intercept)        educ      impinc      female
##  0.02717901  0.97840340  0.87753610  0.75628928
```

Note that we have estimated the exponential AFT model previously, so this time we simply take a negative sign of the coefficient vector, `-(aftexp$coef)`, thus turning raw coefficients from an exponential AFT model into coefficients for an exponential PH regression model. Standard errors still stay the same after such transformation. Thus, all three coefficients in the model remain statistically significant at the .05 level. Exponentiating these coefficients, `exp(phexpcoef)`, we get hazard ratios. Here, we can say that, for each year increase in education, ceteris paribus, we would expect the hazard to decrease by a factor of 0.978; being female as opposed to male decreases the hazard by a factor of 0.756, holding all other variables constant.

Instead of taking the additional turn to get the estimates for an exponential PH regression from its AFT variant, one can use the `phreg` function from the `eha` package to get the same results directly (Broström, 2012),

```
library(eha)
exp.ph <- phreg(Surv(time, death) ~  educ + impinc + female,
              data=mydta, dist="weibull", shape=1)
exp.ph
```

```
## Call:
## phreg(formula = Surv(time, death) ~ educ + impinc + female, data = mydta,
##      dist = "weibull", shape = 1)
##
## Covariate        W.mean      Coef Exp(Coef)  se(Coef)   Wald p
## educ             12.546    -0.022     0.978     0.009    0.020
## impinc           10.070    -0.131     0.878     0.031    0.000
## female            0.561    -0.279     0.756     0.056    0.000
##
## log(scale)                  3.605               0.284    0.000
##
##   Shape is fixed at  1
##
## Events                      1283
## Total time at risk        264931
## Max. log. likelihood     -8093.9
## LR test statistic          55.61
## Degrees of freedom            3
## Overall p-value     5.09903e-12
```

5.5.2   Weibull PH Regression

The Weibull PH regression is among the most popular forms of parametric survival regression models. It assumes that the event time variable, T, follows a Weibull distribution with two parameters, including the shape parameter η and the scale parameter σ, written as $T \sim \text{Weibull}(\eta, \sigma)$ with $\eta, \sigma > 0$. Depending on the parameterization, its probability density function can be written as $f(t) = \eta \sigma^{-\eta} t^{\eta-1} \exp\left(-t/\sigma\right)^{\eta}$, and its cumulative density function is $F(t) = 1 - \exp\left(-t/\sigma\right)^{\eta}$, which then leads to the hazard function,

$h(t) = f(t)/S(t) = \eta\sigma^{-\eta}t^{\eta-1}$. Suppose we have a regression model with a covariate vector, $\mathbf{x}$; if we use the exponential function to estimate σ, such as $\sigma = \exp(\mathbf{x}\theta)$, then we can set $\beta = -\theta\eta$, leading to $\sigma^{-\eta} = \exp(-\mathbf{x}\theta\eta) = \exp(\mathbf{x}\beta)$. Depending on the parameterization, the natural exponential function may or may not contain an intercept term. Instead of setting the baseline hazard function, $h_0(t)$, to be a constant, the Weibull regression sets it to be a function of t, $\eta t^{\eta-1}\exp(\beta_0)$, such that

$$
\begin{aligned}
h(t|\mathbf{x}) &= h_0(t)\exp(\mathbf{x}\beta) \\
&= \eta t^{\eta-1}\exp(\beta_0 + \mathbf{x}\beta)
\end{aligned}
\tag{5.10}
$$

Since the term in front of the exponential function does not involve $\mathbf{x}$, the interpretation of parameter estimates is exactly the same as those from an exponential regression; that is, net of the effects of other variables, for each unit increase in x_k, a generic variable in the $\mathbf{x}$ vector, we would expect the hazard to vary by a factor of $\exp(\beta_k)$. With the relationship established in 5.3, we have $S(t|\mathbf{x}) = \exp(-t^{\eta}\exp(\beta_0 + \mathbf{x}\beta))$.

To illustrate, we can use the same data from the previous sections to run a Weibull PH survival model. Like the exponential survival regression model, the `survreg` function from the `survival` package only provides an accelerated failure time (AFT) version of the model, so we have to turn to the well-established relationship between the Weibull PH and AFT survival models that $\beta_{\text{PHWeib}} = -\eta\beta_{\text{AFTWeib}}$. The `survreg` function reports an estimate of what it calls the scale parameter , $\phi = 1/\eta$. Thus, $\beta_{\text{PHWeib}} = -\beta_{\text{AFTWeib}}/\phi$ after running the `survreg` function to estimate the Weibull AFT regression model.

```
phweibcoef <- -coef(aftweib)/aftweib$scale
weibhr <- exp(phweibcoef)
phexpcoef

## (Intercept)        educ      impinc      female
## -3.60531028 -0.02183322 -0.13063719 -0.27933133

weibhr

## (Intercept)         educ       impinc       female
## 5.799647e-17 9.802430e-01 8.705088e-01 7.163674e-01
```

Alternatively, one can use the `phreg` function from the `eha` package to estimate the same Weibull PH regression. By taking out the "`shape = 1`" option, the function estimates a typical Weibull PH model.

```
library(eha)
phreg(Surv(time, death) ~ educ + impinc + female,
            data=mydta, dist="weibull")
```

Note that the results from our Weibull PH survival regression are similar to those from the exponential PH survival regression model. Again, we simply exponentiate the raw coefficients and get our hazard ratio coefficients. For example, for each year increase in education, we would expect the hazard to change by a factor of 0.98, net of the effects of other covariates; being female as opposed to male decreases the hazard by a factor of 0.716, holding all other variables constant. To collate and compare results, we can use the `stargazer` function from the `stargazer` package to create publication-quality tables, such as the following one (Hlavac, 2018). For the arguments in `stargazer`, one needs to first list the names of estimation objects, and other options that follow are rather self-explanatory.

Table 5.6

Model Comparisons

	Dependent variable:	
	time	
	exponential	*Weibull*
	(1)	(2)
Education	0.022**	0.002**
	(0.009)	(0.001)
Ln(Family Income)	0.131***	0.016***
	(0.031)	(0.004)
Gender(female=1)	0.279***	0.037***
	(0.056)	(0.006)
Constant	3.605***	4.183***
	(0.284)	(0.032)
Observations	3,801	3,801
Log Likelihood	−8,093.931	−6,275.042
χ^2 (df = 3)	55.606***	65.129***
Note:	*p<0.1; **p<0.05; ***p<0.01	

```
require(stargazer)
stargazer(aftexp, aftweib, type='latex', align= T,
        title = "Model Comparisons",
        covariate.labels = c("Education", "Ln(Family Income)", "Gender(female=1)"),
        # colnames=F,
        # column.labels=c("Exponential Reg", "Weibull Reg"),
        keep.stat=c("n", "ll", "chi2", "aic", "bic", "adj.rsq", "scale"))
```

5.6   Cox Regression

The Cox regression model, or sometimes known as the Cox proportional hazard model (Cox, 1972, 1975), is probably the most popular survival regression model in biomedical, epidemiological, and socio-behavioral fields. An interesting anecdote about the discovery of this model recounts that Cox came up with the model while being ill with very high temperature after five years of seemingly fruitless work (Reid, 1994). The Cox model, on one hand, is similar to other PH models previously discussed because its hazard function also follows the form of $h(t) = h_0(t) \exp(\mathbf{x}\beta)$. On the other hand, the Cox model is quite different from other parametric PH models since in Cox regression no distributional assumption is used about the response variable, time T; accordingly, we are not able to derive the functional form for the baseline hazard $h_0(t)$, but the basic relationships still hold among $h(t)$, $f(t)$, and $S(t)$ without relying on any specific distributional form. Cox (1972) reasons that if the quantities of our main interest are regression parameters, or the effects of covariates on the relative risk, such as $h(t|x_i)/h(t|x_j)$, then we can let $h_0(t)$ take any arbitrary form since the baseline hazard cancels and thus drops out in this ratio, and we can simply focus on the second term, $\exp(\mathbf{x}\beta)$. Note that the constant term in $\exp(\mathbf{x}\beta)$ does not exist since it is not identifiable without the functional form of $h_0(t)$ specified and may well be absorbed into $h_0(t)$.

Despite some similarity across all proportional hazard models that hinge upon the hazard function as specified previously, the procedure to estimate the Cox regression is somewhat different in ideas and execution. Instead of constructing a full likelihood for each single observation, the Cox regression constructs its likelihood function using conditional likelihood, or what is later called partial likelihood. Cox (1972) proposes that for an event time, t_i, the probability that an observed failure occurs on a particular individual conditional on all the individuals at risk, or the risk set $R(t_i)$ at that particular time (interval) is given by

$$l_{t_i} = \frac{\exp(\mathbf{x}_i\beta)}{\sum_{j \in R(t_i)} \exp(\mathbf{x}_j\beta)} \tag{5.11}$$

We can then go through all individuals with event time to form an individual likelihood/ probability for each of them and then multiply through all these individual likelihoods. Because each individual likelihood is conditional on a risk set in a certain time interval, the product of all individual likelihood, $L_{t_i} = \prod_{i=1}^{n} l_{t_i}$, is only a partial (conditional) likelihood. To get parameter estimates, we can maximize the partial likelihood, L_{t_i}. Cox (1975) and Andersen and Gill (1982) later show that usual asymptotic properties of ML estimates still hold under partial likelihoods. Note that this model specification assumes an ideal circumstance; that is, there is always a single event (failure) for each discrete time interval. Adjustments need to be made when no event occurs or there are multiple events (tied failure time) occuring.

Using the exact same procedure as we did for the exponential and Weibull PH models, $\frac{h(t|\mathbf{x},x_k+1)}{h(t|\mathbf{x},x_k)} = \exp(\beta_k)$, we can provide a simple interpretation for the effects of covariates. Below, we use the `coxph` function from the `survival` package to estimate a Cox PH model with the same GSS-NDI data for illustration.

```
mycox <- coxph(Surv(time, death) ~ educ + impinc+ female,
        data=mydta)
summary(mycox)

## Call:
## coxph(formula = Surv(time, death) ~ educ + impinc + female, data = mydta)
##
##    n= 3801, number of events= 1283
##    (13 observations deleted due to missingness)
##
##               coef exp(coef)  se(coef)       z Pr(>|z|)
## educ    -0.020755  0.979459  0.009326  -2.225    0.026 *
## impinc  -0.140832  0.868635  0.031218  -4.511 6.44e-06 ***
## female  -0.333588  0.716349  0.056380  -5.917 3.28e-09 ***
## ---
## Signif. codes:  0 '***' 0.001 '**' 0.01 '*' 0.05 '.' 0.1 ' ' 1
##
##         exp(coef) exp(-coef) lower .95 upper .95
```

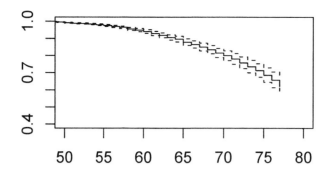

College-Educ Female

Figure 5.8
Cox PH Reg Survival Curve

```
## educ        0.9795        1.021        0.9617        0.9975
## impinc      0.8686        1.151        0.8171        0.9234
## female      0.7163        1.396        0.6414        0.8000
##
## Concordance= 0.566  (se = 0.008 )
## Likelihood ratio test= 66.75  on 3 df,    p=2e-14
## Wald test             = 68.41  on 3 df,    p=9e-15
## Score (logrank) test = 68.57  on 3 df,    p=9e-15
```

Regarding interpretation, we can say, for example, for each year increase in education, one can expect the hazard to change by a factor of 0.979, while holding other covariates constant. Being female vs. male, ceteris paribus, decreases the hazard by a factor of 0.716 .

We can also plot survival curves using the `plot` function from the `survival` package, as shown in Fig. 5.8. Instead of setting predictors to their mean values, we can set them to other substantively meaningful values, for example a female (`female = 1`) with 20 years of education (`educ = 20`) and sample average income (`impinc = mean(mydta$impinc)`).

```
base = data.frame(educ = 20, impinc = mean(mydta$impinc), female=1)
basefit = survfit(mycox, newdata=base)
plot(basefit, newdata=base,xlim=c(50, 80), ylim=c(0.4, 1),
sub="College-Educ Female")
```

5.7   Testing the PH Assumption

The proportional hazard (PH) assumption is probably the single most important assumption in parametric and semi-parametric survival regression models. Take the Cox regression, a survival regression model largely built upon the PH assumption, as an example, the fundamental relationship revolves around the equality about the hazard function, $h(t) =$

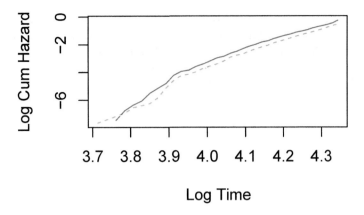

Figure 5.9
Graphical Test of the PH Assumption

$h_0(t) \exp(\mathbf{x}\beta)$. This would imply

$$\ln(-\ln(S(t))) = \ln(-\ln(S_0(t))) + \mathbf{x}\beta \tag{5.12}$$

That is, if the PH assumption holds, we would expect the $\ln(-\ln(S(t)))$ curve shifts parallel by a fixed amount as $\mathbf{x}$ changes.[1] This provides some statistical ground for a graphical check of the model assumption. In the following example, we first fit the survival curve with a single binary covariate, gender (`female`), and then we plot two transformed survival curves against the log of time.

```
kmFit = survfit(Surv(time, death) ~ female, data=mydta)
names(kmFit)

## [1] "n"          "time"       "n.risk"    "n.event"   "n.censor"  "surv"
## [7] "std.err"    "cumhaz"     "std.chaz"  "strata"    "type"      "logse"
## [13] "conf.int"  "conf.type"  "lower"     "upper"     "call"

cumH = -log(kmFit$surv)
plot( log(kmFit$time), log(cumH),
        type="n", xlab = "Log Time", ylab = "Log Cum Hazard")
n = kmFit$strata
lines( log(kmFit$time)[1:n[1]], log(cumH)[1:n[1]], col=2 )
lines( log(kmFit$time)[(n[1]+1):sum(n)], log(cumH)[(n[1]+1):sum(n)],
col=3, lty=2)
```

It can be shown from Fig. 5.9 that the two transformed survival curves are roughly parallel to each other, except that they slightly crisscross at the low end of log time. So there is little evidence of assumption violation with regard to gender, and we should feel comfortable to use a bivariate Cox PH regression. The downside of this graphical method, albeit straightforward, is that it can only work with categorical covariates, not continuous ones.

---

[1] Take the exponential function of both sides of the Eq. 5.12, and use the three equalities, including $h(t) = \frac{f(t)}{S(t)}$, $S(t) = 1 - F(t)$, and $\frac{dS(t)}{dt} = -f(t)$ to get $h(t) = -\frac{d}{dt}\log(S(t))$. Q.E.D.

Below we discuss more formal ways to test the assumption in the Cox PH regression model. In R, after running the `coxph` function from the `survival` package, we can use the `cox.zph` function from the same package to test the PH assumption. The `cox.zph` function conducts a χ^2 test of the association between the scaled Schoenfeld residuals and time for each covariate and jointly (Grambsch and Therneau, 1994).

```
cox.zph(mycox)
```

```
##           chisq df      p
## educ       1.47  1  0.226
## impinc     3.19  1  0.074
## female     1.12  1  0.291
## GLOBAL     8.41  3  0.038
```

Statistically significant results (e.g., $p < 0.05$) from the test suggest clear evidence of violation of the PH assumption. The above results show that the χ^2 tests of the PH assumption for the three variables individually do not appear to show any evidence for the violation of the PH assumption, whereas the global test shows significant results, thus raising a red flag for closer scrunity of this issue.

```
plot(cox.zph(mycox))
```

In addition to the usual statistical tests, we can turn to graphical methods to assess the appropriateness of the PH assumption of our model in a multivariate setting. We can use the `plot` function from the `survival` package to plot the scaled Schoenfeld residuals against transformed time. Any systematic deviation from a horizontal line would suggest some degree of a time-dependent pattern and accordingly a violation of the PH assumption. Again, the first two residual plots in Fig. 5.10 show some evidence of systematic association between time and residual.

Like the parallel lines (PL) assumption of the ordered regression models, the PH assumption allows for a parsimonious, albeit effective, parameterization and is an apparently very strong assumption, often violated in empirical data analysis. When the PH assumption is not met, this does not necessarily imply that the assumption is invalid. Instead, there could be several possibilities, and there is no definitive remedy. But there are still several guidelines that one can follow when the violation occurs. The first step is to avoid the omitted variables problem by including all essential covariates. The second is to create interaction terms between time and the variables for which the PH test fails, or to include some time-varying covariates. If none of these work, then we might consider models that extend beyond the Cox PH regression model.

5.8   Bayesian Approaches to Survival Regression

Bayesian survival analysis follows the general principles of Bayesian estimation laid out in the first two chapters. However, the additional challenges associated with survival analysis, for example censoring, require some special treatment of the data when we use BUGS-like software, such as Stan, to estimate the models. Such (coding) problems are not a concern when we use Bayesian estimation functions from user-written R packages for survival regression models, such as `rstanarm` (Brilleman et al., 2020) and `spBayesSurv` (Zhou

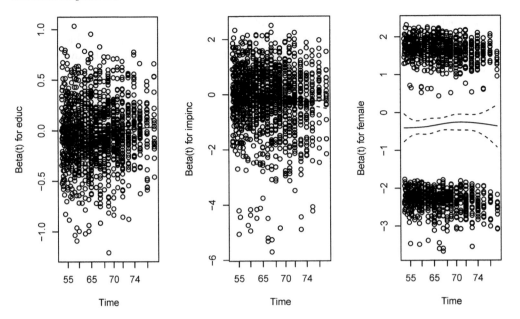

Figure 5.10
Cox PH Reg Residual Plot

et al., 2020). The cost of such ease of using these functions, unsurprisingly, is that we can lose flexibility and power.

5.8.1 Bayesian Estimation of Weibull PH Model Using `rstan`

There are a handful of R packages that can be used for estimating Bayesian survival regression models, including `brms`, `dynsurv`, `rstanarm`, and `spBayesSurv`, with each specialized in certain types of survival models (Bürkner, 2017; Wang et al., 2017; Goodrich et al., 2018; Zhou et al., 2020). But we will first illustrate how to use `rstan`, an R package for interfacing with Stan, to estimate a Bayesian Weibull PH model (Carpenter et al., 2017). Although using `rstan` increases initial cost because we have to supply the likelihood, it does pay off if one wants to estimate some extension to classical survival models and to have more control over the estimation process and post-estimation analysis. Below, we use the ovarian data from the `survival` package to illustrate. The variables used in this example include the number of days from the beginning of observation to death (`futime`) as the event time response variable, censoring status (`fustat`), age (`age`), and residual disease status (`resid.ds`), with a total of 26 observations.

As discussed previously, with `rstan` and Stan, there are several customized sections of codes for 1) managing the data, 2) setting up a Stan model, 3) linking raw data with the Stan model, 4) running the MCMC chains, and 5) diagnosing and interpreting results. For different software applications or packages for estimating Bayesian survival models in R, the ways to approach censoring can be quite different. What is discussed below only pertains to the `rstan` package.

First, we load several R packages to be used for our estimation and post-estimation analysis of the Bayesian Weibull PH model. Then, we put the ovarian data to the R search path using the `attach` function. The following R and Stan codes are based on BUGS and Stan's posted mice example. In this text, we only cover one method with which Stan processes censored cases. For more discussions, please refer to Carpenter et al. (2017). Before separating the data, we use the **as.numeric** function to create a set of dummy variables since all categorical variables are coded as ordinal variables in the small subset of the ovarian data initially.

```
require(survival)
require(rstan)
require(BEST) # for plotting HDI+ROPE

attach(ovarian)
resid1 = as.numeric(resid.ds==1)
rx1    = as.numeric(rx==1)
ecog1  = as.numeric(ecog.ps==1)
```

Using one of the ways Stan deals with censored cases, one needs to turn all data into two sets with the same set of variables. One set has all the uncensored cases and the other censored cases. Below, **nUnc** corresponds to the number of uncensored cases, and **nCen** for the number of censored cases. These two numbers will be communicated to the Stan model so that Stan knows the number of cases for each subset to correctly process. We use the censoring status variable **fustat** to recreate all variables for both subsets, including survival time, **tUnc** for the uncensored subset and **tCen** for the censored subset, age, **ageUnc** for uncensored and **ageCen** for censored, and existence of residual disease, **resUnc** for uncensored and **resCen** for censored.

```
nUnc <- sum(fustat == 1)
nCen <- sum(fustat == 0)
tUnc <- futime[fustat == 1]
tCen <- futime[fustat == 0]
ageUnc <- age[fustat == 1]
ageCen <- age[fustat == 0]
resUnc <- resid1[fustat == 1]
resCen <- resid1[fustat == 0]
```

As Stan prefers the data to be read in matrix format, we use the **cbind** function to column-combine covariates and then turn it into a covariate matrix, one for the uncensored and the other for the censored subset, using the **as.matrix** function. The **xMatUnc** matrix contains all covariates for uncensored cases and **xMatCen** for censored cases. We use the **NCOL** function to find the total number of columns of the covariate matrix, which is equivalent to the total number of covariates. It does not matter whether this data management step is executed before or after the next Stan model setup step, as long as both steps are taken before using the **stan** function from **rstan** to implement the MCMC simulation.

```
xMatUnc = as.matrix(cbind(ageUnc, resUnc))
xMatCen = as.matrix(cbind(ageCen, resCen))
K = NCOL(xMatUnc)
```

There are three Stan code blocks, and each block begins with a key word and all the details are bracketed in between. The first block starting with **data** declares data information. These should be details about the raw data information that will be read into Stan through the **stan** function from the **rstan** package, and they will also be used to construct the likelihood function in the model block that follows. **nUNC**, **nCen**, and **K** are the number of uncensored cases, censored cases, and covariatess respectively. `tUnc` and `tCen` are the response variable names for uncensored and censored cases respectively, with their corresponding column length specified in the bracket right after. `xMatUnc` and `xMatCen` are the matrices of covariates for uncensored and censored cases respectively, with their dimensions specified right before and after. Here we use the same names as those from the real data just for naming convenience and to avoid confusion, but users can choose alternative naming convention, but such changes have to inform `rstan` through the data argument in the `stan` function.[2]

```
# THE MODEL.
modelString = "
data {
int<lower=0> nUnc;
int<lower=0> nCen;
int<lower=2> K;
real<lower=0> tUnc[nUnc];
real<lower=0> tCen[nCen];
row_vector[K] xMatUnc[nUnc];
row_vector[K] xMatCen[nCen];
}
```

The second block that begins with the key word `parameters` is to declare parameters, including the shape parameter `eta`, the constant term `cons`, the β vector for all covariates `b`, in this particular example. As mentioned before, there are several ways to deal with censored cases in Stan. For the method we present below, Stan views the censored time variable as one with missing values and treats it like a parameter, `t2Cen`, to be sampled each time along with other parameters through the MCMC chains.

```
parameters {
real<lower=0> eta;
real cons;
vector[K] b;
real<lower=1> t2Cen[nCen];
}
```

The third block that begins with the key word `model` is to set up the priors and likelihood function to loop through. We use an exponential prior for `eta`, and weakly informative normal for the constant term and the two slopes in the β vector. For the uncensored cases, we loop the likelihood through `nUnc` uncensored cases, and for the censored cases, we loop through the `nCen` censored cases. After the ending quotation of all the Stan codes, we use the `writeLines` function from R base to divert the codes in between the two quotation marks into a file connection, `model.txt`.

---

[2]The `dataList` list after the Stan codes is used specifically to link names used in the raw data that R processes and the names used in the Stan code blocks, and such information is relayed through the `data` argument in the `stan` function.

```
model {
eta ~ exponential(0.001);
cons ~ normal(0, 10);
for (k in 1:K) {
b[k] ~ normal(0, 10);
}
for (n in 1:nUnc) {
tUnc[n] ~ weibull(eta, exp(-(cons + xMatUnc[n]*b) / eta));
}
for (n in 1:nCen) {
t2Cen[n] ~ weibull(eta, exp(-(cons + xMatCen[n]*b) / eta) / tCen[n]);
}
}
" # close quote for modelstring
writeLines(modelString,con="model.txt")
```

If we combine the Stan codes from all three blocks, then it looks like the following,

```
modelString = "
data {
int<lower=0> nUnc;
int<lower=0> nCen;
int<lower=2> K;
real<lower=0> tUnc[nUnc];
real<lower=0> tCen[nCen];
row_vector[K] xMatUnc[nUnc];
row_vector[K] xMatCen[nCen];
}

parameters {
real<lower=0> eta;
real cons;
vector[K] b;
real<lower=1> t2Cen[nCen];
}

model {
eta ~ exponential(0.001);
cons ~ normal(0, 10);
for (k in 1:K) {
b[k] ~ normal(0, 10);
}
for (n in 1:nUnc) {
tUnc[n] ~ weibull(eta, exp(-(cons + xMatUnc[n]*b) / eta));
}
for (n in 1:nCen) {
t2Cen[n] ~ weibull(eta, exp(-(cons + xMatCen[n]*b) / eta) / tCen[n]);
}
}
" # close quote for modelstring
writeLines(modelString,con="model.txt")
```

Once we are done with the Stan model step, we want to link the real data to the data declared in the **data** block of the Stan codes. Below, we create an R list to link these two. We use same names for scalars and matrices from real observed data (on the right of equal signs) and data declared in the **data** block of the Stan model codes (on the left of equal signs) so as to avoid any inadvertent mistake. They do not have to be the same, but one needs to be careful if different names are used. This list needs to be passed on to the **data** argument of the **stan** function to make this communication valid.

```
dataList = list(
        nUnc = nUnc,
        nCen = nCen,
        K = K ,
        tUnc = tUnc,
        tCen = tCen,
        xMatUnc = xMatUnc,
        xMatCen = xMatCen )
```

Afterwards, we specify values for various MCMC options to be used in the **stan** function. In this case, we request to have summary statistics about the constant term (**cons**), the vector for parameter estimates (**b**), and **eta**, the shape parameter. Other parameter values are for the calibration of MCMC sampling.

```
parameters = c("cons", "b", "eta")
adaptSteps = 500
burnInSteps = 500
nChains = 3
numSavedSteps=6000
thinSteps=1
nPerChain = ceiling( ( numSavedSteps * thinSteps ) / nChains )
```

Because the whole MCMC sampling process can be quite time-consuming, we can also initiate a time monitoring system. We use the **proc.time** function to begin the timing process.

```
time.used <- proc.time()
```

Once all these house-keeping issues are taken care of, we can then start the MCMC process using the **stan** function, the main workhorse of our Bayesian estimation of the Weibull PH model. There are several arguments in the **stan** function. First, **model_code** communicates the model setup to Stan. Since all the information has been saved to a string object, we simply specify the object name, **modelString**. The **data** argument is to supply the raw data to be read into Stan. To make our results reproducible, we also set the **seed** to a specific number so that the results will stay the same everytime we run the same codes. The **pars** option is to request which parameters are to be selected for presentation from the MCMC chains.

```
mcmcSamples <- stan(model_code=modelString,
                    data=dataList,
                    seed = 47306,
                    pars=parameters,
                    chains=nChains,
```

```
                                  iter=nPerChain,
                      warmup=burnInSteps ) # init=initsChains
```

After the MCMC step is implemented, we can use the `proc.time` function again to record current time and then substract the start time recorded previously and get the total time taken to run the MCMC chain. If it takes too much time, then we may consider changing some of our modeling or sampling strategies to speed up the MCMC process.

```
proc.time() - time.used
```

Right after recording the time for the MCMC procedure, we can also take a look at the summary results of our MCMC sampling

```
#summary(mcmcSamples)
print(mcmcSamples, digits=3)

## Inference for Stan model: anon_model.
## 3 chains, each with iter=2000; warmup=500; thin=1;
## post-warmup draws per chain=1500, total post-warmup draws=4500.
##
##              mean se_mean    sd     2.5%      25%      50%      75%     97.5%
## cons      -17.385   0.100 3.794  -25.359  -19.784  -17.252  -14.738  -10.460
## b[1]        0.120   0.001 0.038    0.049    0.092    0.119    0.146    0.196
## b[2]       -1.121   0.016 0.774   -2.697   -1.617   -1.097   -0.590    0.291
## eta         1.580   0.009 0.352    0.946    1.334    1.562    1.800    2.340
## survp100    0.984   0.000 0.017    0.936    0.979    0.989    0.994    0.999
## survp200    0.958   0.001 0.033    0.870    0.945    0.967    0.981    0.994
## survp500    0.851   0.002 0.083    0.651    0.804    0.866    0.913    0.970
## survp800    0.722   0.003 0.133    0.435    0.634    0.736    0.821    0.935
## survp1200   0.554   0.004 0.184    0.195    0.420    0.561    0.693    0.881
## lp__     -111.965   0.106 3.714 -120.157 -114.307 -111.674 -109.326 -105.683
##          n_eff Rhat
## cons      1435 1.000
## b[1]      1860 1.000
## b[2]      2243 1.000
## eta       1385 0.999
## survp100  2695 1.000
## survp200  2747 1.000
## survp500  2634 1.000
## survp800  2309 1.000
## survp1200 1881 1.000
## lp__      1230 1.001
##
## Samples were drawn using NUTS(diag_e) at Mon Apr 04 13:04:48 2022.
## For each parameter, n_eff is a crude measure of effective sample size,
## and Rhat is the potential scale reduction factor on split chains (at
## convergence, Rhat=1).

#traceplot(mcmcSamples)
#stan_plot(mcmcSamples)
```

As how the variables are entered the covariate matrix, `b[1]` and `b[2]` in the printed summary results of the MCMC sampling chains correspond to parameter estimates for the coefficients of age and residual disease status respectively. As discussed in previous chapters,

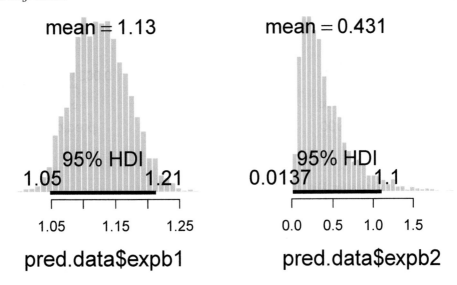

Figure 5.11
Posterior Distributoins of Hazard Ratios

we can directly use a probabilistic statement to describe the uncertainty associated with our parameter estimates. Again, one usually uses the 2.5% and 97.5% columns in the results section to provide the 95% "credible intervals" of the parameter estimates. For example, we can say there is a 0.95 probability that the parameter for age lies in between 0.049 and 0.196. With these parameter estimates, we can do many things; for example, we can provide substantive interpretation of these estimates by exponentiating them. To derive the probability distribution of quantities of our interest, such as the hazard ratio or survival probabilities, we can convert the Stan object into a data frame. We first use the `as.matrix` function to convert mcmcSamples, the Stan estimation object, into a matrix and then apply the `data.frame` function to turn the matrix into a data frame. Next, we use the `exp` function to exponentiate the raw coefficients and turn them into hazard ratio coefficients. We can use various R functions to get the usual statistics of such distributions of hazard ratios, for example, mean, median, some quantile, or variance and standard deviation. Below we use the `plotPost` function from the BEST package to plot the distributions of two hazard ratios, including their 95% HDI ranges (Kruschke, 2013). It can be shown from the graph on the left of Fig. 5.11 that the mean of the hazard ratio for age is 1.13 and its 95% HDI is from 1.05 to 1.21. As such, we can safely reject the null that age is not related to the event time response variable (i.e., excluding 1) at the conventional credible level. On the right, we can see that the 95% HDI of the distribution of the hazard ratio for residual disease ranges from 0.01 to 1.1, thus providing empirical support for the null that the effect of this variable is probably not statistically significant (i.e., including 1). But note that such Bayesian assessment of null values is not very robust since it does not involve ROPEs.

```
# turning Stan fit object into a data frame and columns into variables
# b[1] and b[2] are turned into b.1. and b.2. automatically in R
pred.data = data.frame(as.matrix(mcmcSamples))
pred.data$expb1 = exp(pred.data$b.1.)
pred.data$expb2 = exp(pred.data$b.2.)
pred.data$expc = exp(pred.data$cons)
par(mfrow=c(1,2))
plotPost(pred.data$expb1)
```

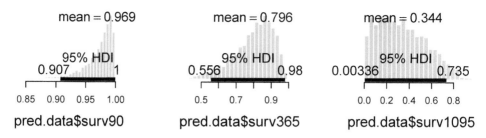

Figure 5.12
Survival Probabilities for Different Time

```
plotPost(pred.data$expb2)
```

We can calculate survival probabilities given some typical values of the predictors. In the following example, we set age to 63, the age at or above which usually half of the ovarian cancer cases are found, and the residual disease status to be 1. Then we vary the time to be 90 (three months), 365 (one year), and 1095 (three years) days. We use the formula, $S(t|\mathbf{x}) = \exp\left(-t^\eta \exp\left(\beta_0 + \mathbf{x}\beta\right)\right)$, for calculating the survival probabilities, and then use the `plotPost` function to plot these probability distributions.

```
pred.data = data.frame(as.matrix(mcmcSamples))
# 63 (half of ovarian cancer found) and with residual disease
pred.data$surv90 =
        exp(-exp(pred.data$cons + 63*pred.data$b.1. + pred.data$b.2.)
                        *90^(pred.data$eta))
pred.data$surv365 =
        exp(-exp(pred.data$cons + 63*pred.data$b.1. + pred.data$b.2.)
                        *365^(pred.data$eta))
pred.data$surv1095 =
        exp(-exp(pred.data$cons + 63*pred.data$b.1. + pred.data$b.2.)
                        *1095^(pred.data$eta))
par(mfrow=c(1,3))
plotPost(pred.data$surv90)
plotPost(pred.data$surv365)
plotPost(pred.data$surv1095)
```

As can be shown from the three sub-graphs in Fig. 5.12, as the survival time increases, the average probability of survival decreases from 0.969 for about three months, to 0.796 for about a year, and 0.344 for roughly three years.

5.8.2 Bayesian Estimation of Survival Models Using spBayesSurv

Using `rstan` along with Stan in the background to estimate survival models has clear advantages and limitations. With `rstan`, one has to code the Stan model and supply the likelihood function, and it may take much longer to get convergence if some of the estimation conditions are not set within their proper ranges. Nonetheless, once the MCMC chains are obtained, one can conduct various post-estimation analyses. In addition, one is able to code some uncommon models that functions in user-written packages cannot estimate. But if one has typical Bayesian survival models to estimate, then using readily made R functions would be a better alternative. In this section, we illustrate how to use some of these functions to estimate popular Bayesian survival regression models. First, the same Weibull PH that we looked at in the previous section can be estimated using the `survregbayes` function in the `spBayesSurv` package (Zhou et al., 2020). The syntactic structure of `survregbayes` is quite similar to other survival regression functions discussed previously. Other than the typical model/formula argument, one needs to specify model type (`survmodel= "PH"`) and the distributional assumption for the error term (`dist="weibull"`). The `mcmc` argument is to specify all usual MCMC sampling option values.

```
require(spBayesSurv)
nburn=500
nsave=6000
nskip=0
mcmc=list(nburn=nburn, nsave=nsave, nskip=nskip, ndisplay=1000)
prior = list(M=10, r0=1)
set.seed(47306)
bayesWeib = survregbayes(formula = Surv(futime, fustat) ~ age + resid1,
                                       survmodel="PH",
                                       dist="weibull",
                    mcmc=mcmc) # prior=prior,
```

We can then use the `summary` function to list results and produce model fit statistics. Since the results are similar in nature (though they do differ somewhat probably due to different priors and other MCMC conditions) to those discussed previously, we will not get into detailed discussion.

```
summary(bayesWeib)

## Proportional hazards model:
## Call:
## survregbayes(formula = Surv(futime, fustat) ~ age + resid1,
##      survmodel = "PH",
##      dist = "weibull", mcmc = mcmc)
##
## Posterior inference of regression coefficients
## (Adaptive M-H acceptance rate: 0.269):
##          Mean      Median   Std. Dev.  95%CI-Low  95%CI-Upp
## age     0.14366   0.14188   0.04288    0.06358    0.23230
## resid1  -0.99063  -0.97183  0.76010    -2.60230   0.38991
##
## Log pseudo marginal likelihood: LPML=-93.44227
## Deviance Information Criterion: DIC=184.4266
## Watanabe-Akaike information criterion: WAIC=186.1609
## Number of subjects: n=26

#names(bayesWeib)
bayesWeib$WAIC

## [1] 186.1609
```

Bayesian Cox PH model is not that easy to code in Stan because it involves non-failures, tied failures, and censoring. One possible workaround is to use the Poisson distribution to code the model. There are several customized R functions that are able to estimate it. Below, we show how to use the `indeptCoxph` function from the same `spBayesSurv` package to estimate the Bayesian Cox PH regression model. We can supply the same formula as we use throughout this section, and let the function set MCMC options to their default values.

One nice feature of the `spBayesSurv` package is that it provides a post-estimation plotting function. In this case, we want to plot five survival curves, corresponding to ages 40, 50, 60, 70, and 80 with residual disease status all set to 1. Then we turn these five pairs of hypothetical values, each representing a hypothetical individual, into a data frame. Once the new data frame is created, we pass it onto the `plot` function and plot it.

```
bayesCox = indeptCoxph(Surv(futime, fustat) ~ age + resid1)

summary(bayesCox)

## Cox PH model with piecewise constant baseline hazards
## Call:
## indeptCoxph(formula = Surv(futime, fustat) ~ age + resid1)
##
## Posterior inference of regression coefficients
## (Adaptive M-H acceptance rate: 0.1385):
##          Mean     Median   Std. Dev.  95%CI-Low   95%CI-Upp
## age      0.15731  0.15225  0.05898    0.06233     0.27648
## resid1  -0.99810 -0.85538  0.92786   -2.77711     0.61689
##
## Log pseudo marginal likelihood: LPML=-102.2295
## Number of subjects: n=26

setx = data.frame(age=c(40, 50, 60, 70, 80),
                  resid1=c(1, 1, 1, 1, 1),
                  row.names = c("age=40", "age=50", "age=60",
                                "age=70", "age=80"))


plot(bayesCox, xnewdata=setx, CI=0.001, cex=0.5)
```

It can be easily shown from Fig. 5.13 that the survival curves for older individuals have lower survival probabilities throughout all time points from the very beginning of the observation. In addition, the survival curves for the older ones decline much faster than those of the younger ones, especially starting from age 60.

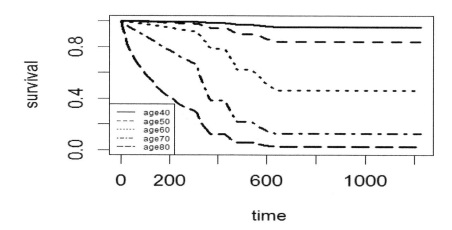

Figure 5.13
Survival Curves for Different Age Groups

6

Extensions

> To govern a vast expanse is to concoct a simple dish.
>
> in Chapter 60 of *Dao De Jing* by Laozi

6.1 Multilevel Regression

Multilevel regression has been widely used in major fields across behavioral, biomedical, social, and statistical sciences. It is also called mixed models in (bio-)statistics, hierarchical models in education, multilevel models in political science and sociology (Gelman and Hill, 2007; Rabe-Hesketh and Skrondal, 2012; Raudenbush and Bryk, 2002). Previous chapters focus exclusively on units of analyses at the same level, while largely disregarding possible nested or clustered data structures in empirical analysis. For example, individual health conditions are not only affected by individual chances and conducts but also community and national characteristics; individual educational outcomes result from the synergy of individual traits, family milieu, and school as well as community environment. The impact of environment or context is not some novelty discovered recently; instead, it has long existed in traditional wisdom. For example, it is recorded that the mother of Mencius, a Confucian saint idolized for thousands of years in China, moved her dwelling three times to a place that can provide the best educational environment for Mencius.

In general, for some data point i nested within an aggregate unit a, one has

$$\eta_{ia} = \sum_{k=0}^{K} x_{kia}\beta_{ka} \tag{6.1}$$

where x_k is a generic predictor and β_k is its corresponding slope coefficient (x_{0ia} is a unit vector and β_{0a} is an intercept) ; η_{ia} is usually called the linear predictor. For an observed response variable, y_{ia}, it is usually assumed that $\mathrm{E}(y_{ia}|\mu_{ia}) = \mu_{ia}$. Generally, one can relate y_{ia} to η_{ia} by having

$$\eta_{ia} = g(\mu_{ia}) \tag{6.2}$$

where $g(\mu_{ia})$ is a canonical link function (Raudenbush and Bryk, 2002). For linear regression models, one usually has $g(\mu_{ia}) = \mu_{ia}$, and for binary logit models, $g(\mu_{ia}) = \ln\frac{\mu_{ia}}{1-\mu_{ia}}$. In general, if this first-level model is a generalized linear model, then the multilevel model is called generalized mixed models or generalized linear mixed models (GLMMs). If the aggregate level is ignored, all the models in previous chapters can be re-parameterized in a similar manner and grouped under what is usually called generalized linear models. Without

DOI: 10.1201/9780429056468-6

loss of generality, a two-level analysis usually has,

$$\beta_{ka} = \gamma_{k0} + \sum_{h=1}^{H_k} \gamma_{kh} Z_{ha} + u_{ka} \qquad \mathbf{u} \sim N\left(\mathbf{0}, \mathcal{V}\right) \tag{6.3}$$

where β_{ka} is an intercept (for x_{kia} as a column vector of ones) or slope associated with a level-one predictor, x_{kia}, Z_{ha} is an aggregate level variable featuring some characteristic of an aggregate unit a, and the disturbance vector $\mathbf{u}$ is a $(K+1) \times 1$ matrix containing disturbance terms with its kth element equal to u_{ka}. Note that $\mathbf{u}$ is usually assumed to follow a multivariate-normal distribution with its mean set to $\mathbf{0}$ and its variance-covariance matrix set to $\mathcal{V}$; $\mathcal{V}$ is a $(K+1) \times (K+1)$ matrix with its (k,k) diagonal element containing the variance component for u_{ka} and (k,j) off-diagonal element containing the covariance between u_{ka} and u_{ja} for $j \neq k$. Note that the estimation of parameters in this matrix, especially the off-diagonal elements, often pose great challenges to model convergence. To give a concrete example, y_{ia} can be a binary indicator of self-rated health for individuals with one denoting poor health and zero good health, and x_{kia} can be a measure of individual education or income; individuals can be nested within an administrative unit, such as county, so Z_{ha} may correspond to some county-level characteristic, such as the percentage of college graduates. This example is for a two-level binary regression model, and one can move to a higher level analysis if the data structure allows for it. Below, the individual-level data from the 2010 Behavioral Risk Factor Surveillance System along with county data from the 2010 Census data and the American Community Survey are used to estimate two-level binary (logit) regression and count regression models for illustrative purposes.

6.1.1 Multilevel Logit Regression

In two-level binary regression models, the link function is set to be $F^{-1}(\mathbf{x}\beta)$, where $F(\mathbf{x}\beta)$ is usually a cumulative density function, such as that of the standard normal or logistic distribution. For a two-level logit model, the link function is $g(\mu_{ia}) = \ln \frac{\mu_{ia}}{1-\mu_{ia}}$ at level one. For the level-2 equation 6.3, one can begin with the simplest random intercept model, in which the only level-two equation is $\beta_{0a} = \gamma_{00} + u_{0a}$, without any aggregate predictors but an intercept term γ_{00} and a disturbance term u_{0a}. In the example below, the binary response variable is `heart`, denoting whether one has any of the three major types of heart disease, including heart attack, angina, and stroke. Its predictors include age (in decades), gender (male = 1), race/ethnicity, and education levels. Race/ethnicity is measured with five binary indicators, including Asian (`rAsian`), Black (`rBlack`), Hispanic (`rHispn`), and other races (`rOther`), with white used as the reference category. Education is also measured with four dummy variables, including less than high school (`lthischl`), high school (`highschl`), and some college (`somecolg`), with college and beyond (`colggrad`) being the reference category. The R code chunk below first reads in the data, re-scales the age variable, and then randomly samples 5,000 cases from the 2010 BRFSS database for later analysis.

```
require(foreign)
readin <- read.dta("data/indcntdta01.dta", convert.factor = F)
# create list of variable names used for variable selection below
usevar <- c("heart", "genhlthR", "physhlthR", "menthlthR", "ageR", "male",
    "rAsian", "rBlack", "rHispn", "rOther", "rWhite", "lthischl", "highschl",
    "somecolg", "colggrad", "ctyfips", "ctymdinK", "ctygini")
usedta <- subset(readin[complete.cases(readin[usevar]), ], select = usevar)
set.seed(47306)
sampData <- usedta[sample(1:nrow(usedta), size = 5000, replace = T), ]
```

```
sampData$age10 = sampData$ageR/10
```

There are several R packages frequently used for estimating multilevel linear and generalized linear models, including **brms** (Bürkner, 2017). **lme4** (Bates et al., 2015), and **nlme** (Pinheiro et al., 2020). The examples below use both the **glmmPQL** function from the **MASS** package and the **glmer** function from the **lme4** package for estimation and post-estimation analysis.

One of the most widely used R package to estimate multilevel linear and generalized linear (mixed effect) models is the **lme4** package. Its **glmer** function is the workhorse for estimating multilevel generalized linear models via maximum likelihood. The syntactic structure of **glmer** is very similar to that of the **glm** function, except that **glmer** also allows users to specify random effects. The R code chunk below estimates an intercept-only random effects model, wherein only the random and fixed effects for the intercept are estimated. This model is often called the null or naive model, used as the baseline for later model comparison and selection. After the response variable (**heart**) and the unit vector or intercept ("1") are specified and delimited by the tilde sign (~) in between, users need to add a parenthesis afterwards, within which the random effects are specified first and then the level-2 (aggregate level) identifier, separated by a vertical line. The unit vector ("1"), the first element within the parenthesis, corresponds to the intercept that is assumed to vary across counties, denoted by the county identifier, **ctyfips**. Since the intercept is the only random component specified before the vertical line within the parenthesis, this is a random-intercept model.

```
library(lme4)
mod0 <- glmer(heart ~ 1 + (1 | ctyfips),
                data = sampData, family = binomial)
summary(mod0)

Generalized linear mixed model fit by maximum likelihood (Laplace
  Approximation) [glmerMod]
 Family: binomial  ( logit )
Formula: heart ~ 1 + (1 | ctyfips)
   Data: sampData

     AIC      BIC   logLik deviance df.resid
  3657.3   3670.3  -1826.6   3653.3     4998

Scaled residuals:
    Min      1Q  Median      3Q     Max
-0.3742 -0.3681 -0.3668 -0.3651  2.7759

Random effects:
 Groups  Name        Variance Std.Dev.
 ctyfips (Intercept) 0.008881 0.09424
Number of obs: 5000, groups:  ctyfips, 625

Fixed effects:
            Estimate Std. Error z value Pr(>|z|)
(Intercept) -2.00159    0.04517  -44.31   <2e-16 ***
---
Signif. codes:  0 '***' 0.001 '**' 0.01 '*' 0.05 '.' 0.1 ' ' 1
```

Based on results from this model, one can calculate the intra-class correlation (ICC), or sometimes called variance partition coefficient (VPC),

$$\rho = \frac{\tau^2}{\tau^2 + \sigma^2} \tag{6.4}$$

wherein $\tau^2 = V(\beta_{0a})$ and σ^2 denote between- and within-group variance respectively, and as such ρ is a measure of the proportion of the between-group variance out of the total variance (Raudenbush and Bryk, 2002, pp. 36,74). In intercept-only random-effects two-level logistic regression models, σ^2 is the variance of the unobserved latent variable y^*, and $\sigma^2 = \frac{\pi^2}{3}$ as one of the model assumptions. One can then use the output from the `glmer` function to calculate ρ, or one can turn to the `icc` function from the **performance** package to do the same (Lüdecke et al., 2020),

```
library(performance)
icc(mod0)

# Intraclass Correlation Coefficient

    Adjusted ICC: 0.003
  Conditional ICC: 0.003
```

It appears that the between-group variance only accounts for a very small amount, which is 0.3%, of the total variation. Such a small ICC is indicative of very limited level-2 variation, suggesting that single-level analysis may suffice. One can continue building the model despite the small ICC by adding level-1 predictors, such as age (`age10`), gender (`male`), race (`rBlack`), and education (`lthischl` and `highschl`) like the following,

```
mod1 <- glmer(heart ~ age10 + male +  rBlack + lthischl + highschl +
              (1 | ctyfips), data = sampData, family = binomial)
summary(mod1)

Generalized linear mixed model fit by maximum likelihood (Laplace
  Approximation) [glmerMod]
 Family: binomial  ( logit )
Formula: heart ~ age10 + male + rBlack + lthischl + highschl + (1 | ctyfips)
   Data: sampData

     AIC      BIC   logLik deviance df.resid
  3245.7   3291.3  -1615.8   3231.7     4993

Scaled residuals:
    Min      1Q  Median      3Q     Max
-1.3829 -0.3918 -0.2705 -0.1730  8.5367

Random effects:
 Groups  Name        Variance Std.Dev.
 ctyfips (Intercept) 0.01722  0.1312
Number of obs: 5000, groups:  ctyfips, 625

Fixed effects:
            Estimate Std. Error z value Pr(>|z|)
```

```
(Intercept) -5.90046    0.23583 -25.020  < 2e-16 ***
age10        0.56982    0.03326  17.132  < 2e-16 ***
male         0.62152    0.09366   6.636 3.22e-11 ***
rBlack       0.43571    0.14713   2.961  0.00306 **
lthischl     0.70840    0.15003   4.722 2.34e-06 ***
highschl     0.10933    0.10504   1.041  0.29795
---
Signif. codes:  0 '***' 0.001 '**' 0.01 '*' 0.05 '.' 0.1 ' ' 1

Correlation of Fixed Effects:
         (Intr) age10  male   rBlack lthsch
age10    -0.941
male     -0.304  0.104
rBlack   -0.142  0.073  0.086
lthischl -0.098 -0.008  0.045 -0.125
highschl -0.080 -0.091  0.091 -0.038  0.239
```

The interpretation of results from multilevel logistic regression models is not that different from the usual practice for single-level models except that one needs to be cautious about the distinction between uni-specific and population-average models, which is beyond the scope of this text. For example, in `mod1`, one can say that net of the effect of all other variables as well as unobserved county-level heterogeneities accounted for in the model (unit-specific model), each ten-year increase in age would increase the odds of having any of the three heart diseases by a factor of $\exp(0.57) = 1.768$. Our text focuses on unit-specific models, without getting into nuanced differences between population-average and uni-specific models in the GLMM framework. One can then compare the two models previously estimated using the `anova` function,

```
anova(mod0, mod1)

Data: sampData
Models:
mod0: heart ~ 1 + (1 | ctyfips)
mod1: heart ~ age10 + male + rBlack + lthischl + highschl + (1 | ctyfips)
     npar    AIC    BIC  logLik deviance  Chisq Df Pr(>Chisq)
mod0    2 3657.3 3670.3 -1826.6   3653.3
mod1    7 3245.7 3291.3 -1615.8   3231.7 421.55  5  < 2.2e-16 ***
---
Signif. codes:  0 '***' 0.001 '**' 0.01 '*' 0.05 '.' 0.1 ' ' 1
```

And the results show that the second model that has level-1 predictors clearly outperforms the intercept-only random effects model. One can further allow for more random effects. For example, the race coefficient (`rBlack`) can also be set to vary randomly across counties by adding the variable within the parenthesis before the county identifier separated by the "|" sign. Without any adjustment in some of the arguments, this model cannot converge because of the additional layer of model complexity, including two additional parameters in the variance-covariance matrix of the random effects. The addition of covariance and, to a lesser degree, variance component of random effects can largely increase the computational cost and make the convergence difficult. Thereby, the `nAGQ`'s default option (1 for the Laplace approximation) is reset to zero, denoting an estimation method that is less exact but easier to converge in estimating the parameters [1].

```
mod3a <-glmer(heart ~ age10 + male +  rBlack + lthischl + highschl
                + (1 + rBlack | ctyfips),
                nAGQ=0, data = sampData, family = binomial)
```

One can get similar results from the `glmmPQL` function from the MASS package. Unlike `glmer`, `glmmPQL` uses penalized quasi-likelihood to fit generalized linear mixed models.[2]

```
require(MASS)
mod3b <- glmmPQL(heart ~ age10 + male +  rBlack + lthischl + highschl,
                random = ~ 1 + rBlack | ctyfips, data = sampData,
                family = binomial)
```

One can also have a slope-as-outcome model by including a level-2 predictor, for example, county median household income (`ctymdinK`), and an interaction term between black and county median household income. As a result, there are two equations at the second (county) level, including an intercept-as-outcome equation and a slope-as-outcome equation (see Eq. 6.3 , using county median household income as the sole predictor.

```
mod4a <-glmer(heart ~ age10 + male +  rBlack + lthischl + highschl
                + ctymdinK + ctymdinK:rBlack
                + (1 + rBlack | ctyfips),
                nAGQ=0, data = sampData, family = binomial)
```

After estimation, we can then compare `mod3a` with `mod4a`, which adds a level-2 predictor, `ctymdinK`, to mod3a.

```
# results suppressed
anova(mod3a, mod4a)

Data: sampData
Models:
mod3a: heart ~ age10 + male + rBlack + lthischl + highschl
        + (1 + rBlack | ctyfips)
mod4a: heart ~ age10 + male + rBlack + lthischl + highschl +
        ctymdinK + ctymdinK : rBlack + (1 + rBlack | ctyfips)
        npar  AIC  BIC logLik deviance Chisq Df Pr(>Chisq)
mod3a     9 3249 3308  -1616  .  3231
mod4a    11 3244 3315  -1611     3222 9.75  2     0.0076 **
```

---

[1]Note that by so doing, we only force the `glmer` function to produce results without explicit errors; its estimation of the variance-covariance components of the random effects, however, is still problematic. We still proceed to illustrate the R codes to show what is possible, but the results are suppressed and have to be read with caution.

[2]Penalized quasi-likelihood (PQL) is a commonly used estimation and inference technique in the GLMMs. Quasi-likelihoods do not use real likelihood functions that are based on probability distributional functions; instead, a quasi-likelihood function establishes a relationship between the mean and variance of (usually conditionally) observed response data, and the function has properties similar to real likelihood functions. In PQL, a second-order (penalized) Taylor approximation (Laplace's method) of the quasi-likelihood function is used. PQL was initially devised as a procedure for approximate likelihood inference in generalized mixed models (Wedderburn, 1974; McCullagh and Nelder, 1989; Breslow and Clayton, 1993).

For an illustration of the interpretation of binary indicators, for example in `mod4a`, one can say that holding all other variables constant (including unobserved county characteristics captured in the model), being male vs. female increases the odds of having any of three heart diseases by a factor of $\exp(0.637) = 1.891$ (results not shown here, and the raw coefficient for male is 0.637). One can also calculate predicted probabilities and graph them using functions from the `ggeffects` package (Lüdecke, 2018),

```
library(ggeffects)
library(ggplot2)
maleProb = ggpredict(mod4a, "male")
maleProb

# Predicted probabilities of heart

male | Predicted |       95% CI
--------------------------------
  0 |       0.07 | [0.06, 0.08]
  1 |       0.12 | [0.11, 0.14]

Adjusted for:
*     age10 =   5.61
*    rBlack =   0.10
* lthischl =   0.08
* highschl =   0.27
* ctymdinK =  54.77
*   ctyfips = 0 (population-level)

# plot(ggpredict(mod4a, "age10"))
# plot(ggpredict(mod4a, c("ctymdinK[all]", "rBlack")))
plot(ggpredict(mod4a, c("ctymdinK[all]", "rBlack")), colors = "bw") +
labs(shape = "Male")
```

The results show that the predicted probability of having heart disease for a male with average sample characteristics is 0.124, and the predicted probability for an otherwise similar female is 0.07. Figure 6.1 illustrates the associations among county median income, race (black vs. non-black), and the predicted probabilities of having heart disease; as county median income is in its low end, there is a large gap between blacks and non-blacks in their likelihood of having heart disease, with blacks clearly being above non-blacks. As the county-level income increases, the gap narrows and even disappears roughly above the $100,000 income mark. The graph also shows that the variance of the predictions among blacks is much greater than that of non-blacks.

6.1.2 Multilevel Count Regression

As discussed previously, when the response variable of interest is a count variable, usually count regression should be used. When the count data follow a clustered, hierarchical, or nested structure, one may have to turn to multilevel count regression models. Their model setup follows that of GLMMs, except that the link function is specifically devised for count data. For Poisson count regression models, the link function is commonly defined as $\eta_{ia} = \ln(\mu_{ia})$.

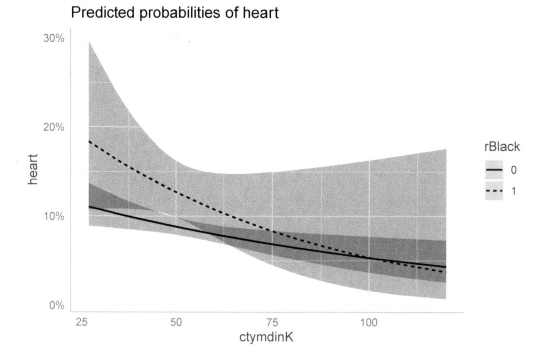

Figure 6.1
Predicted Probabilities of Heart Disease by Race and Count Median Income

Below, a two-level Poisson regression model is estimated using the same data and predictors but with a different response variable, `physhlthR` (the number of days that the respondents do not feel very well about their physical health during the past 30 days).

```
summary(pois <- glmer(physhlthR ~ age10 + male + rBlack + lthischl + highschl
                      + (1|ctyfips), nAGQ=0, data=sampData, family = poisson))

Generalized linear mixed model fit by maximum likelihood (Adaptive
  Gauss-Hermite Quadrature, nAGQ = 0) [glmerMod]
 Family: poisson  ( log )
Formula: physhlthR ~ age10 + male + rBlack + lthischl + highschl + (1 |
    ctyfips)
   Data: sampData

     AIC      BIC   logLik deviance df.resid
 58109.3  58154.9 -29047.6  58095.3     4993

Scaled residuals:
   Min     1Q Median     3Q    Max
-5.285 -1.945 -1.326 -0.031 26.560

Random effects:
 Groups Name        Variance Std.Dev.
 ctyfips (Intercept) 1.66     1.288
Number of obs: 5000, groups:  ctyfips, 625
```

```
Fixed effects:
              Estimate Std. Error z value Pr(>|z|)
(Intercept) -0.003626   0.062253  -0.058 0.953557
age10        0.143647   0.004753  30.224  < 2e-16 ***
male        -0.158389   0.015684 -10.099  < 2e-16 ***
rBlack       0.097391   0.027012   3.606 0.000312 ***
lthischl     0.688569   0.023514  29.283  < 2e-16 ***
highschl     0.343957   0.016521  20.819  < 2e-16 ***
---
Signif. codes:  0 '***' 0.001 '**' 0.01 '*' 0.05 '.' 0.1 ' ' 1

Correlation of Fixed Effects:
         (Intr) age10  male   rBlack lthsch
age10    -0.448
male     -0.106  0.035
rBlack   -0.053  0.034  0.028
lthischl -0.061 -0.017  0.023 -0.074
highschl -0.057 -0.106  0.042 -0.037  0.275
```

Again, the interpretation of results from multilevel models is very similar to that in single-level models, except that users need be careful about the distinction between unit-specific and population-average models for nonlinear models. For example, for the effect of education, one can say that having less than high school education, as opposed to college college and above, increases the predicted self-reported poor physical health days by a factor of $\exp(0.689) = 1.991$, while holding all other variables, including the unobserved county-level heterogeneities, constant. If there is any indication of over-dispersion or a large number of zeros, one can turn to the multilevel negative binomial regression by using the `glmer.nb` function, also from the `lme4` package,

```
negb <- glmer.nb(physhlthR ~ age10 + male + rBlack + lthischl + highschl
                 + (1|ctyfips), nAGQ=0, data=sampData)
negb@beta

[1]  0.41065934  0.16152191 -0.21020718  0.09386091  0.68579233  0.37824640
```

Then one can compare the two models to test for over-dispersion like what is usually done in single-level models,

```
anova(pois, negb)

Data: sampData
Models:
pois: physhlthR ~ age10 + male + rBlack + lthischl + highschl + (1 | ctyfips)
negb: physhlthR ~ age10 + male + rBlack + lthischl + highschl + (1 | ctyfips)
     npar   AIC   BIC  logLik deviance Chisq Df Pr(>Chisq)
pois    7 58109 58155 -29047.6    58095
negb    8 19260 19313  -9622.2    19244 38851  1  < 2.2e-16 ***
---
Signif. codes:  0 '***' 0.001 '**' 0.01 '*' 0.05 '.' 0.1 ' ' 1
```

The results show that the multilevel negative binomial model outperforms its Poisson counterpart, and one may feel comfortable to go with the two-level negative binomial model for analysis and reporting. Note that the discussion about multilevel count regression models is to illustrate the estimation of GLMMs beyond a two-level binary regression models. The model complexity and challenge for convergence can increase exponentially by adding random effects and/or moving onto more complicated level-1 models. But the general logic and procedure for model comparison and selection remain roughly the same as illustrated in this and previous sections.

6.1.3 Bayesian Multilevel Regression

The Bayesian paradigm appears to dovetail with the multilevel regression framework naturally. With Bayesian methods, parameters are assumed to follow chosen prior distributions. In multilevel models, the parameters in the level-1 equation, as described in Eq. 6.3, can come from similar or same prior distributions, such as normal. If preferred in either practice or theory, one can have hierarchical (or hyper-) parameters of the parameters in the prior distributions (or hyperpriors). For example, if one assumes that the slopes at the level-1 equation are from a normal distribution, it can be further assumed that the mean in such normal distributions comes from a hierarchical (weakly informative) prior of another normal. Recall that the traditional frequentist approach to estimating multilevel generalized linear models, as illustrated in previous sections, sometimes is not very effective especially when it comes to models with a complex covariance structure of random effects. Bayesian methods nonetheless perform relatively well in such cases where the traditional approach fails. Below, we use the `stan_glmer` function from the `rstanarm` package (Goodrich et al., 2018) to estimate the same model specified in `mod4a`. The general syntactic structure of `stan_glmer` is similar to that of `glmer` except that `stan_glmer` has an additional set of arguments for Bayesian estimation and MCMC sampling, such as priors (e.g., priors for the covariance structure of random effects) and sampling conditions (e.g., iterations and chains).

```
library(rstanarm)
```

```
# iter = 5000, chains = 3, warmup = 500, seed = 47306
stanmod01 <- stan_glmer(heart ~ age10 + male + rBlack + lthischl + highschl
+ ctymdinK + ctymdinK : rBlack + (1 + rBlack | ctyfips), data = sampData,
family = binomial)
```

By default, the `stan_glmer` function executes MCMC sampling with four chains, each with 2000 iterations, and half of them are used in the warm-up period and discarded. Once the MCMC simulation concludes, we can use the `print` function to request summary results of our simulation,

```
print(stanmod01, detail=T, digits=3)
stan_glmer
family:        binomial [logit]
formula:       heart ~ age10 + male + rBlack + lthischl + highschl
               + ctymdinK + ctymdinK : rBlack + (1 + rBlack | ctyfips)
observations: 5000
------
               Median MAD_SD
(Intercept)    -5.342  0.329
age10           0.574  0.034
male            0.639  0.092
```

```
rBlack            0.751  0.670
lthischl          0.657  0.148
highschl          0.077  0.103
ctymdinK         -0.011  0.004
rBlack:ctymdinK  -0.007  0.013

Error terms:
 Groups   Name        Std.Dev. Corr
 ctyfips  (Intercept) 0.1678
          rBlack      0.3349   -0.17
Num. levels: ctyfips 625
------
```

These summary statistics provide concise and important information about the posterior, including the medians of the simulated distributions for the intercept and slopes, their associated standard deviations, and estimates for the variance-covariance components.[3] Note that MAD—the second column of the results section for parameter estimates—denotes median absolute deviation. If one is interested in having customized results of the posterior distribution, then the following R code chunk can be used, for example, to construct 95% credible intervals of selected parameters,

```
summary(stanmod01,
        pars = c("(Intercept)", "age10", "rBlack", "ctymdinK",
                "Sigma[ctyfips:(Intercept),(Intercept)]",
                "Sigma[ctyfips:rBlack,rBlack]",
                "Sigma[ctyfips:rBlack,(Intercept)]"),
        probs = c(0.025, 0.50, 0.975), digits = 3)
# 95% credible interval
posterior_interval(stanmod01)
```

Once the preliminary examination of Bayesian estimation results concludes, we can move onto model diagnostics and interpretation. For example, we can investigate a trace plot of the posterior sampling like before using the mcmc_trace function from the bayesplot package (Gabry and Mahr, 2021) as follows,

```
require(bayesplot)
mcmc_trace(stanmod01, pars = "age10", size = 0.5) +   xlab("Iteration")
```

It can be shown in the trace plot Fig. 6.2 that roughly, all four chains converge on the same posterior (i.e., sampled values). We can also take a further step to calculate out-of-sample predictions, for example, the predicted probability of having a heart condition for a 35-year-old non-black female with a county-level median household income in a particular county,

```
# Lake County of Indiana
filter <- subset(sampData, ctyfips==18089, select=c("ctymdinK"))
ctyinc = mean(filter$ctymdinK)
```

---

[3]The standard deviations under the error terms section are calculated as the square root of the means of their corresponding (diagonal) elements in the variance-covariance matrix of the group-specific deviations from the common parameters from the posterior.

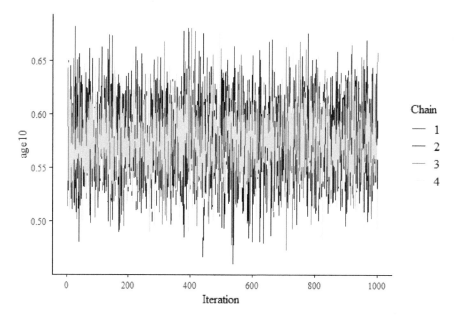

Figure 6.2
Trace Plot of Sample Values for Age

```
x.vector <- data.frame( age10  = 3.5, male   = 0, rBlack = 0,
                        lthischl = 0, highschl = 0,   ctymdinK = ctyinc,
                        ctyfips  = 18089 )
```

Then this new data vector can be relayed to the `posterior_epred` function from the `rstanarm` package to construct the empirical distribution of this predicted probability,

```
pred.prob = posterior_epred(stanmod01, newdata=x.vector)
mean(pred.prob)
```

```
## [1] 0.02123976
```

```
(ci95 = quantile(pred.prob, probs=c(0.025, 0.5, 0.975)))
```

```
##       2.5%        50%       97.5%
## 0.01404520 0.02078748 0.03140672
```

It can be shown from the results that the median predicted probability of the previously characterized individual is 0.021, and there is a 0.95 probability that the predicted probability lies between 0.014 and 0.031.

6.2   Causal Inference

The term causal inference, along with several other technical expressions such as propensity score analysis, treatment effect (analysis), and observational study, refers to roughly the same toolkit of techniques to mine observational or survey data to approximate randomized experimentation for finding causation (Rubin, 1973; Rosenbaum and Rubin, 1983; Rosenbaum, 2010; Guo and Fraser, 2015; Morgan and Winship, 2015). Similar to methods used for dealing with missing data, this ensemble of techniques is probably the best choice when one does not have good choices. The origin of causal inference can date back to the 1920-1930s when experimental design was first devised and introduced in agricultural statistics (Splawa-Neyman et al., 1990; Fisher, 1935). Investigating the effect of some treatment of interest, be it a drug, a social program, or distinct life experience, is often the primary goal of behavioral, health, and social sciences. The golden standard in such research is to apply randomized experimentation, with which subjects can be randomly assigned to the treatment group that receives the treatment of interest, and concurrently to one or multiple control/reference groups, in which the subjects usually do not receive the treatment (e.g., placebo) or receive a different treatment for reference. Then one compares the outcome of interest between subjects with and without the treatment. In many cases, such experimentation that originated in engineering and natural science can be quite unrealistic in social science research for multiple reasons, such as cost or ethical concerns. Thus researchers frequently have to use observational or even survey data for finding causation.

Problems similar to the challenges in causal inference can also come up in issues such as endogeneity and sample selection, and the three could overlap or coincide sometimes. In econometrics, endogeneity refers to the scenario in which independent variables are correlated with the error term in a regression equation. Endogeneity can arise in different circumstances, such as measurement errors (with independent variables), omitted variables, and simultaneity, most of which are (or partially) related to issues with independent variables (Wooldridge, 2010). Sample selection is mostly about the limited observability of response variables, and those interested please refer to the chapter on survival regression that has cogent discussion about the two most commonly occurring issues with sample selection, censoring and truncation.

6.2.1 Average Treatment Effects

6.2.1.1 Average Treatment Effects

While making causal inference using observational or survey data, one usually aims to compare the outcome of some response variable for subjects across multiple groups that are formed mainly to sort (e.g., admission to a GT program or college) or evaluate programs/treatments (e.g., the effectiveness of an antiviral drug such as Remdesivir for treating COVID-19 or a means-tested welfare program such as WIC). This is usually done when randomized trials are not available for various reasons. So suppose one is interested in the outcome of a continuous response variable, y, for subjects in two groups, denoted by a binary treatment indicator, T, without loss of generality; when $T = 1$, subjects are in the treatment group, and when $T = 0$, the subjects are in the non-treatment or control group. If one is interested in the partial effect of a binary treatment indicator variable on some response variable of interest for a population under study, then the average treatment effect (ATE) is usually the target quantity to be estimated,

$$\text{ATE} = \quad E\left(y^1 - y^0\right) \tag{6.5}$$

where y^1 and y^0 denote the response values for the treated ($T = 1$) and non-treated ($T = 0$) respectively. In theory, ATE is calculated as the expected difference in the response outcome for subjects randomly assigned to the treatment or control group. If the random assignment assumption holds, then the estimation of ATE becomes quite straightforward, since $E\left(y^1 - y^0\right) = E\left(y^1\right) - E\left(y^0\right)$ (Wooldridge, 2010). In practice, however, the assignment processes in observational studies or surveys are frequently non-random and may operate through other mechanisms, such as self-selection, administrative selection, or other biased procedures introduced either explicitly or inadvertently in these studies, such that the two groups for comparison are not similar for an array of attributes. If the latter is true, which probably holds in most cases, then adjustments have to be made to compute ATE or related quantities of interest, such as the average treatment effect on the treated (ATT). Although there are other types of treatment effects, such as local average treatment effect (LATE) and intent-to-treat effect (ITTE), the following discussion focuses on the two most frequently estimated quantities of interest in treatment effects, ATE and ATT. It is also of importance to note that our discussion in this section only pertains to data that satisfy the stable unit treatment value assumption (SUTVA) that (non-)treatment of subjects only affect their own outcomes, not those of others.

6.2.1.2   Average Treatment Effects for the Treated

In practice, researchers are frequently interested in the treatment effects for those who qualify for or choose to participate in a program or treatment, or it may be that the program or treatment of interest targets only some segment of the population such as WIC or Head Start participants; that is, researchers usually want to know if a treatment specifically designed for some targeted (sub-)population with certain attributes is effective. Under such circumstances, one needs to calculate the average treatment effect for the treated,

$$
\begin{aligned}
\text{ATT} = \ & E\left(y^1 - y^0 | T = 1\right) \\
= & E\left(y^1 | T = 1\right) - E\left(y^0 | T = 1\right)
\end{aligned}
\tag{6.6}
$$

This quantity cannot be estimated directly because the second term, $E\left(y^0 | T = 1\right)$ or the non-treatment effect for the treated, is unobservable since the treated *is* treated. This challenge actually stays persistently in causal inference and observational studies that the treatment of interest is NOT randomized across subjects, and subjects are either in treatment or not. In essence, one can argue that this is a problem of missing data. Thus, the problem about causal inference, on this account, is similar to that of missing data; and the solutions—imputation of missing data—may also be similar, to some extent. The challenge or point of divergence is only about how to impute the missing data.

6.2.1.3   (Strong) Ignorability of Treatment Assumption

A variety of serious biases can arise when subjects are not randomly assigned to groups, making most estimation efforts futile. Rosenbaum and Rubin (1983) propose the ignorability of treatment assumption to purportedly address this issue in a general regression framework. The ignorability assumption states that T, the binary indicator variable for treatment assignment, is independent of $y(y^0, y^1)$, while conditioning on $\mathbf{x}$,

$$
E\left(y^{0,1} | \mathbf{x}, T\right) = \ E\left(y^{0,1} | \mathbf{x}\right)
\tag{6.7}
$$

That means, if one can control for sufficient amount of information from an array of exogenous variables, then the way how subjects are assigned to one group instead of the other (or others) has trivial effect on the expected outcome of interest across groups, or the effect of assignment process on the average outcome can be ignored. In essence, what causal inference does, as manifested in existing literature, is to extract information from such exogenous variables to find a non-treatment sub-sample matching the treatment sub-sample for sufficient attributes. There is an array of methods that can be used to estimate treatment effects given the ignoribity assumption, and the following discussion focuses on propensity score analysis.

6.2.2 Propensity Score Analysis

According to (Guo and Fraser, 2015, p. 1), propensity score analysis may include a wide variety of related, albeit distinct, techniques for estimating treatment effects, including exclusively regression-based modeling techniques such as the Heckman sample selection model and instrumental variables, a combination of regression modeling, predicting, and matching such as propensity score matching, and a few other propensity score based methods such as sub-classification, weighting, and non-parametric regression. As causal inference and propensity score analysis is a general analytic framework, this section focuses on matching methods with discussion of other techniques, wherever appropriate. Note that sample selection models or some regression adjustments, such as instrumental variables, use parametric regression techniques to take both the non-random assignment process and other relationships (e.g., control variables and the response outcome) into account, either simultaneously or with two-step procedures. These methods deserve some chapter-length treatment and thus are excluded from our discussion below.

Analysts are usually advised to follow a four-step procedure for implementing matching methods (Stuart, 2010, p. 5). The first step is to find a "distance" measure for matching; that is, one needs to select the measure/s used for finding matches in the treatment and control groups so that the two groups are balanced in all important attributes for meaningful comparisons. Other "closeness" measures are possible, such as exact and linear propensity score, but the two most commonly used measures are propensity score and the Mahalanobis distance. The second step is to select matching methods, which may include nearest neighbor, optimal, and ratio, among others. The third is to appraise if the matched samples are balanced with regard to $\mathbf{x}$ or observed attributes; this step is quite similar to diagnostics under the general regression framework. Usually, some calibration of covariates, distance measures, matching methods, or some combination of the previous three, may have to be made in order to find the best-matching samples. The last step is to analyze the outcome variable and estimate the treatment effect. It is of note that one can have multiple options in each step, and the total number of combinations can grow rapidly. Thus, a systemic selection of several combinations, usually comprising most popular/reasonable methods in each step and depending on the research areas and questions at hand, is recommended for triangulation.

6.2.2.1 Propensity Score Matching

Despite some recent criticism of the method (King and Nielsen, 2019), propensity score matching remains the most popular matching method for analyzing treatment effects (Rosenbaum and Rubin, 1983; Stuart, 2010; Guo and Fraser, 2015; Morgan and Winship, 2015). Propsensity score is commonly defined as $\pi(\mathbf{x}) \equiv P(T = 1|\mathbf{x})$, or the probability of being in the treatment group, conditional on covariates. It is usually obtained by first running a binary regression, such as a logit or probit model, of the binary treatment indicator variable on a set of control variables. Depending on the link function used for the binary regression, the propensity score (i.e., the conditional probability) is calculated differently. For example, if a logit model is chosen, then $P(T = 1|\mathbf{x}) = \frac{\exp(\mathbf{x}\beta)}{1+\exp(\mathbf{x}\beta)}$. Afterwards, one can calculate differences in propensity scores to measure and compare the "distance," and select cases for estimating treatment effects. With propensity scores, one can calculate $D_{ij} = |P_{T_i} - P_{C_j}|$ as a distance measure, where P_{T_i} and P_{C_j} denote propensity scores for cases i and j from the treatment (T) and control (C) groups respectively.

The next step is to select a matching method from a variety of options, including exact (naive), nearest neighbor (greedy), optimal, ratio (caliper) matching, subclassification, and weighting, each of which has its advantages as well as limitations. Since a careful and thorough elaboration on how to select a matching method out of an array of candidates and to triangulate results from different methods requires a multi-chapter treatment, this section focuses on the most popular one, the nearest neighbor matching, with occasional discussion of other methods where appropriate. In its simpliest form, nearest neighbor matching selects the case in the control group that has the smallest distance from a target case given some caliber, and it is frequently used for estimating ATT.

Below, propensity score matching is illustrated using data from the General Social Survey cumulative file. Let us assume that the quantity of interest is the treatment effect of college education on income for the general population (hence ATE). The data used for propensity score matching include the response variable of my interest, `qrincome`, the recoded (approximately numericalized from interval data) individual income variable measured in 100 dollars. We also include several exogeneous variable, including age (`age`), gender (`female`), and race (`white`). In this example and what follows, education (`college`) is used as the binary treatment indicator variable. When `college` is one, that means the respondent received a college-level or above education; otherwise, college is coded as zero. Below, we first read in the data and provide summary statistics for each variable by college.

```
readin <- read.dta("data/gssCum7212Teach.dta", convert.factor=F)
# create list of variable names used for variable selection below
usevar <- c("qrincome", "age", "college", "female", "white")
# select variables and eliminate missing cases
usedta <- subset(readin[complete.cases(readin[usevar]),],
                        select=c(qrincome, age, college, female, white))
# set up an x matrix to be fed into, for example, Mahalanobis distance
# matching
X = cbind(usedta$age, usedta$female, usedta$white)
# check data
tapply(usedta$qrincome, usedta$college, summary)
```

```
## $`0`
##    Min. 1st Qu.  Median    Mean 3rd Qu.    Max.
##     6.6    75.0   125.0   184.2   375.0   375.0
##
## $`1`
##    Min. 1st Qu.  Median    Mean 3rd Qu.    Max.
##     6.6   175.0   375.0   282.4   375.0   375.0
```

```
tapply(usedta$age, usedta$college, mean)
```

```
##          0          1
## 40.24852 41.40438
```

```
tapply(usedta$female, usedta$college, mean)
```

```
##          0          1
## 0.4910962 0.4953917
```

```
tapply(usedta$white, usedta$college, mean)
```

```
##          0          1
## 0.8001583 0.8571429
```

The results from our descriptive analysis of income and other covariates by college show that the two groups, people with college education and those without, are unbalanced in all covariates, especially race (`white`).

```
# Difference without Adjustments
meanT = mean((usedta[usedta$college==1,])$qrincome)
meanC = mean((usedta[usedta$college==0,])$qrincome)
meanT - meanC

## [1] 98.18734
```

We can also calculate the naive treatment effect without adjustment by taking a means difference as shown in the R code chunk above, and the results show that the income difference is 98.187 ($\times 100$) dollars. In reality, we know that only a select segment of the high-school student population goes to college, and this process involves both self-selection (i.e., decision to go to college) and institutional selection (e.g., certain standards have to be met). So to explore the causal relationship between college education and income, one may consider this selection process.

Below, we illustrate a simple example of one-to-one nearest matching using propensity score matching. First, we use a parametric model to produce propensity scores, and to that end, we choose to run a binary logit model (`logitPS`) using `college` as the binary treatment indicator variable, and `age`, `female`, and `white` as covariates.

```
logitPS <- glm(college ~ age + female + white,
    data=usedta, family=binomial(link="logit"))
```

Then we invoke the `Match` function from the `Matching` package (Sekhon, 2011). By default, the `Match` function performs the one-to-one nearest neighbor matching. The `Y`, `Tr`, and `X` arguments are to specify the response variable (treatment outcome), the treatment indicator variable, and the covariates or the propensity score to match on respectively. In this case, propensity score is used (`X=logitPS$fitted`). Since the quantity of interest is ATE, the `estimand` argument is set to ATE; if ATT is specified instead, then an estimate of the average treatment effect for the treated is calculated instead.

```
require(Matching)
# Average Treatment Effect on the Treated
ATEPS = Match(Y=usedta$qrincome, Tr=usedta$college, X=logitPS$fitted,
estimand="ATE") summary(ATEPS)

##
## Estimate...  88.517
## AI SE......  5.7481
## T-stat.....  15.399
## p.val......  < 2.22e-16
##
## Original number of observations..............  3395
## Original number of treated obs..............  868
## Matched number of observations..............  3395
## Matched number of observations  (unweighted).  34973
```

The results show that the estimated average treatment effect is 88.517 ($\times 100$) dollars and the difference is statistically significant at the conventional level. Next, one needs to check if the matched samples are balanced. There are multiple statistical measures serving this purpose. The first is called the standardized mean difference (SMD), which is equivalent of the test statistic from a mean difference test for two independent samples in a one-to-one matching scenario (Linden and Samuels, 2013). So for a generic covariate x_k,

$$\text{smd}_k = \frac{\overline{x}_{Tk} - \overline{x}_{Ck}}{\sqrt{\frac{S^2_{Tk} + S^2_{Ck}}{2}}} \tag{6.8}$$

where $\overline{x}_{Tk}$ and $\overline{x}_{Ck}$ are sample means of x_k in the treatment (T) and control (C) groups respectively. This statistic should be examined without referencing to the response variable. The idea behind this balance statistic is straightforward. The statistical decision behind this measure, however, is somewhat opaque, and scholars propose different threshholds for this standardized difference, from 0.10, 0.20, to 0.25 (Rubin, 2001; Linden and Samuels, 2013). But the rule of thumb is that the higher the value for the standardized mean difference, the more unbalanced the two samples are. When multiple covariates are involved, one needs to use a summary measure of imbalance, $\text{SMD} = \frac{1}{K}\sum_{k=1}^{K}\text{smd}_k$.

It is well known that a sample mean only characterizes the central tendency of a distribution, while potentially leaving out all other features and nuances of the distribution. Two distributions with very similar means (expected values) could have drastically different shapes and higher order moments. Thus, it is also helpful to compare sample variances. Rubin proposes to check the ratio of the two variances from the treatment and control groups,

$$\text{VR}_k = \frac{S^2_{Tk}}{S^2_{Ck}} \tag{6.9}$$

For subjects from two balanced samples, the value of this statistic should be close to one. Rubin also suggests that a ratio outside the range of 0.5 to 2 is indicative of imbalance (Rubin, 2001). Similar to the sample mean difference, when multiple covariates are tested, one usually uses a summary measure, the geometric mean of variance ratio, $\text{GMVR} = \left(\Pi_{k=1}^{K}\text{VR}_k\right)^{\frac{1}{K}}$.

To check balance, one can use the `MatchBalance` function from the same `Matching` package. The following R code chunk and results show how it is typically done. The first argument of `MatchBalance` is to specify the formula for balance. To the left of the tilde sign, "~", is the treatment indicator variable, and to the right includes the covariates for which the balance check statistics are calculated. In this case, we include only `age` to simplify presentation. It is advised that more polynomials terms (higher-order terms and interactions) than those in matching itself should be used in balance check. The `match.out` argument specifies the matching output object, and the `nboots` option requests the number of replications for the Kolmogorov-Smirnov (KS) test, a non-parametric test of equality of probability distribution for continuous variables (Massey, 1951), and it is suggested that the number of bootstrap samples be set to at least 500 to be acceptable and 1000 preferably (Sekhon, 2011).

```
MatchBalance(college ~ age, match.out=ATEPS, nboots=1000, data=usedta)
```

```
##
## ***** (V1) age *****
##                        Before Matching    After Matching
## mean treatment........    41.404          40.789
## mean control..........    40.249          40.532
## std mean diff.........    9.9853          1.9927
##
```

```
## mean raw eQQ diff.....        2.3341           0.21151
## med  raw eQQ diff.....             2                 0
## max  raw eQQ diff.....            10                10
##
## mean eCDF diff........     0.034509           0.00315
## med  eCDF diff........     0.023645        0.00037172
## max  eCDF diff........      0.10702           0.03171
##
## var ratio (Tr/Co).....      0.69644           0.93798
## T-test p-value........     0.016164        5.9304e-08
## KS Bootstrap p-value.. < 2.22e-16        < 2.22e-16
## KS Naive p-value...... 7.4868e-07        1.1102e-15
## KS Statistic..........      0.10702           0.03171
```

The results above show that some balance check statistics are within acceptable range, and we are unsure about the degree to which the two groups are matched with regard to age. Although the mean difference, differences in empirical-QQ plots, and variance ratio show clear signs of balance improvement, the t test and the Kolmogorov-Smirnov test indicate otherwise. Usually, one needs to go through several iterations back and forth from balance check to the parametric model for matching or vice versa to make sure that the treatment and control groups are sufficiently matched on major attributes.

6.2.2.2 Mahalanobis Distance Matching

The Mahalanobis distance can be broadly defined as a scale-free dissimilarity index between points/vectors in a multi-dimensional space (Guo and Fraser, 2015). It was first proposed in 1936 by an Indian statistician, P. S. Mahalanobi, who is often viewed as the father of modern statistics in India (Mahalanobis, 1936). The Mahalonobis distance or metric is a multi-dimensional generalization of the standardized score in a univariate case, or the number of standard deviations one realized/hypothetical value of a distribution deviates from the mean (expected value) of that distribution. When this measure is applied to the distance of two random vectors in a multi-dimensional space, it is calculated as,

$$\text{MD}_{i,j} = \left(\mathbf{x}_{Ti} - \mathbf{x}_{Cj}\right)' Q^{-1} \left(\mathbf{x}_{Ti} - \mathbf{x}_{Cj}\right) \tag{6.10}$$

where $\mathbf{x}_{Ti}$ and $\mathbf{x}_{Cj}$ are vectors for a treated subject i and a control subject j, and Q is a $K \times K$ sample variance-covariance matrix of the K covariates in $\mathbf{X}$ (i.e., excluding the response and treatment variables) in the full control group if one is to estimate ATT, or the variance-covariance matrix of $\mathbf{X}$ in the pooled treatment and full control groups if ATE is the quantity of interest (Rubin, 1980; Stuart, 2010). Usually, this distance measure works best when covariates are continuous. Below, the same `Match` function is used to execute the Mahalanobis Distance matching with the `Weight` argument set to 2 for using the Mahalanobis distance measure.

```
ATTMM = Match(Y=usedta$qrincome, Tr=usedta$college, X=X,
              replace = TRUE, ties = TRUE, Weight = 2)
summary(ATTMM)

##
## Estimate...  81.317
## AI SE......   4.926
## T-stat.....  16.508
## p.val......  < 2.22e-16
```

```
##
## Original number of observations.............  3395
## Original number of treated obs..............  868
## Matched number of observations..............  868
## Matched number of observations  (unweighted).  16796
```

The results show that the ATT is 81.317, and is statistically significant. Afterwards, one can use the `MatchBalance` function again to conduct balance diagnostics to see if the two matched samples are balanced on covariates. This time, the race variable, `white`, is selected just for illustrative purposes.

```
MatchBalance(college ~ white, match.out=ATTMM,
                nboots=1000, data=usedta)
```

```
##
## ***** (V1) white *****
##                           Before Matching      After Matching
## mean treatment........      0.85714          0.85714
## mean control..........      0.80016          0.85714
## std mean diff.........      16.275                 0
##
## mean raw eQQ diff.....      0.057604               0
## med   raw eQQ diff.....            0               0
## max   raw eQQ diff.....            1               0
##
## mean eCDF diff........      0.028492               0
## med   eCDF diff........      0.028492               0
## max   eCDF diff........      0.056985               0
##
## var ratio (Tr/Co).....      0.76634                1
## T-test p-value........  7.0484e-05                 1
```

The balance statistics show that the race variable is balanced across the two groups.

6.2.2.3   Genetic Matching

Genetic matching could be viewed as an extension of the Mahalanobis distance matching by adding a weight matrix in the distance measure. Instead of having a simple normalizing variance-covariance matrix, the distance metric is refined as

$$\text{GMD}_{i,j} = \quad (\mathbf{x}_{Ti} - \mathbf{x}_{Cj})' \left(Q^{-1/2}\right)' \left(Q^{-1/2}\right) (\mathbf{x}_{Ti} - \mathbf{x}_{Cj}) \qquad (6.11)$$

where W is a $K \times K$ weight matrix with its off-diagonal elements set to zero for computational convenience (e.g., computational power and estimability of the optimization problem) (Sekhon, 2011) and $Q^{-1/2}$ is the Cholesky factorization of Q, the variance-covariance matrix of the covariate matrix $\mathbf{X}$. One could also include the propensity score P as a covariate, thus turning $\mathbf{X}$ into $\mathbf{Z} = [\mathbf{X}, P]$. If a weight of one is given to all the diagonal elements, then we have the usual Mahalanobis distance matching. If, however, a weight of one is given to P, and all other covariates receive a weight of zero, then the genetic matching reduces to propensity score matching. The genetic matching method described here uses a genetic search algorithm to search for the weights in the weight matrix, W, that optimizes (usually minimizes) a user-specified loss function in measuring covariates balance. This algorithm first proposes an initial batch of weights and iteratively improves weights that can minimize the loss function (Sekhon, 2011; Diamond and Sekhon, 2013).

The `GenMatch` function from the same `Matching` package creates the weight matrix, `weightMat` (W in 6.11), to be used in the `Match` function. By default, `GenMatch` iteratively searches for solutions to the diagonal elements in W that would minimize the largest observed individual discrepancy in covariates for matching at every iteration (generation) (Sekhon, 2011). The `X` argument specifies the covariates to match on, and it may include propensity scores, if so chosen. The `BalanceMatrix` argument specifies the variables that one intends to achieve balance on. For both arguments, covariates need to be collected in a matrix for use. The `pop.size` argument controls the number of individuals the evolutionary algorithm uses to solve the optimization problem, and can largely determine the amount of time the `GenMatch` function takes to find the solution. Here `pop.size` is set to 10 for computational convenience,

```
# genetic matching
weightMat = GenMatch(Tr=usedta$college, X=X, BalanceMatrix=X, pop.size=10)
```

After the weight matrix is created, one can use it in the `Match` function to implement genetic matching. Everything else follows as usual.

```
geneticMatch = Match(Y=usedta$qrincome, Tr=usedta$college, X=X,
Weight.matrix=weightMat)
summary(geneticMatch)

##
## Estimate...  81.317
## AI SE......  4.926
## T-stat.....  16.508
## p.val......  < 2.22e-16
##
## Original number of observations..............  3395
## Original number of treated obs..............   868
## Matched number of observations..............   868
## Matched number of observations  (unweighted).  16796
```

Up until now, the `Matching` package has been used exclusively in this section. There are several other packages that can perform propensity score analysis, for example, `MatchIt` (Ho et al., 2011), `optmatch` (Hansen and Klopfer, 2006), and `twang`, of which `MatchIt` is another popular package (Ho et al., 2011). Below are a few lines of R codes to show how the same genetic matching can be implemented with functions from the `MatchIt` package. The `matchit` function is the workhorse to implement the requested matching method. The `method` argument in the `matchit` function is set to `genetic` for genetic matching, which calls the `Match` function of the `Matching` package. The method can also be set to a variety of other options, for example, `nearest` (default), `exact`, `optimal`, and `subclass` (subclassication).

```
library(MatchIt)
m.out <- matchit(college ~ age + female + white,
                 data = usedta, method = "genetic")
summary(m.out)
plot(m.out)
m.data <- match.data(m.out)
```

The `summary` and `plot` functions present balance statistics, and the `match.data` function produces matched data, with which one can proceed for subsequent analysis.

6.2.2.4   Coarsened (Exact) Matching

Unlike the previously discussed matching methods, coarsened exact matching (CEM) is a member of the "Monotonic Imbalance Bounding" (MIB) class of matching methods (Iacus et al., 2012). The statistical theory and proof behind CEM can be convoluted, but the general idea is quite straightford and similar to that of the randomized full block design. In such experimental research designs, nuisance factors, or the variables of little interest to experimenters (i.e., control variables), are held constant in multidimensional blocks. Within each block, the values of nuisance factors are fixed for all subjects therein, whereas the factor of interest (treatment factor) varies across subjects, thus making it possible to examine the effect of this particular factor.

So if one has an array of control variables and can create structured blocks such that each treated subject is paired with one or multiple subjects from the control group, then balance can be easily obtained. The reality, however, is more complicated when one has continuous variables and more than a few covariates. If researchers use the measurements exactly as they are from empirical data, the curse of dimensionality will almost surely lead to insufficient blocks/cells with matched pairs. What coarsened exact matching does in essence is, as its name suggests, to coarsen the measurement of the data and combine values that are substantively indistinguishable with fewer categories (coarsened bins) such that the number of blocks reduce to manageable ones with sufficient matched subjects for estimating causal effects. So when it comes down to execution, coarsened exact matching becomes finding histogram bins for each covariate and then creating a multidimensional histogram space, in which each treated and non-treated subject can be mapped onto one unique block in that space. To check imbalance, Iacus et al. (2012) propose the $\mathcal{L}_1$ statistic that sums across the differences in the relative frequencies of the treated and control groups for all blocks in the multidimensional hisrogram space,

$$\mathcal{L}_1(t, c; H) = \frac{1}{2} \sum_{m_1...m_K \in H(\mathbf{X})} |t_{m_1...m_K} - c_{m_1...m_K}| \tag{6.12}$$

in which $H(\mathbf{X})$ denotes a multidimensional histogram (bin) space, compartmentalized with the Cartesian product $H(x_1) \times ... \times H(x_K)$, $t_{m_1...m_K}$ and $c_{m_1...m_K}$ are empirical relative frequencies of observations for the block/cell with its Cartesian coordinates being $m_1...m_K$ (Iacus et al., 2011, p. 352). This statistic is bound between zero and one, and its value increases as the degree of imbalance rises, and its evaluation is contingent upon data and selected covariates. For example, if $\mathcal{L}_1 = 0.8$, then that means only 20% of the two empirical multidimensional distributions (from the two groups) overlap.

Below, functions from the `cem` package are used to execute automated coarsened exact matching (Iacus et al., 2009). After calling the package, the first few lines of R codes look at descriptive statistics of the treated (college) and control (non-college) groups, and compare the pre-matching means of the two groups. Below, we show a slightly different method to compute the same statistic as we did earlier, by extracting case IDs first.

```
library("cem")
cemdat <- data.frame(na.omit(usedta))
# identify cases in the treated and control groups
treat <- which(cemdat$college == 1)
control <- which(cemdat$college == 0)
# count the number of cases in the treated group
```

```
treatN <- length(treat)
# count the number of cases in the control group
controlN <- length(control)
# naive mean difference
meandiff = mean(cemdat$qrincome[treat])-mean(cemdat$qrincome[control])
meandiff
```

```
## [1] 98.18734
```

A naive method for computing the treatment effect is to simply take a difference in the means of the two groups, which turns out to be 98.187, the same as we computed previously. Because the process of getting into college is not through randomized assignment, the covariates can differ significantly in their distributional properties and sample statistics. One can use the `imblance` function to check pre-matching balance statistics. The first column in the results section of univariate imbalance statistics, "`statistics`", computes the observed mean differences between the two groups (i.e., college vs. non-college).

```
covars <- c("age", "female", "white")
preMat = imbalance(group = cemdat$college, data = cemdat[covars])
preMat
```

```
##
## Multivariate Imbalance Measure: L1=0.175
## Percentage of local common support: LCS=76.4%
##
## Univariate Imbalance Measures:
##
##             statistic    type           L1 min 25% 50% 75% max
## age     1.155861853 (diff) 0.0007914523   5   3   2  -1 -10
## female  0.004295544 (diff) 0.0042955436   0   0   0   0   0
## white   0.056984567 (diff) 0.0569845667   0   0   0   0   0
```

In this case, the omnibus imbalance measure $L1 = 0.175$. The numbers from the "`min`" to "`max`" columns report the differences in percentiles of corresponding empirical (observed) distributions of the covariates for the two groups. For example, the age difference between the treated and control groups is 1.156, and the difference in the 50th percentiles (medians) for age of the two groups is 2.

Next, one can use the `cem` function to implement CEM. Below, an automated (bin sizes are automatically set by the function) CEM is used. Alternatively, users can supply user-defined coarsened cutpoints or groupings. The `drop` argument excludes the response variable, `qrincome`, from the matching process, and other variables in the `cemdat` data frame are used as covariates to be matched on between the treated and control groups. The `eval.imbalance` option is set to T so that an array of balance statistics can be displayed in the results.

```
# automated coarsening
matAutCoarsen <- cem(treatment = "college", data = cemdat,
    drop = "qrincome", eval.imbalance=T)
```

```
##
## Using 'college'='1' as baseline group
```

```
matAutCoarsen
```

```
##              G0  G1
## All        2527 868
## Matched    2485 868
## Unmatched    42   0
##
##
## Multivariate Imbalance Measure: L1=0.048
## Percentage of local common support: LCS=89.4%
##
## Univariate Imbalance Measures:
##
##            statistic    type            L1 min 25% 50% 75% max
## age     1.013065e-01 (diff) 0.000000e+00   5   0   0   0  -1
## female  5.551115e-17 (diff) 2.775558e-17   0   0   0   0   0
## white   0.000000e+00 (diff) 0.000000e+00   0   0   0   0   0
```

By comparing the two sets of results from running the `imbalance` function before and after CEM, it can be shown that the level of imbalance has reduced significantly. If such level of imbalance reduction is satisfactory, then one can proceed to estimate the treatment effect by running a (weighted) difference in means test, a bivariate, or a multiple regression. Below, a bivariate linear regression is used.

```
estAutCoarsen <- att(matAutCoarsen, qrincome ~ college, data = cemdat)
estAutCoarsen
```

```
##
##              G0  G1
## All        2527 868
## Matched    2485 868
## Unmatched    42   0
##
## Linear regression model on CEM matched data:
##
## SATT point estimate: 83.069314 (p.value=0.000000)
## 95% conf. interval: [73.045466, 93.093162]
```

The results show that college graduates can earn 83.069×100 more dollars than the annual income that they would have received had they not gone to college. In this example, a complete or near-complete balance is assumed, so the treatment variable, `college`, is the only explanatory variable. If there is still some indication of imbalance for other covariates, then one can run a multiple regression instead by including those covariates.

6.3   Machine Learning

Machine learning is usually viewed as a sub-field of or alternative (insomuch as artificial intelligence was not very successful in its formative years) to artificial intelligence (AI), and it has been around for at least seven decades without a very clear trail of its official debut. This, in and of itself, is probably indicative of the interdisciplinary nature of this field. The original idea about machine learning is to let machine learn, think, and behave just like or to approximate human intelligence. From the initial idea about machine learning (i.e., the game of checkers) to the most recent admirable accomplishments (e.g., Deep Blue's victory over Kasparov in chess in 1997 and AlphaGo's defeat of Lee Sedol in go in 2016 and then Ke Jie in 2017) of machine learning, board games have been the prototypical human intelligence activity for a machine to emulate. AlphaGo's defeat of Lee and especially the world number-one ranked go player, Ke Jie, can be viewed as watershed events in the world of machine learning and AI, since go is usually viewed as the epitome of human intelligence that integrates art, mathematics, logic, and human intuition and counter-intuition.

Arguably being the one who coined and popularized the term machine learning, Samuel (1959) in "Some Studies in Machine Learning Using the Game of Checkers," for the first time, proposed two machine-learning procedures to program a computer to learn playing checkers and potentially defeat human beings. In it, Samuel (1959) probably gave the initial definition of machine learning, that "[p]rogramming computers to learn from experience should eventually eliminate the need for much of this detailed programming effort" (p. 535). Although Samuel might be the first one who explicitly used the term machine learning and gave a formal definition, the ideas and some basic models of machine learning probably have earlier origins. For example, Hebb (1949, p. 62) proposed Hebb's postulate (A Neurophisilogical Postulate) in his seminal work, *Organization of Behavior*, that laid the foundation for neural network in machine learning. While some argue that in his ground-breaking work that bridges neural science with psychology, Hebb probably proffered the prototypical ideas of machine learning by specifying a model for brain cell interaction, others indicate that it is Arthur Samuel, a pioneer in computer gaming and artificial intelligence, coined the term machine learning while developing a computer program for playing checkers. Probably except for historians of science, it is not worth prodigious efforts to map out all the branches of the idea tree about machine learning as no branch or field in science arises out of nowhere, and they all, to some extent, are fireworks ignited by one or multiple fuses of brilliance of great thinkers of previous generations.

In broad terms, machine learning can be defined as an interdisciplinary science to devise algorithms to identify patterns in usually massive data either through prescriptive procedures, and more importantly, through automating this whole process by minimizing human intervention. It is a new and important area that cross-fertilizes among artificial intelligence, computer science, statistics, and other related fields.

6.3.1 Basic Concepts

There is a new collection of terms used in machine learning, some of which are similar or almost identical to expressions already existing in traditional statistical science, whereas others are relatively new concepts and ideas. Below, we briefly discuss the differences between machine learning and statistical learning, between supervised learning and unsupervised learning, between regression and classification, and between training and validation.

6.3.1.1 Machine Learning and Statistical Learning

The relationship between machine learning (and possibly data science) and statistics could be compared to how the GLM (generalized linear models) framework is associated with specific categorical and limited response variable models, insomuch as machine learning is conceptualized as a general paradigm to synthesize several related areas, such as computer science and statistics. According to James et al. (2014), "Statistical learning refers to a set of tools for modeling and understanding complex data sets. It is a recently developed area in statistics and blends with parallel developments in computer science and, in particular, machine learning" (p. vii). Murphy (2012, p. 1) defines "...machine learning as a set of methods that can automatically detect patterns in data, and then use the uncovered patterns to predict future data, or to perform other kinds of decision making under uncertainty." Based on these two definitions and others, one can see although there is a considerable overlap between the two areas, statistical learning aims to model and understand data, whereas machine learning is to automatically detect patterns and to predict future data. In practice, one would find that machine and statistical learning are almost identical in some areas but diverge in others, especially with regard to functional emphasis (e.g., explanation vs. prediction).

6.3.1.2 Supervised Learning and Unsupervised Learning

Supervised and unsupervised learning are two commonly used terms in machine and statistical learning. Supervised learning is similar to statistical analysis or modeling that has response variables so that predictions from our models/hypotheses can be checked against values from observed response variables, and this machine learning process is called supervised learning. In contrast, unsupervised learning is a learning process that does not involve response variables for post-estimation predictive diagnostics. Since this text primarily focuses on regression analyses that usually have responses variables, techniques for data reduction (unsupervised learning) are not included in the discussion of the following sections.

6.3.1.3 Regression and Classification

In machine and statistical learning, regression usually refers to a model with a continuous response variable, and classification, on the other hand, entails models with non-continuous and usually categorical response variables, such as logistic and multinomial logit models.

6.3.1.4 Training, Validation, and Test

Since much of the goal of machine learning is to build a predictive model, investigating the predictive power of a model or algorithm can be essential. The data analysis process in machine learning usually involves first randomly partitioning the initial data into several parts with a priori proportions. The part that is used for initial data analysis and model building is called training set, and the part (usually parts) that is used for tuning parameters and selecting models is called the validation set. The testing set is usually used to estimate the performance of the chosen model. To improve accuracy and reduce variability, one usually uses multiple sets for validation, and the results are combined to give an overall assessment (test) of the predictive power of the algorithm and model of our interest.

6.3.2 Supervised Learning

For supervised learning, usually analysts have both predictors and their response variables. The goal is to link the response variable to the set of, usually a large number of, predictors available at hand so that one can either make predictions of the response variable for future observations or gain a sophisticated understanding of the relationship between the response variable and its predictors.

6.3.2.1 Regularization: Ridge and Lasso Regression

Model selection is an enduring theme in classical statistical literature, and it has become even more so under the framework of machine and statistical learning that usually serves to detect patterns in massive data and make accurate prediction about future in-coming data points. When researchers have a large number of both observations and variables, it becomes almost imperative to sift through noise and find the best predictive pattern. Ridge and lasso regressions are designed to reduce model complexity and alleviate the problem of over-fitting commonly seen in models with many predictors and higher-order terms. When over-fitting occurs, the model may achieve its optimal explanatory power with the training data (or data at hand), but can fit poorly the test data or data to be investigated in the future; in other words, over-fitting can lead to a mismatch between high explanatory but low predictive power.

Ridge Regression

Under the linear regression framework, the least squares method estimates $\hat{\beta} = \left(\beta_0, \widetilde{\beta}\right)$ that minimizes, the residual sum of squares, $RSS = \sum_{i=1}^{N}\left(y - \beta_0 - \mathbf{x}\widetilde{\beta}\right)^2$, wherein β_0, by convention, denotes the intercept, $\widetilde{\beta}$ is the slopes vector without the intercept term, and $\mathbf{x}\widetilde{\beta} = \sum_{k=1}^{K}\beta_k x_k$; or, RSS can be re-expressed as $e'e$ in the matrix form, where $e = y - \beta_0 - \mathbf{x}\widetilde{\beta}$. Ridge regression is similar to least squares, except that the coefficients are estimated by minimizing the sum of RSS and a shrinkage penalty, so the loss function [4] becomes

$$\sum_{i=1}^{N}\left(y - \beta_0 - \mathbf{x}\widetilde{\beta}\right)^2 + \lambda\widetilde{\beta}'\widetilde{\beta} = RSS + \lambda\sum_{k=1}^{K}\beta_k^2 \qquad (6.13)$$

where $\widetilde{\beta}$ is a column vector of slopes excluding the intercept β_0, the tuning parameter $\lambda \geq 0$ adjusts the level of penalty for having additional parameters (James et al., 2014), $\lambda\sum_{k=1}^{K}\beta_k^2$ is called a shrinkage penalty, and β_k is estimated through such shrinkage and thus called a shrinkage estimator. The term, $\sum_{k=1}^{K}\beta_k^2$, is the squared l_2 norm (also known as the Euclidean norm or distance), which can be expressed as $\|\beta\|_2 = \left(\sum_{k=1}^{K}\beta_k^2\right)^{\frac{1}{2}}$ in this case. Since the main goal of ridge regression is to solve the problem of over-fitting, or regularization, it can be said that ridge regression applies an l_2 regularization.

One can use multiple R packages to estimate regularized (penalized) regressions or regressions with shrinkage, including `caret`(Kuhn, 2022), `elasticnet`(Zou and Hastie, 2020), `glmnet`(Friedman et al., 2010), `h2o`(LeDell et al., 2022), `lars`(Hastie and Efron, 2013), and `penalized`(Goeman et al., 2018), just to name a few. Below, the same data used in the previous section are recycled to estimate a ridge regression, with two additional predictors, including marital status (`married`), and level of happiness (`happy`). Note that the number of predictors in the examples illustrated below is atypical and much fewer than what one would usually have with a large database, or "big data," so they only serve illustrative purposes. The following few lines of R codes manage and prepare data,

```
require(foreign)
readin <- read.dta("data/gssCum7212Teach.dta", convert.factor = F)
# create list of variable names used for variable selection
usevar <- c("qrincome", "hlthc2", "age", "educ", "female",
    "white", "married", "happy")
# subset the data (select variables)
usedta <- subset(readin[complete.cases(readin[usevar]), ],
    select = c(qrincome, hlthc2, age, educ, female, white,
```

---

[4]Loss or cost function, $L(y, f(X))$, refers to a target function used for estimating model parameters usually via penalizing prediction errors. For example, the absolute error (L_1) loss and squared error (L_2) loss.

```
          married, happy))
x = as.matrix(usedta)[, -1:-2]
y1 = log(usedta$qrincome)
y2 = usedta$hlthc2
```

To divide the raw data into several parts and use them for model building, validation, and testing is a common practice in machine learning. Below, we randomly split the data into two parts with a split ratio of 70/30 only for illustrative purposes. The `set.seed` function is used so that each time the data are randomly split, they will be split in the exact same way for reproducibility. The `sample` function randomly generates a list of numbers stored in `part01` as case index.

```
set.seed(473062000)
# rndnum <- .Random.seed
part01 = sample(1:nrow(usedta), nrow(usedta)*0.7)
part02 = (-part01)
# x needs to stay as a matrix for predict.glmnet to work
x01 = as.matrix(usedta)[part01, -1:-2]
# y can be kept as a numeric column
y01 = log(usedta[part01, 1])
```

Then we can estimate a ridge regression with the `glmnet` function from the `glmnet` package. The first two arguments in the `glmnet` function include an input (usually predictors) matrix and a response (column) vector. When the `alpha` argument is set to zero, then the model imposes a ridge penalty. By default, the `nlambda` argument is set to be 100. The `lambda` argument requires a sequence supplied; by default, the program provides a sequence valued from high to low, based on `nlambda` and `lambda.min.ratio`, the latter of which is the ratio of the minimum to the maximum value of the sequence. [5] According to the `glmnet` introduction manual, users are advised to use the sequence provided by the function (i.e., without specifying the `lambda` argument) and need to be cautious when user-supplied lambda sequence is used instead. But there are also concerns about blindly accepting the default lambda sequence, and that users should come up with their own choices and finer-tune the λ's. Below, we request to experiment with 10 λ's for illustrative ease.

```
# lamseq = 2^seq(9, -3, length = 100)
# ridge.fit = glmnet(x, y1, alpha = 0, nlambda = 100, lambda = lamseq)
require(glmnet)
# alpha = 0 for ridge
ridge.reg = glmnet(x01, y01, alpha = 0, nlambda = 10)
```

Once the results are stored, we can request to investigate them in various ways, for example, beginning with the dimension of the resultant coefficient matrix and then the summary results of our ridge regression,

```
dim(coef(ridge.reg))
```

```
## [1]   7 10
```

---

[5] According to Friedman et al. (2010, p. 7), $\lambda_{\max}$ is calculated as the minimum of λ when all the parameter estimates are zero. Then `lambda.min.ratio` is used to get $\lambda_{\min}$ such that $\lambda_{\min} = \sigma\lambda_{\max}$, and the middle numbers of this sequence are filled out in descending order on the log scale.

```
print(ridge.reg)
```

```
##
## Call:  glmnet(x = x01, y = y01, alpha = 0, nlambda = 10)
##
##      Df   %Dev Lambda
## 1     6   0.00 337.80
## 2     6   0.35 121.40
## 3     6   0.94  43.63
## 4     6   2.46  15.68
## 5     6   5.81   5.64
## 6     6  11.25   2.02
## 7     6  16.52   0.73
## 8     6  19.13   0.26
## 9     6  19.84   0.09
## 10    6  19.96   0.03
```

The results show that `coef(ridge.reg)` is a 7 x 10 numeric matrix. The few columns in the results section correspond to degrees of freedom (`Df`), percent of null devianced explained (`%Dev`), and λ values (`Lambda`). The last column shows the 10 λ's chosen by the function. When $\lambda = \lambda_{\max} = 337.809$, the estimation procedure, as designed, suppresses all slope coefficients (excluding the intercept) to approximate zero; and when $\lambda = \lambda_{\min} = 0.034$, there is minimum shrinkage, and the estimated slopes should be very similar to those from a traditional least squares,

```
cbind(coef(ridge.reg)[, 1], coef(ridge.reg)[, 10])
```

```
##                        [,1]          [,2]
## (Intercept)   4.924560e+00   3.030085848
## age           1.141046e-38   0.011922980
## educ          1.139406e-37   0.117842943
## female       -5.255006e-37  -0.513136423
## white         1.758237e-37   0.033650227
## married       2.211232e-37   0.117255410
## happy        -1.155037e-37   0.001836924
```

We can also visually investigate how the magnitude of the coefficients change with the λ's by graphing a coefficient profile plot as shown in Fig. 6.3.

```
plot(ridge.reg, xvar=c("lambda")) # "dev", "norm"
```

Once the estimates are obtained, one can then make various predictions, beginning with the predicted values for the responose variable while choosing $\lambda=121.402$, for example, from the λ sequence, using the `predict.glmnet` function,

```
x01.pred = predict(ridge.reg, s=ridge.reg$lambda[2], newx=x01)
```

In the R code chunk above, the `newx` argumet is to supply a new data matrix with the same number of columns. In this case, we still use x01 to request predictions of the response variable in the initial estimation sample. One can take a step further to calculate mean squared error (MSE), a commonly used measure to assess model fit and predictive

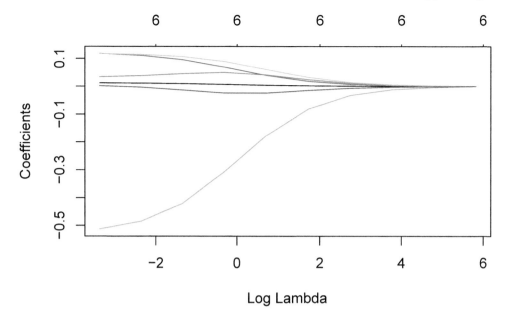

Figure 6.3
Coefficient Profile Plot for Ridge Regression

power especially in regression models predicting continuous response variables. Note that $\text{MSE} = \frac{1}{N} \sum_{i=1}^{N} (y_i - \widehat{y}_i)^2$, $\widehat{y}_i = \widehat{f}(x_i)$, where the form of $f(x_i)$ varies, depending on the specified statistical model. In this example, $f(x_i) = x\beta$. We can use the following R code chunk to compute the MSE associated with our initial estimation data and chosen λ,

```
mean((x01.pred - y01)^2)
```

```
## [1] 1.125526
```

Albeit details absent, it can be easily shown that $\lambda_{\min}$ from our data for initial modeling yields the smallest MSE. The limitation to go with $\lambda_{\min}$ or virtually $\lambda = 0$ is that it is likely over-fitting the current data while yielding poor predictive power with future data. The following R code chunk uses the estimates from the ridge regression to predict the response variable in the hold-out data previously split out.

```
x02 = as.matrix(usedta[part02,-1:-2])
y02 = log(usedta[part02, 1])
# predict for all lambda's with x02
ridge.pred=predict(ridge.reg, s=ridge.reg$lambda, newx=x02)
mse.test = as.matrix(apply((ridge.pred-y02)^2, 2, mean))
min(mse.test)
```

```
## [1] 0.8922916
```

```
(answer <- which(mse.test ==  min(mse.test), arr.ind=T))
```

```
##    row col
## s8   8   1
```

It can be shown from the results that the λ yielding the minimum MSE of the hold-out data is λ_8, not λ_{min}. One can also use the `predict.glmnet` function to compute ridge regression coefficients for $\lambda = 5$ instead of the λ's supplied either by users or the `glmnet` function.

```
predict(ridge.reg, s = 5, type = "coefficients")[1:7, ]
```

The naive training and testing procedures are used to illustrate the basic ideas of regularization (penalized linear regression), and they may seem ad-hoc and even primitive. For empirical research, one is usually advised to follow well-established procedures and methodically analyze the data.

Next, we use functions from the `caret` package to showcase how machine and statistical learning is performed in a systematic way. Note that other R packages can accomplish many or even all of the following tasks, some with even more efficient and effective ways, for example, the `cv.glmnet` function from the `glmnet` package and the `elasticnet` package. The `caret` package nonetheless is a mega package that absorbs probably the largest number of packages for streamlining the process for classification and regression training, including decision trees. As any experienced data analyst would expect, breadth and sophistication also come with its cost that sometimes it can be hard for users find specific information for a particular learning algorithm.

The R code chunk below first splits the original data into a training and a test set to illustrate the process of creating a regression model and assessing model accuracy using validation under the general framework of machine/statistical learning. Eighty percent of the cases are randomly selected from the original data to be used as the training set, and the rest 20% is used as the test set for an unbiased assessment of the proposed model[6]. The `set.seed` function specifies the seed for random sampling so that the results are reproducible. For a numeric vector y (the first argument), the `createDataPartition` function (in the `caret` package) first stratifies the data into different sections based on the percentiles of y, and then random sample from these sections.

```
library(elasticnet)
library(caret)
set.seed(47306)
trainIdx <- createDataPartition(
      y = y1,      # the outcome data are needed
   times = 1, # the number of partition
      p = .80,    # percent for training set
   list = FALSE # results not in list
)
```

---

[6]No concensus has been reached about how to split the initial coming data into training, validation, and test dataset. Suggestions have been made to have, for example, 60/20/20, 70/15/15/, and 80/10/10. The choices depend on initial data size and related consideration.

Although it is informative to know the model accuracy/test error associated with specific λ's, it would be most fruitful to obtain the optimal λ value that produces the smallest MSE using cross-validation (cross-validation test error). There is a variety of cross-validation methods, including the leave-one-out (LOO) cross-validation and k-fold cross-validation. In LOO, one observation is left out to be used as the validation (testing) set, and the remaining observations compose the training set, until all N observations are used as the validation set. Then one can estimate the test MSE as the average of these N test errors, $\mathrm{CV}_{(N)} = \frac{1}{N} \sum_{i=1}^{N} \mathrm{MSE}_i$. For k-fold cross-validation, the data are divided into k non-overlapping sets, or folds. Then each fold takes turn to be used as the validation set, and the remaining is used as the training set. In k-fold cross-validation, the test MSE is the average of the k test errors, $\mathrm{CV}_{(k)} = \frac{1}{k} \sum_{q=1}^{k} \mathrm{MSE}_q$ (James et al., 2014).

Below, we use the `createFolds` function from the `caret` package to create 10-fold data for cross-validation,

```
# create 10-folds data for reproducibility set.seed(47306)
folds.k <- createFolds(y1[trainIdx], k = 10)
```

Next, we create lists of options to be used later. Note that `trainControl` adjusts various training options, and `expand.grid` supplies tuning parameters. For the training control options, 10-fold (`number = 10`) cross-validation (`method = "cv"`) is selected. For tuning parameters, we set alpha to be zero and use the old lambda sequence (`lambda = ridge.reg$lambda`) automatically chosen by the ridge regression previously estimated using the `glmnet` function,

```
ridge.control <- trainControl( method = "cv",
                               number = 10,
                               index = folds.k) # can add seeds option
ridge.grid <- expand.grid( alpha = 0,
                           lambda = ridge.reg$lambda)
```

Then we can use the `train` function from the `caret` package to estimate the model and perform the 10-fold cross-validation,

```
ridge.train <- train ( x = x[trainIdx,],
                       y = y1[trainIdx],
                       method = "glmnet",
                       trControl = ridge.control,
                       tuneGrid = ridge.grid )
```

Once the training and cross-validation is completed, one can request to show the summary results,

```
ridge.train

## glmnet
##
## 1834 samples
##    6 predictor
##
## No pre-processing
## Resampling: Cross-Validated (10 fold)
```

```
## Summary of sample sizes: 183, 184, 183, 183, 184, 182, ...
## Resampling results across tuning parameters:
##
##     lambda         RMSE       Rsquared    MAE
##     0.03378094     0.9545184  0.1780900   0.7221273
##     0.09399749     0.9536723  0.1780700   0.7219145
##     0.26155359     0.9545334  0.1777367   0.7247716
##     0.72778839     0.9653501  0.1762917   0.7385099
##     2.02511443     0.9914436  0.1731443   0.7638963
##     5.63500121     1.0193714  0.1697571   0.7893401
##    15.67972559     1.0365824  0.1676509   0.8062920
##    43.62976787     1.0443704  0.1666953   0.8139200
##   121.40242080     1.0474312  0.1663195   0.8168511
##   337.80944743     1.0488958  0.1641699   0.8182347
##
## Tuning parameter 'alpha' was held constant at a value of 0
## RMSE was used to select the optimal model using the smallest value.
## The final values used for the model were alpha = 0 and lambda = 0.09399749.
```

There are four columns in the results section, corresponding to the chosen λ sequence and its associated root mean square error (RMSE), R-squareds, and mean absolute error (MAE). Note that RMSD is the square root of MSE discussed previously, and MAE $= \frac{1}{N} \sum_{i=1}^{N} |y_i - \widehat{y}_i|$ is another measure of model accuracy. One can also graph the tuning parameter profile plot to see how λ varies with RMSE, as shown in Fig. 6.4.

```
plot(ridge.train)
```

Based on the results and the diagnostics performed by the **train** function, $\lambda=0.094$ provides the best model fit with its associated parameter estimates, and thus is chosen as the final model,

```
# show the coefficients associated with the best lambda
coef(ridge.train$finalModel, ridge.train$bestTune$lambda)
```

```
## 7 x 1 sparse Matrix of class "dgCMatrix"
##                       s1
## (Intercept)  3.122129180
## age          0.013254334
## educ         0.110184580
## female      -0.441291261
## white        0.009966035
## married      0.086312812
## happy       -0.012852290
```

Next, we can use the final model chosen by our cross-validation analysis for the test set and calculate prediction error,

```
ypred <- predict(ridge.train, type = "raw", newdata = x[-trainIdx, ])
(mse = mean((y1[-trainIdx]-ypred)^2))
```

```
## [1] 0.978222
```

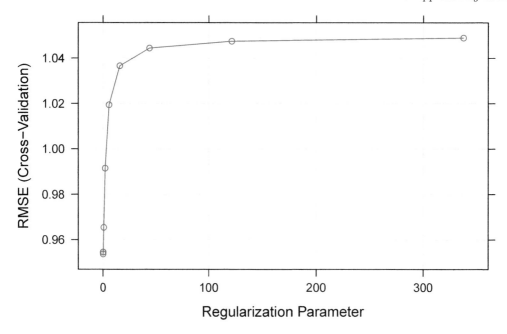

Figure 6.4
Regularization Parameter Profile for Ridge Regression

Without such "learning" process, how would a normal linear regression model perform? Below, we simply estimate an OLS regression using the training set, and then use the estimates to calculate predictions and their associated MSE to assess model accuracy,

```
x.old = x[trainIdx,]
y.old = y1[trainIdx]
x.new = as.data.frame(x[-trainIdx,])
data.old = as.data.frame(cbind(y.old, x.old))
lm.train = lm(y.old~., data=data.old)
lm.pred = predict(lm.train, newdata=x.new)
(lm.mse = mean((y1[-trainIdx]-lm.pred)^2))
```

```
## [1] 0.9837906
```

It is not surprising to see that the MSE calculated based on the single OLS regression model is $\text{MSE}_{\text{lm}} = 0.984$, greater than the $\text{MSE}_{\text{cv}} = 0.978$ produced with cross-validation. Thus, it can be shown in comparing the two MSEs, first using the ridge regression with $\lambda_{\min}$ (i.e., the λ in the chosen or supplied sequence that gives the smallest value of mean cross-validation error) and then the least squares method, that ridge regression with a carefully chosen λ outperforms the classical linear regression in improving model predictive accuracy.

Lasso Regression

In lasso (least absolute shrinkage and selection operator), the shrinkage penalty uses $|\beta_k|$ as opposed to β_k^2 in the ridge regression, so the loss function becomes

$$\sum_{i=1}^{N}\left(y - \beta_0 - \mathbf{x}\widetilde{\beta}\right)^2 + \lambda\widetilde{\mathbf{1}}|\widetilde{\beta}| = \quad RSS + \lambda\sum_{k=1}^{K}|\beta_k| \qquad (6.14)$$

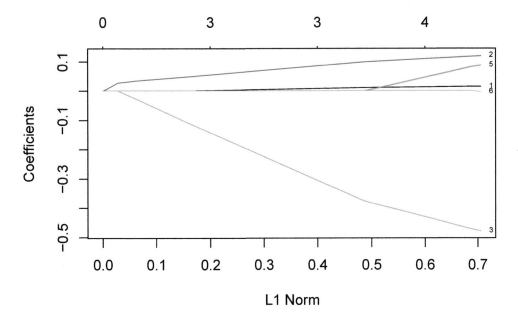

Figure 6.5
Lasso Coefficients Path Graph

wherein $\widetilde{\mathbf{1}}$ is a unit row vector with k elements, and $\widetilde{\mathbf{1}}|\widetilde{\beta}| = \sum_{k=1}^{K} |\beta_k|$ is called a l_1 norm (or Manhattan distance) and can be written as $\|\beta\|_1 = \sum_{k=1}^{K} |\beta_k|$ in this case. Related to—yet somewhat distinct from—ridge regression, lasso is another regularization technique that also performs variable selection so that only a subset of the coefficients are retained and the others are dropped (i.e., $\beta_k = 0$). Below, the `glmnet` function is used again to estimate a lasso regression with the same predictors and response variable used previously. The `alpha` argument in the `glmnet` function is set to zero by default, corresponding to a ridge regression. By resetting it to one, we can then run a lasso regression.

```
lasso.train = glmnet(x[trainIdx,], y1[trainIdx], alpha = 1)
plot(lasso.train, label=T)
```

Fig. 6.5 shows the variation paths of the coefficients as the l_1 norm of the whole coefficient vector increases. To save didactic illustrations similar to those for ridge regression, the following R code chunk directly uses cross-validation to choose the optimal tuning parameter λ and plot the CV error as a function of λ,

```
set.seed(47306)
lasso.cv = cv.glmnet(x[trainIdx,], y1[trainIdx], alpha=1)
plot(lasso.cv)
```

The two dotted vertical lines in Figure 6.6 mark on the scale of natural log of $\lambda_{\min}$, the λ that yields the minimum cross-validation error, and λ_{1se}, the largest λ value that gives a CV error that is roughly one standard error from the minimum. The numbers along the top graph area line denote the numbers of non-zero coefficient estimates for their corresponding λ values. Once $\lambda_{\min}$ is obtained, we can calculate its associated MSE in the test set,

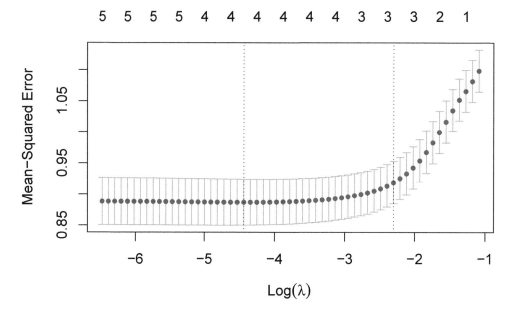

Figure 6.6
Plot of Tuning Parameters and MSE for Lasso Regression

```
lam.best= lasso.cv$lambda.min
lasso.pred=predict(lasso.cv, s=lam.best, newx=x[-trainIdx,])
mean((lasso.pred-y1[-trainIdx])^2)
```

```
## [1] 0.9804707
```

Based on the few MSEs calculated using different regularization parameterizations, including the one for classical least squares by setting λ to zero, it appears that the ridge and lasso regression models using optimal tuning parameter estimates outperform their traditional alternatives, with a small but clear advantage with the ridge regression obtaining the smallest MSE, given the atypical data size and number of predictors in the example used.

Note that when the model accuracy measure is about the same, lasso has a significant advantage with a simpler model having fewer parameter estimates, compared with ridge; lasso uses a sparse matrix to perform the computation and suppresses some parameters to be zero,

```
lasso.est  = glmnet(x, y1, alpha=1)
best.coef = predict(lasso.est, type="coefficients", s=lam.best)
best.coef
```

```
## 7 x 1 sparse Matrix of class "dgCMatrix"
##                     s1
## (Intercept)  3.12778868
## age          0.01256144
## educ         0.11188403
## female      -0.45583564
## white        .
## married      0.07027643
```

happy

It can be shown from the results that both parameters for `white` and `happy` are set to zeroes. If models have about comparable predictive power, then Occam's razor (the principle of parsimony) should be invoked to prefer the simpler one.

6.3.2.2 Penalized Binary Logit

In machine and statistical learning, the process to predict response variables that are categorical and noncontinuous is called classification. Note that the regularization techniques discussed in previous sections can also be applied to binary regression models by adding the l_1 or l_2 norm to the loss function. Building on the usual log likelihood function for a binary logit, the objective (loss function and regularization) function for a penalized logit model can be defined as,

$$-\left\{\frac{1}{N}\sum_{i=1}^{N}\left[y\cdot\left(\beta_0+\mathbf{x}\widetilde{\beta}\right)-\ln\left(1+\exp\left(\beta_0+\mathbf{x}\widetilde{\beta}\right)\right)\right]\right\}+\lambda\left[(1-\alpha)\,||\beta||_2^2/2+\alpha||\beta||_1\right]$$

$$(6.15)$$

in which the term within the first pair of curly brackets corresponds to the log likelihood function in a classical binary logit model, $||\beta||_2^2=\sum_{k=1}^{K}\beta_k^2$, and $||\beta||_1=\sum_{k=1}^{K}|\beta_k|$; $\lambda\geq 0$ is a complexity parameter, and $0\leq\alpha\leq 1$ is a mixture parameter; when $\alpha=0$, one has ridge, and when $\alpha=1$, then the lasso version of the model is used instead. If however α is set to be between zero and one, then one has a mixture of lasso and ridge, or elastic net (Friedman et al., 2010).

The following R code chunk illustrates how to estimate a penalized binary logit model using the `glmnet` function again. The binary response variable of this model is self-rated health (SRH), with one denoting fair or poor health and zero for good or excellent health, with the same set of predictors used in the previous examples. The same data random split scheme is used as in the ridge cross-validation example, with 80% of the cases for the training and 20% for the test set. First, a lassso logit regression of SRH is run using the training set.

```
binreg.lasso = glmnet(x[trainIdx,], y2[trainIdx], family = "binomial")
# alpha = 1
plot(binreg.lasso, xvar = "dev")
```

Like what was done previously for the lasso linear regression, one usually uses cross-validation to find the optimal λ, in search of the smallest cross-validation error rate. The same `cv.glmnet` function can be used to achieve this goal, and one can also plot a similar lambda-against-MSE graph.

```
set.seed(47306)
binreg.cv=cv.glmnet(x[trainIdx,], y2[trainIdx], family = "binomial")
# alpha = 1    plot(binreg.cv)
```

After the results from the cross-validation are obtained, one can use the $\lambda_{\min}$ to re-estimate the model and calculate the error rate using the test set. Below, the `predict.glmnet` function sets the `type` argument to `response` to calculate the predicted probabilities for all cases in the test set,

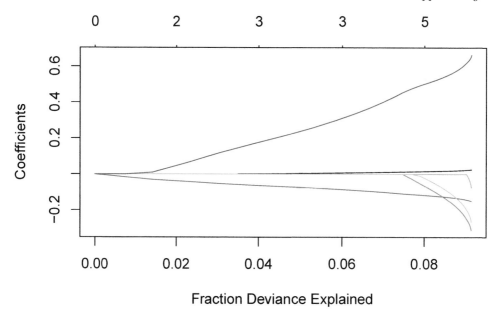

Figure 6.7
Lasso Logit Coefficients Path Graph

```
# obtain the lambda.min from binreg.cv
brlam.best = binreg.cv$lambda.min
cvbinreg.pred = predict(binreg.cv, s=brlam.best, newx=x[-trainIdx,],
 type="response")
head(cvbinreg.pred)
```

```
##               s1
## 5   0.24454454
## 10  0.16642790
## 11  0.12813897
## 30  0.26986223
## 49  0.20043299
## 65  0.08936801
```

After the predictions are obtained, one can set the threshold to be .5 to classify predictions into one and zero, and compare the predicted class with observed,

```
n.test = length(y2[-trainIdx])
brlasso.pred=rep("0goodHlth", n.test)
brlasso.pred[cvbinreg.pred>.5] ="1poorHlth"
brlasso.tab = table(brlasso.pred, y2[-trainIdx])
brlasso.tab
```

```
##
## brlasso.pred    0    1
##    0goodHlth  381   66
##    1poorHlth    5    5
```

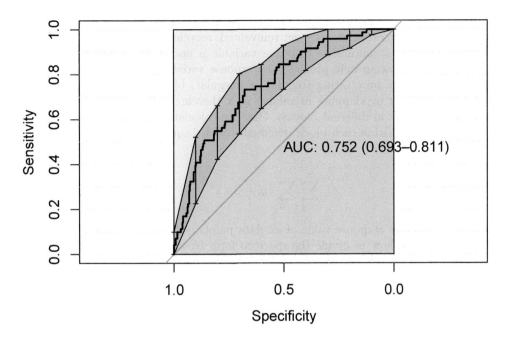

Figure 6.8
Plot of ROC Curve for CV Lasso Logit

```
(accuracy.rate = (brlasso.tab[1,1] + brlasso.tab[2,2])/sum(brlasso.tab))
## [1] 0.8446389
```

It turns out that our predictive accuracy rate is about 0.845. Using .5 as the only cut-off point to dichotomize the predicted probabilities and re-classify cases can be arbitrary, so the ROC curve can be used as an omnibus predictive measure of the model predictive power. Below, we use functions from the pROC package (Robin et al., 2011) to plot an ROC curve using the cvbinreg.pred prediction object obtained previously, It can be shown from Fig. 6.8 that the area under the curve is roughly 0.752, indicative of an acceptable level of model discrimination, based on Hosmer et al. (2013, p. 177).

6.3.2.3 Decision/Regression Trees

Like many great ideas in statistics as well as data science broadly defined, decision tree is underlaid by a seemingly simple yet fascinating idea. The tree methods were initially developed by Leo Breiman and Jerome H. Friedman, independently of each other, in the early 1970s. As stated by the few authors of the first and probably still the deinitive work in this area, *Classification and Regression Trees* (hence the origin of CART), the implementation of these methods had been invincible before the computer age (Breiman et al., 1984, p. viii). With the exponential growth of computing power almost religiously observed by the Moore's law (the number of transistors in a microchip doubles every two years, and so is predicted about the speed of computers) in the development of semiconductors, the use of tree methods nowadays has become as easy as other commonly used techniques that were initially able to be carried out with paper and pencil. It is also of note that the Bayes rule, introduced in the first chapter of this text, is implicitly used in and serves as a major theoretical foundation for the tree methods (Breiman et al., 1984). So brilliant ideas usually do not die, nor do they get conceived anymore; they only get reinvented, recycled, or revived.

Assuming there is a continuous response variable y and a set of predictors, $\mathbf{x}$, and the primary goal is to accurately predict this response variable, then one can construct a multi-dimensional space by creating (regular or irregular) blocks [7] that have data points similar (i.e., close to or overlapping in values) in $\mathbf{x}$ therein, relative to other data points outside these blocks (or in different blocks). One can calculate the summed squared errors of the response values within each block (residual summed squares or RSS) and sum across all blocks as follows,

$$\sum_{s=1}^{S} \sum_{i \in B_j} (y_i - \widehat{y}_{B_s})^2 \tag{6.16}$$

where $\hat{y}_{B_s}$ is the average response value of all data points in block B_s (James et al., 2014). The main challenge is how to divide the space to form blocks that are (usually) mutually exclusive. Since the partitioning strategy closely follows decision trees that probably got the "tree" part in its name from graph theory, the techniques discussed in this section are generally called classification and regression trees. Usually, the computational cost is considerably high to consider all possible partitions in the multidimensional space to find the tree structure with the minimum residual summed squares. Regression trees use different methods to map out such tree structures, and the most commonly used one is recursive binary splitting (i.e., two branches at each split point) that begins with a monolithic block with all data points contained therein and then scour the best split of nodes at each step without any prospective or retrospective consideration (greedy) (James et al., 2014, p. 306). This algorithm searches for the best binary split at each candidate node by considering all possible predictors (e.g., by finding the predictor that is most strongly associated with the response variable at the node of interest) and splitting points that would lead to the greatest reduction in RSS for regression trees or other selection criteria for other types of decision trees (e.g., highest impurity reduction).

There are multiple R packages for decision trees, including `ipred` (Peters and Hothorn, 2021), `randomForest` (Liaw and Wiener, 2002), `rpart` (Therneau and Atkinson, 2019), `tree` (Ripley, 2019), and the meta package `caret` (Kuhn, 2022). The `tree` package was written to conduct common classification as well as regression tree analyses (Ripley, 2019), and its syntactic structure is quite similar to that of `glm`. If the example in the previous section is recycled for regression tree here, then one could simply invoke the `tree` function from the `tree` package, specify the response variable `qrincome` to the left of a tilde sign, "~", and put all predictors to the right. The `data` argument is to specify data source. One can request a summary of the results by using the `summary` function, or directly call the tree analysis object to examine the tree structure presented in a textual format.

```
require(tree)
tree.reg <- tree(log(qrincome) ~ age + educ , data = usedta)
summary(tree.reg)

##
## Regression tree:
## tree(formula = log(qrincome) ~ age + educ, data = usedta)
## Number of terminal nodes:  7
```

---

[7]This idea is similar to the multidimensional histogram bin space used in the coarsened matching discussed in the section on causal inference. Such blocks, in essence, are rectangular room if there are three orthogonal predictors or multidimensional enclosed space with more predictors.

```
## Residual mean deviance:  0.869 = 1985 / 2284
## Distribution of residuals:
##    Min. 1st Qu.  Median   Mean 3rd Qu.    Max.
## -3.5420 -0.3346  0.2028  0.0000  0.4974  2.1490
```

```
tree.reg
```

```
## node), split, n, deviance, yval
##        * denotes terminal node
##
##  1) root 2291 2525.00 4.942
##     2) age < 23.5 190   242.20 3.890 *
##     3) age > 23.5 2101 2053.00 5.038
##       6) educ < 13.5 1165 1199.00 4.786
##        12) educ < 11.5 369   420.40 4.504
##          24) age < 62.5 316   317.30 4.626 *
##          25) age > 62.5 53    70.53 3.778 *
##        13) educ > 11.5 796   735.10 4.917
##          26) age < 38.5 369   377.90 4.720 *
##          27) age > 38.5 427   330.30 5.088 *
##       7) educ > 13.5 936   689.50 5.350
##        14) age < 27.5 113   140.70 4.772 *
##        15) age > 27.5 823   505.90 5.430 *
```

One can also request to draw the tree to illustrate all the thresholds and nodes using the `plot` and `tree` functions as shown in Fig. 6.9,

```
plot(tree.reg, type=c("proportional")) # uniform
text(tree.reg)
```

Alternatively, one can graph the above partition in a two-dimensional feature space. The blocks demarcated by the dotted lines in Fig.6.10 are the ones found through a tree regression of the income variable on age and education. The averages of income within each block are believed to produce the smallest RSS, given the default estimation options.

Just like a real tree with many branches and leaves growing uncontrollably that requires pruning for aesthetic reasons, a decision tree also needs pruning so as to avoid over-fitting and present a clear pattern. To prune, one usually uses recursive binary splitting to grow a very large tree, and then implements cost complexity pruning. To accomplish that, one first needs to define a cost-complexity function,

$$R_\alpha = R(T) + \alpha |\tilde{T}| \tag{6.17}$$

where $R(T)$ is the loss function of the tree, $|\tilde{T}|$ is the complexity measure, or the total number of terminal nodes in a tree, and α is the complexity/tuning parameter (Breiman et al., 1984, p. 66). In regression trees, the previous equation can be re-expressed as,

$$\sum_{s=1}^{|\tilde{T}|} \sum_{i:x_i \in B_s} (y_i - \hat{y}_{B_s})^2 + \alpha |\tilde{T}| \tag{6.18}$$

where the double summation term produces RSS, B_s denotes the block corresponding to the sth terminal node, and $\hat{y}_{B_s}$ is the predicted response in B_s. This is similar to the loss

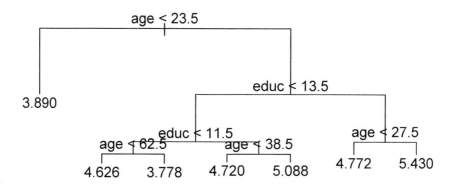

Figure 6.9
Regression Tree

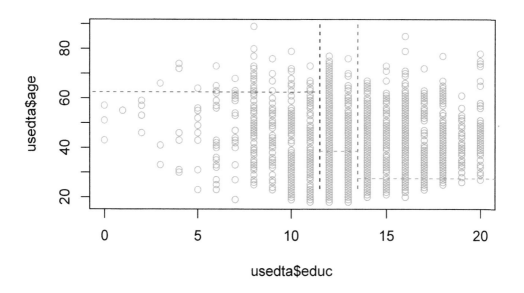

Figure 6.10
Partition Plot of Two-Dimensional Feature Space

function of the ridge or lasso formulation. The `rpart` function in the `rpart` package performs k-fold (10) cross-validation by default and provides rich features in growing trees,

```
require(rpart)
rpart.full <- rpart(log(qrincome) ~ age + educ + female + white + married
+ happy, method = "anova", data = usedta) # control=rpart.control(xval=0)
rpart.full
```

```
## n= 2291
##
## node), split, n, deviance, yval
##         * denotes terminal node
##
##   1) root 2291 2524.76900 4.942342
##     2) age< 23.5 190   242.24010 3.889991 *
##     3) age>=23.5 2101 2053.08600 5.037510
##       6) educ< 13.5 1165 1198.52500 4.786254
##        12) female>=0.5 560   638.67600 4.501170
##          24) educ< 11.5 148   172.87950 4.111714 *
##          25) educ>=11.5 412   435.28470 4.641071 *
##        13) female< 0.5 605   472.20850 5.050134
##          26) educ< 9.5 124   142.95800 4.608220
##            52) age>=62.5 25    38.75000 3.706840 *
##            53) age< 62.5 99    78.76651 4.835841 *
##          27) educ>=9.5 481   298.79190 5.164058 *
##       7) educ>=13.5 936   689.47640 5.350237
##        14) age< 27.5 113   140.70240 4.772471 *
##        15) age>=27.5 823   505.87400 5.429565 *
```

It is of note that the cost function that `rpart` uses is a unitless updated version of Eq. 6.17, $R_{cp} = R(T) + cp \cdot |\tilde{T}| \cdot R(T_1)$, in which cp is the complexity parameter and $R(T_1)$ is the loss function for the tree without splits (Therneau and Atkinson, 2022, p. 24). One can then proceed to plot the tree using the `rpart.plot` function from the `rpart.plot` package (Milborrow, 2021),

```
library(rpart.plot)
rpart.plot(rpart.full, digits=5, roundint=F)
```

Fig. 6.11 shows how the decision nodes grow, where the "leaves" (terminal nodes) situate, how splits are made, and the proportions at each split, in a visually effective manner. One can also invoke the `printcp` function in the `rpart` package to display the complexity parameters and their associated model prediction errors,

```
printcp(rpart.full)
```

```
##
## Regression tree:
## rpart(formula = log(qrincome) ~ age + educ + female + white +
##     married + happy, data = usedta, method = "anova")
##
## Variables actually used in tree construction:
```

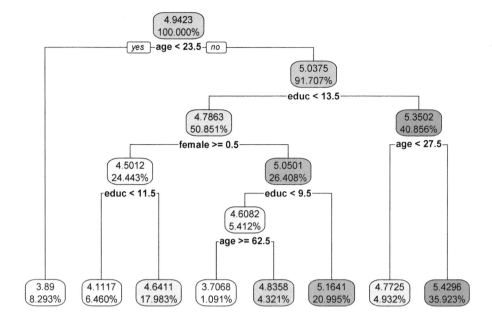

Figure 6.11
CV Tree Plot

```
## [1] age     educ    female
##
## Root node error: 2524.8/2291 = 1.102
##
## n= 2291
##
##          CP nsplit rel error   xerror      xstd
## 1 0.090877      0   1.00000  1.00077  0.036133
## 2 0.065386      1   0.90912  0.93124  0.035333
## 3 0.034712      2   0.84374  0.86964  0.034357
## 4 0.016992      3   0.80902  0.81913  0.032957
## 5 0.012085      4   0.79203  0.80735  0.032526
## 6 0.012064      5   0.77995  0.80545  0.032497
## 7 0.010077      6   0.76788  0.79692  0.031979
## 8 0.010000      7   0.75781  0.78983  0.031805
```

Note that in the results section, the first to the fifth columns correspond to CP (complexity parameter), nsplit (number of split points), rel error (summed squared error or $1-R^2$), xerror (the cross-validation error), and xstd (standard error of cross-validation errors). One can also plot the results using the plotcp function,

```
plotcp(rpart.full)
```

and Fig. 6.12 shows how the cross-validation relative error changes as the values of CP vary that in general as CP goes down, so does the error. In tree methods, the 1 SE rule is commonly used to select the tree with the right size and account for variability in cross-validation error (xerror) so as to avoid possible model over-fitting. To implement, one can first find the tree with the miminum cross-validation error, and then add its corresponding

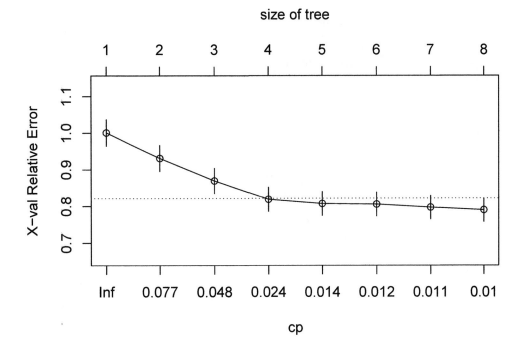

Figure 6.12
CV Tree CP Plot

standard error (deviation). The tree model with the largest cross-validation error within that one standard error range should be chosen. Below, we use this rule to select that tree model with its corresponding CP,

```
xerror.min = min(rpart.full$cptable[,"xerror"])
xstd.min = rpart.full$cptable[which.min(rpart.full$cptable[,"xerror"]),"xstd"]
xerror.best = xerror.min + xstd.min
xerror.all = rpart.full$cptable[,"xerror"]
cp.which = which(abs(xerror.all - xerror.best)==min(abs(xerror.all - xerror.best)))
cp.best =( rpart.full$cptable[cp.which,"CP"])
```

Once the CP (0.016992) is selected, we can use the **prune** function to prune or simplify the tree,

```
rpart.prune<- prune(rpart.full, cp= cp.best)
# printcp(rpart.prune)
```

and then plot the pruned tree,

```
rpart.plot(rpart.prune, digits=3)
```

It can be shown from the results that a tree with three splits and four terminal nodes is chosen as the optimal structure for this tree regression model.

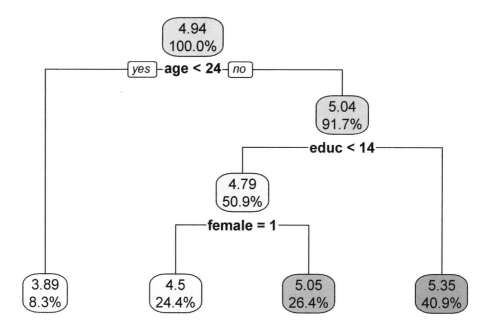

Figure 6.13
CV Tree Plot after Prune

6.3.2.4 Bagging and Random Forests

One major challenge in using normal decision tree in machine learning (i.e., training, validation, and test) is its high variability; that is, if one randomly divides the data into multiple
parts for training and validation purposes, and even if random sampling with the same size
and replacement is implemented, one would frequently find quite different tree structures.
Bagging (Bootstrap AGGregatING) and random forests are two similar methods to address
the problem of high variance, and bagging can be viewed as a special case of random forests.
Both bagging and random forests are viewed as ensemble algorithms/models that create
multiple models to improve predictive accuracy. Although decision tree is a relatively new
technique in data analysis and prediction, the approach to solving the problem of high variance is built upon classic statistical ideas, such as bootstrap and averaging (James et al.,
2014). The bagging technique re-samples from the original data with replacement and usually with the same size, build the tree from top down using the same algorithm described in
the previous section, and repeat this process for R times and average the predictions from
all R trees.

To estimate a tree regression with bagging, one can use the `randomForest` function from
the `randomForest` package (Liaw and Wiener, 2002), the oldest and probably the most
popular one that implements both bagging and random forests. Below, the original data
is randomly split into halves. Bagging is first implemented on the training set, and MSE
is calculated using the validation/holdout data. By setting the `mtry` argument to the total
number of predictors/features, one can run a regression tree with bagging.

```
library(randomForest)
# bagging
set.seed(47306)
bag.train <- randomForest(log(qrincome)~age + educ + female + white + married + happy,
                          data=usedta, subset=trainIdx, mtry=6, importance = TRUE)
```

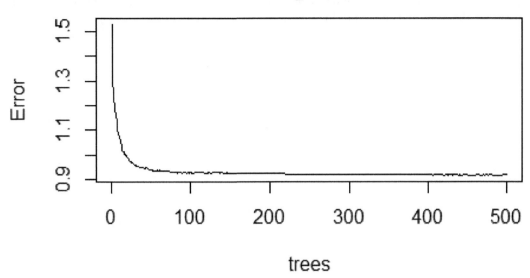

Figure 6.14
Bagging for Reducing Error

```
plot(bag.train)
```

```
bag.yhat <- predict(bag.train, newdata=usedta[-trainIdx,])
# test error
(bag.mse <- mean((bag.yhat - y1[-trainIdx])^2))
```

```
## [1] 1.48966
```

Fig. 6.14 shows that the error rate reduces dramatically as the number of trees increases approximately from 0 to 60–70.

Random forest is an ensemble learning technique generalized from the bagging method. Generally, this method randomly selects predictors at each split piont so as to minimize the correlations among different trees. The number of predictors randomly selected is usually set to be $m = K/3$ for regression and $m \simeq \sqrt{K}$ for classification (usually rounded down), where K is the total number of predictors available. If we are to estimate a random forest analysis of the same variables, then the `mtry` argument of the `randomForest` function should be set to 2 $(6/3 = 2)$.

Below, we use functions from the `caret` package to run a random forest with cross-validation. As usual, we can set up the training and tuning options. The argument "`expand.grid(.mtry = (1:6))`" is to supply a vector of integers from one to six, denoting the experimented number of predictors randomly chosen at each split point for growing the tree.

```
library(caret)
rf.control <- trainControl(method = "cv", number = 10)
rf.grid <- expand.grid(.mtry = (1:6))
```

Then we can run a CV random forest with the previously prescribed training and tuning parameters. The `createFolds` function in the `caret` package is to index the randomly chosen cases for different folds for reproducibility.

```
set.seed(47306)
folds.k <- createFolds(y1[trainIdx], k = 10)
rf.cv <- train(  x = x[trainIdx,], y = y1[trainIdx],
                 method = "rf", tuneGrid = rf.grid,
                 trControl = rf.control, index=folds.k)
print(rf.cv)

Random Forest

1834 samples
   6 predictor

No pre-processing
Resampling: Cross-Validated (10 fold)
Summary of sample sizes: 1650, 1652, 1650, 1650, 1651, 1651, ...
Resampling results across tuning parameters:

mtry  RMSE       Rsquared   MAE
1     0.930614   0.272747   0.717543
2     0.893942   0.275220   0.674116
3     0.913607   0.253332   0.681021
4     0.931192   0.237496   0.692614
5     0.941746   0.228051   0.699413
6     0.945921   0.224929   0.703260

RMSE was used to select the optimal model using the smallest value.
The final value used for the model was mtry = 2.
```

The results show that `mtry` = 2, or choosing 2 predictors at each split point yields the best fitting model. One can proceed to use this optimal model to calculate MSE for the test set. The results show that random forest outperforms bagging in predictive accuracy, as is usually the case.

```
rf.yhat <- predict(rf.cv, newdata = x[-trainIdx,], type = "raw")
(rf.mse <- mean((rf.yhat - y1[-trainIdx])^2))
[1] 0.836068
```

6.3.2.5   Classification Trees

Similar to what regression tree is for continuous response variables, classification tree is specifically designed for categorical response variables. Just like how the average is computed for each block in a regression tree, we first calculate the proportion of each response category (i.e., the categories of the response variable) in each block, and predict that each observation falls under the most frequently occurring class in that particular block. In growing classification trees, one can use either of the two counterparts of RSS in regression trees, including the Gini index/impurity and the information entropy measure. The Gini index (of diversity) is used frequently in classification and regression tree (CART) analysis and in this context is defined as

$$G = 1 - \sum_{c=1}^{C} \widehat{p}^2(c|B) \qquad (6.19)$$

where $\widehat{p}(c|B)$ denotes the proportion of cases in block B that are from class c (Breiman et al., 1984, p. 103), and $\widehat{p}^2(c|B)$ can be understood as the probability of having two random draws from the same space that happen to be from the same class. Then G is the complement of the probability of all possible two random draws from the same class, thus measuring the level of diversity. When one has a class that dominates the classification and accordingly have a $\widehat{p}^2(c|B)$ close to one, G takes on a relatively small value.

The information entropy measure is usually used in other types of decision trees (James et al., 2014), such as iterative dichotomizer 3 (ID3) and C4.5 (an extension of ID3), and is defined as

$$E = -\sum_{c=1}^{C} \widehat{p}(c|B) \log\left[\widehat{p}(c|B)\right] \qquad (6.20)$$

Developed by Shannon (1948, p. 12) as the underpinning of information theory, the information entropy measure may be inspired by the concept of entropy in statistical thermodynamics (e.g., the second law of thermodynamics). In this case, $\log\left[\widehat{p}^{-1}(c|B)\right] = -\log\left[\widehat{p}(c|B)\right]$ is the "entropy" of $\widehat{p}(c|B)$, and it increases as $\widehat{p}(c|B)$ decreases. E is the average entropy (weighted by corresponding probabilities) for all $\widehat{p}(c|B)'s$.

Below, we use the `rpart` function in the `rpart` package to estimate a classification tree (Therneau and Atkinson, 2019). The response variable is a binary indicator of self-reported health, with one denoting poor health and zero for good health. Note that in the R code chunk below, we set the `method` argument to `class` to request a classification tree. The `cp` option in the `rpart.control` argument is to set a benchmark complexity parameter to discern splits with trivial model fit improvement (as a factor of the number supplied for `cp`).

```
library(rpart)
tree.class <- rpart(hlthc2 ~ age + educ + female + white + married + happy,
            data = usedta, subset=trainIdx, method = "class",
       control = rpart.control(cp = 0.005)) #
```

After constructing the tree in the background, one can use the `printcp` function to request printing the complexity parameters and their associated model accuracy statistics,

```
printcp(tree.class, digits=3)

##
## Classification tree:
## rpart(formula = hlthc2 ~ age + educ + female + white + married +
##     happy, data = usedta, subset = trainIdx, method = "class",
##     control = rpart.control(cp = 0.005))
##
## Variables actually used in tree construction:
## [1] age   educ   happy
##
## Root node error: 304/1834 = 0.166
##
```

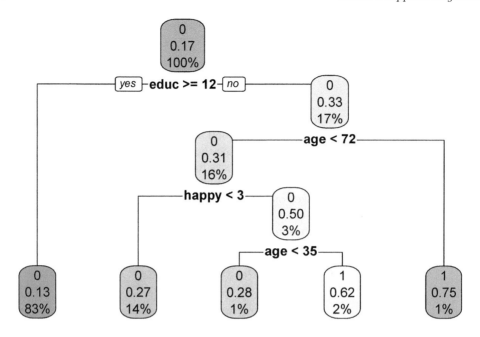

Figure 6.15
Classification Tree Plot

```
## n= 1834
##
##        CP nsplit rel error xerror   xstd
## 1 0.00987      0    1.000   1.00 0.0524
## 2 0.00500      4    0.954   1.01 0.0526
```

It can be shown from the results that the three variables, age, educ, and happy, are retained to construct the tree. Note that in our example, education is very powerful in predicting self-reported health, and it dominates the tree construction and classification process. Thereby, there is not much room to grow and prune the tree or to try additional CPs to further improve the model fit. We can also plot the final tree, as shown in Fig. 6.15,

```
rpart.plot(tree.class)
```

Next step, one can check the model fit by computing predictions using the test set,

```
class.class = predict(tree.class, newdata=as.data.frame(x[-trainIdx,]),
                      type = c("class"))
```

and use both predicted and observed classes for the response variable in the test set to check the model accuracy.

```
(tab = table(class.class, usedta$hlthc2[-trainIdx]))

##
## class.class   0   1
##           0 380  64
##           1   6   7
```

```
(accuracy = (tab[1,1] + tab[2,2])/sum(tab))
```

```
## [1] 0.8468271
```

The results show that the predictive accuracy of this classification tree is 0.847. Note that it is atypical to have so few variables to construct a classification tree, we only use three variables to contruct the tree, and education also largely governs this process. So it is likely that this classification tree model fares poorly, compared with its alternatives using lasso or ridge logit.

Bibliography

Agresti, A. (2010). *Analysis of Ordinal Categorical Data*. John Wiley & Sons, Inc, Hoboken, NJ.

Agresti, A. (2013). *Categorical Data Analysis*. John Wiley & Sons, Hobeken, New Jersey.

Ahn, J., Mukherjee, B., Banerjee, M., and Cooney, K. A. (2009). Bayesian inference for the stereotype regression model: Application to a case-control study of prostate cancer. *Statistics in Medicine*, 28:3139–3157.

Aitchison, J. and Silvey, S. D. (1957). The generalization of probit analysis to the case of multiple responses. *Biometrika*, 44(1/2):131–140.

Akaike, H. (1974). A new look at the statistical model identification. *IEEE Transactions on Automatic Control*, AC-19(6):716–723.

Aldrich, J. (1997). R. A. Fisher and the making of maximum likelihood 1912-1922. *Statistical Science*, 12(3):162–176.

Allison, P. D. (1999). Comparing logit and probit coefficients across groups. *Sociological Methods & Research*, 28(2):186–208.

Allison, P. D. (2014). *Event History and Survival Analysis*. Quantitative Applications in the Social Sciences. SAGE, Los Angeles.

Allison, P. D. (2013, February 13). What's the best R-squared for logistic regression? https://statisticalhorizons.com/r2logistic/.

Andersen, P. K. and Gill, R. D. (1982). Cox's regression model for counting processes: A large sample study. *The Annals of Statistics*, 10(4):1100–1120.

Anderson, J. A. (1984). Regression and ordered categorical variables. *Journal of the Royal Statistical Society. Series B (Methodological)*, 46(1):1–30.

Andrich, D. (1978). A rating formulation for ordered response categories. *Psychometrika*, 43(4):561–573.

Arrow, K. J. (1951). *Social Choice and Individual Values*. Wiley, New York, 1st edition.

Arrow, K. J. (1959). Rational choice functions and orderings. *Economica*, 26(102):121–127.

Arrow, K. J. (1963). *Social Choice and Individual Values*. John Wiley & Sons, Inc, New York, 2nd edition.

Axler, S. (2015). *Linear Algebra Done Right*. Undergraduate Texts in Mathematics. Springer, New York.

Bagozzi, B. E. and Mukherjee, B. (2012). A mixture model for middle category inflation in ordered survey response. *Political Analysis*, 20:369–386.

Bagozzi, B. E., Hill, D. W., Jr, Moore, W. H., and Mukherjee, B. (2015). Modeling two types of peace: The zero-inflated ordered probit (ZiOP) model in conflict research. *Journal of Conflict Resolution*, 59(4):728–752.

Bagozzi, B. E. (2016). The baseline-inflated multinomial logit model for international relations research. *Conflict Management and Peace Science*, 33:174–197.

Bates, D., Mächler, M., Baolker, B., and Walker, S. (2015). Fitting linear mixed-effects models using lme4. *Journal of Statistical Software*, 67(1):1–48.

Bayes, T. and Price, R. (1763). An essay towards solving a problem in the doctrine of chances. by the late Rev. Mr. Bayes, F. R. S. communicated by Mr. Price, in a letter to John Canton, A. M. F. R. S. *Philosophical Transactions (1683-1775)*, 53:370–418.

Berkson, J. (1944). Application of the logistic function to bio-assay. *Journal of the American Statistical Association*, 39(227):357–365.

Berndt, E. K., Hall, B. H., Hall, R. E., and Hausman, J. A. (1974). Estimation and inference in nonlinear structural models. *Annals of Economic and Social Movement*, 3/4:653–665.

Bliss, C. I. (1934). The method of probits. *Science*, 79(2037):38–39.

Bloch, D. A. and Watson, G. S. (1967). A Bayesian study of the multinomial distribution. *The Annals of Mathematical Statistics*, 38:1423–1435.

Blume, J. D., Greevy, R. A., Welty, V. F., Smith, J. R., and Dupont, W. D. (2019). An introduction to second-generation p-values. *The American Statistician*, 73(sup1):157–167.

Bosak, D. (2022). logr: Creates log files. R package version 1.2.9. https://CRAN.R-project.org/package=logr.

Boswell, M. T. and Patil, G. P. (1970). Chance mechanisms generating the negative binomial distributions. In G. P. Patil (Ed.) *Random Counts in Scientific Work, Vol. 1: Random Counts in Models and Structures*, pp. 3–22. Pennsylvania State University Press, University Park, PA.

Box, G. E. P. (1976). Science and statistics. *Journal of the American Statistical Association*, 71(356):791–799.

Breiman, L., Friedman, J. H., Olshen, R. A., and Stone, C. J. (1984). *Classification and Regression Trees*. Chapman & Hall/CRC, New York.

Breslow, N. E. and Clayton, D. G. (1993). Approximate inference in generalized linear mixed models. *Journal of the American Statistical Association*, 88(421):9–25.

Brilleman, S. L., Elci, E. M., Novik, J. B., and Wolfe, R. (2020). Bayesian survival analysis using the rstanarm R package. arXiv: Computation. https://arxiv.org/abs/2002.09633

Broström, G. (2012). *Event History Analysis with R*. Chapman & Hall/CRC, Boca Raton.

Brown, S., Harris, M. N., and Spencer, C. (2020). Modelling category inflation with multiple inflation processes: Estimation, specification and testing. *Oxford Bulletin of Economics and Statistics*, 82(6):1342–1361.

Broyden, C. G. (1967). Quasi-Newton methods and their application to function minimisation. *Mathematics of Computation*, 21:368–381.

Bürkner, P.-C. (2017). brms: An R package for Bayesian multilevel models using Stan. *Journal of Statistical Software*, 80(1):1–28.

Buse, A. (1982). The likelihood ratio, Wald, and Lagrange multiplier tests: An expository note. *The American Statistician*, 36(3):153–157.

Cai, T., Xia, Y., and Zhou, Y. (2021). Generalized inflated discrete models: A strategy to work with multimodal discrete distributions. *Sociological Methods & Research*, 50(1):365–400.

Cameron, A. C. and Trivedi, P. K. (1990). Regression-based tests for overdispersion in the Poisson model. *Journal of Econometrics*, 46(3):347–364.

Cameron, A. C. and Trivedi, P. K. (1998). *Regression Analysis of Count Data*. Econometric Society Monographs. Cambridge University Press, Cambridge, UK.

Cameron, A. C. and Trivedi, P. K. (2005). *Microeconometrics: Methods and Applications*. Cambridge University Press, New York.

Canty, A. and Ripley, B. D. (2019). boot: Bootstrap R (S-Plus) functions. R package version 1.3-22. https://cran.r-project.org/web/packages/boot/.

Carpenter, B., Gelman, A., Hoffman, M., Lee, D., Goodrich, B., Betancourt, M. D., Brubaker, M., Guo, J., Li, P., and Riddell, A. (2017). Stan: A probabilistic programming language. *Journal of Statistical Software*, 76(1):1–32.

Carroll, N. (2018). oglmx: Estimation of ordered generalized linear models. R package version 3.0.0.0. https://CRAN.R-project.org/package=oglmx.

Cheng, S. and Long, J. S. (2007). Testing for IIA in the multinomial logit model. *Sociological Methods & Research*, 35(4):583–600.

Choirat, C., Honaker, J., Imai, K., King, G., and Lau, O. (2018). Zelig: Everyone's statistical software. R package version 5.1.6.1. https://zeligproject.org/.

Chow, G. C. (1960). Tests of equality between sets of coefficients in two linear regressions. *Econometrica*, 28(3):591–605.

Christensen, R. H. B. (2019). ordinal: Regression models for ordinal data. R package version 2019.12-10. https://CRAN.R-project.org/package=ordinal.

Chukhrova, N. and Johannssen, A. (2019). Fuzzy regression analysis: Systematic review and bibliography. *Applied Soft Computing*, 84:105708.

Clark, T. G., Bradburn, M. J., Love, S. B., and Altman, D. G. (2003). Survival analysis part I: Basic concepts and first analyses. *British Journal of Cancer*, 89:232.

Cleves, M., Gould, W. W., and Marchenko, Y. V. (2016). *An Introduction to Survival Analysis Using Stata*. Stata Press, College Station, Texas, 3rd revised edition.

Clogg, C. C. and Shihadeh, E. S. (1994). *Statistical Models for Ordinal Variables*. Advanced Quantitative Techniques in the Social Sciences Series. SAGE Publications, Thousand Oaks, CA.

Clogg, C. C., Petkova, E., and Haritou, A. (1995). Statistifcal methods for comparing regression coefficients across models. *American Journal of Sociology*, 100(5):1261–1293.

Cohen, J. (1988). *Statistical Power Analysis for the Behavioral Sciences*. Lawrence Erlbaum Associates, Hillsdale, NJ, 2nd edition.

Cook, R. D. (1977). Detection of influential observation in linear regression. *Technometrics*, 19(1):15–18.

Cox, D. R. (1972). Regression models and life-tables. *Journal of the Royal Statistical Society. Series B (Methodological)*, 34(2):187–220.

Cox, D. R. (1975). Partial likelihood. *Biometrika*, 62(2):269–276.

Cragg, J. G. and Uhler, R. S. (1970). The demand for automobiles. *The Canadian Journal of Economics / Revue canadienne d'Economique*, 3(3):386–406.

Cramer, J. (2003). The origin and development of the logit model. In J. C. Cramer (Ed.), *Logit Models from Economics and Other Fields* (pp.149–157). Cambridge University Press, New York.

Croissant, Y. (2020). Estimation of random utility models in R: The mlogit package. *Journal of Statistical Software*, 95(11):1–41.

Davidon, W. C. (1991). Variable metric method for minimization. Report, Argonne National Laboratory, SIAM Journal on Optimization, 1(1):1-17.

Davidson, J. (2000). *Econometric Theory*. Blackwell Publishers Inc., Malden, MA.

Davison, A. C. and Hinkley, D. V. (1997). *Bootstrap Methods and their Applications*. Cambridge University Press, Cambridge, UK.

de Valpine, P., Turek, D., Paciorek, C., Anderson-Bergman, C., Temple Lang, D., and Bodik, R. (2017). Programming with models: Writing statistical algorithms for general model structures with NIMBLE. *Journal of Computational and Graphical Statistics*, 26:403–413.

Diamond, A. and Sekhon, J. S. (2013). Genetic matching for estimating causal effects: A general multivariate matching method for achieving balance in observational studies. *The Review of Economics and Statistics*, 95(3):932–945.

Dodge, Y. (2008). *The Concise Encyclopedia of Statistics*. Springer-Verlag, New York.

Edwards, J. R. and Berry, J. W. (2010). The presence of something or the absence of nothing: Increasing theoretical precision in management research. *Organizational Research Methods*, 13(4):668–689.

Eisenpress, H. (1962). Note on the computation of full-information maximum-likelihood estimates of coefficients of a simultaneous system. *Econometrica*, 30(2):343–348.

Eisenpress, H. and Greenstadt, J. (1966). The estimation of nonlinear econometric systems. *Econometrica*, 34(4):851–861.

Engle, R. F. (1984). Wald, likelihood ratio, and Lagrange multiplier tests in econometrics. In Z. Briliches and M. D. Intriligator (Eds.), *Handbook of Econometrics*, Vol. 2 (pp. 776–826). Elsevier, Amsterdam, The Netherlands.

Fagerland, M. W. and Hosmer, D. W. (2013). A goodness-of-fit test for the proportional odds regression model. *Statistics in Medicine*, 32(13):2235–2249.

Fagerland, M. W. and Hosmer, D. W. (2017). How to test for goodness of fit in ordinal logistic regression models. *The Stata Journal*, 17(3):668–686.

Fernihough, A. (2019). mfx: Marginal effects, odds ratios and incidence rate ratios for GLMs. R package version 1.2-2. https://CRAN.R-project.org/package=mfx.

Fienberg, S. E. (1977). *The Analysis of Cross-Classifed Data.* M.I.T. Press, Cambridge, MA.

Fienberg, S. E. and Mason, W. M. (1979). Identification and estimation of age-period-cohort models in the analysis of discrete archival data. *Sociological Methodology*, 10(1):1–67.

Fink, D. (1997). A compendium of conjugate priors. https://www.researchgate.net/ publication/ 238622435_A_Compendium_of_Conjugate_Priors.

Fisher, R. A. (1912). On an absolute criterion for fitting frequency curves. *Messenger of Mathematics*, 41:155–160.

Fisher, R. A. (1922). On the mathematical foundations of theoretical statistics. *Philosophical Transactions of the Royal Society A: Mathematical, Physical & Engineering Sciences*, 222(309-368).

Fisher, R. A. (1934). *Statistical Methods for Research Workers.* Oliver & Boyd, Edinburgh, UK, 5th edition.

Fisher, R. A. (1935). *The Design of Experiments.* Oliver and Boyde, Edingurgh.

Fletcher, R. (1970). A new approach to variable metric algorithms. *The Computer Journal*, 13(3):317–322.

Fletcher, R. and Powell, M. J. D. (1963). A rapidly convergent descent method for minimization. *The Computer Journal*, 6(2):163–168.

Fox, J. (2008). *Applied Regression Analysis and Generalized Linear Models.* SAGE, Thousand Oaks, CA.

Fox, J. (2017). *Using the R Commander: A Point-and-Click Interface for R.* Chapman & Hall/CRC Press, Boca Raton, FL.

Fox, J. and Weisberg, S. (2019). *An R Companion to Applied Regression.* Sage, Thousand Oaks CA, 3rd edition.

Friedman, J., Hastie, T., and Tibshirani, R. (2010). Regularization paths for generalized linear models via coordinate descent. *Journal of Statistical Software*, 33(1):1–22.

Fry, T. R. L. and Harris, M. N. (1998). Testing for independence of irrelevant alternatives: Some empirical results. *Sociological Methods & Research*, 26(3):401–423.

Fu, V. (1998). Estimating generalized ordered logit models. *Stata Technical Bulletin*, 8(44):27–30.

Fullerton, A. S. (2009). A conceptual framework for ordered logistic regression models. *Sociological Methods & Research*, 38(2):306–347.

Fullerton, A. S. and Xu, J. (2016). *Ordered Regression Models: Parallel, Partial, and Non-Parallel Alternatives.* Chapman & Hall Statistics in the Social and Behavioral Sciences. Chapman & Hall, New York.

Gabry, J. and Mahr, T. (2021). bayesplot: Plotting for Bayesian models. R package version 1.8.1. https://mc-stan.org/bayesplot/.

Gauss, C. F. (1963 [1809]). *Theoria Motus Corporum Coelestium in Sectionibus Conicis Solem Ambientium (Theory of the Motion of Heavenly Bodies Moving about the Sun in Conic Sections)*. Dover, New York.

Gelman, A. (2002). Prior distribution. In A. H. El-Shaarawi and W. W. Piegorsch (Eds.), *Encyclopedia of Environmetrics*, Vol. 2 (1634–1637). John Wiley & Sons, Chichester, UK, volume 3, pages 1634–1637. John Wiley & Sons, Chichester, NH.

Gelman, A. (2019). Prior choice recommendations. https://github.com/stan-dev/stan/wiki/Prior-Choice-Recommendations.

Gelman, A., Carlin, J. B., Stern, H. S., Dunson, D. B., Vehtari, A., and Rubin, D. B. (2014). *Bayesian Data Analysis*. CRC Press, Boca Raton, FL, 3rd edition.

Gelman, A. and Hill, J. (2007). *Data Analysis Using Regression and Multilevel/Hierarchical Models*. Cambridge University Press, Cambridge, NY.

Gelman, A., Jakulin, A., Pittau, M. G., and Su, Y.-S. (2008). A weakly informative default prior distribution for logistic and other regression models. *The Annals of Applied Statistics*, 2(4):1360–1383.

Gelman, A. and Su, Y.-S. (2020). arm: Data analysis using regression and multilevel/hierarchical models. R package version 1.11-2. https://CRAN.R-project.org/package=arm

Gelman, A. and Su, Y.-S. (2020b). arm: Data analysis using regression and multilevel/hierarchical models.

Goeman, J. J., Meijer, R. J., and Chaturvedi, N. (2018). penalized: L1 (lasso and fused lasso) and L2 (ridge) penalized estimation in GLMs and in the Cox model. R package version 0.9-51. https://cran.r-project.org/web/packages/penalized/index.html.

Goldfarb, D. (1970). A family of variable-metric methods derived by variational means. *Mathematics of Computation*, 24:23–26.

Goodman, L. A. (1979). Simple models for the analysis of association in cross-classifications having ordered categories. *Journal of the American Statistical Association*, 74(367):537–552.

Goodrich, B., Gabry, J., Ali, I., and Brilleman, S. (2018). rstanarm: Bayesian applied regression modeling via Stan. R package version 2.17.4. https://mc-stan.org/rstanarm.

Grambsch, P. M. and Therneau, T. M. (1994). Proportional hazards tests and diagnostics based on weighted residuals. *Biometrika*, 81(3):515–526.

Graunt, J. (1662). *Natural and Political Observations Made Upon the Bills of Mortality*. John Martyn, London, 2nd edition.

Greene, W. (2000). *Econometric Analysis*. Prentice Hall, Upper Saddle River, New Jersey, 4th edition.

Greene, W. (2007). *Econometric Analysis*. Prentice-Hall, Upper Saddle River, New Jersey, 6th edition.

Greene, W., Harris, M. N., Srivastava, P., and Zhao, X. (2018). Misreporting and econometric modelling of zeros in survey data on social bads: An application ot cannabis consumption. *Health Economics*, 27:372–389.

Greenwood, M. and Yule, G. U. (1915). The statistics of anti-typhoid and anti-cholera inoculations, and the interpretation of such statistics in general. *Proceedings of Royal Society of Medicine*, 8:113–94.

Guo, S. and Fraser, M. W. (2015). *Propensity Score Analysis*. SAGE, Los Angeles.

Gurland, J., Lee, J., and Dahm, P. (1960). Polychotomous quantal response in biological assay. *Biometrics*, 16(3):382–398.

Hald, A. (1990). *A History of Probability and Statistics and Their Applications before 1750*. Wiley Series in Probability and Mathematical Statistics. John Wiley & Sons, New York, NY.

Hald, A. (1998). *A History of Mathematical Statistics from 1750 to 1930*. John Wiley & Sons, New York, NY.

Hald, A. (2000). *A History of Parametric Statistical Inference from Bernoulli to Fisher, 1713-1935*. Sources and Studies in the History of Mathematics and Physical Sciences. Springer, New York.

Hald, A. (2003). *A History of Probability and Statistics and Their Applications before 1750*. Wiley Series in Probability and Mathematical Statistics. John Wiley & Sons, New York, NY.

Hansen, B. B. and Klopfer, S. O. (2006). Optimal full matching and related designs via network flows. *Journal of Computational and Graphical Statistics*, 15(3):609–627.

Harrell, Frank E., J. (2015). *Regression Modeling Strategies: With Applications to Linear Models, Logistic and Ordinal Regression, and Survival Analysis*. Springer, New York, 2nd edition.

Harrell Jr, F. E. (2021). rms: Regression modeling strategies. R package version 6.2-0. https://CRAN.R-project.org/package=rms.

Harrell Jr, F. E. (2021). Hmisc: Harrell miscellaneous. R package version 4.5-0. https://CRAN.R-project.org/package=Hmisc.

Harris, M. N. and Zhao, X. (2007). A zero-inflated ordered probit model, with an application to modelling tobacco consumption. *Journal of Econometrics*, 141:1073–1099.

Hastie, T. and Efron, B. (2013). lars: Least angle regression, lasso and forward stagewise. R package version 1.2. https://CRAN.R-project.org/package=lars.

Hausman, J. and McFadden, D. (1984). Specification tests for the multinomial logit model. *Econometrica*, 52(5):1219–1240.

Hayes, B. (2013). First link in the Markov chain. *American Scientist*, 101(2):92.

Hayes, B. (2017). *Foolproof, and Other Mathematical Meditations*. The MIT Press, Cambridge, MA.

Hebb, D. O. (1949). *The Organization of Behavior: A Neurophychological Theory*. John Wiley & Sons, New York.

Hilbe, J. M. (2008). *Negative Binomial Regression*. Cambridge University Press, New York.

Hirk, R., Hornik, K., and Vana, L. (2020). mvord: An R package for fitting multivariate ordinal regression models. *Journal of Statistical Software*, 93(4):1–41.

Hlavac, M. (2018). stargazer: Well-formated regression and summary statistics tables. Central European Labour Studies Institute (CELSI), Bratislava, Slovakia. R package version 5.2.2. https://CRAN.R-project.org/package=stargazer.

Ho, D. E., Imai, K., King, G., and Stuart, E. A. (2011). MatchIt: Nonparametric preprocessing for parametric causal inference. *Journal of Statistical Software*, 42(8):1–28.

Hosmer, D. W., Lemeshow, S. (1980). Goodness of fit tests for the multiple logistic regression model. Communication in Statistics, 9(10):1043-1069.

Hosmer, D. W., Lemeshow, S., and May, S. (2008). *Applied Survival Analysis*. Wiley Series in Probability and Statistics. John Wiley & Sons, Hoboken, New Jersey, 2nd edition.

Hosmer, D. W., Lemeshow, S., and Sturdivant, R. X. (2013). *Applied Logistic Regression*. Wiley Series in Probability and Statistics. John Wiley & Sons, Hoboken, New Jersey, 3rd edition.

Iacus, S. M., King, G., and Porro, G. (2009). cem: Software for coarsened exact matching. *Journal of Statistical Sofware*, 30(9):1–27.

Iacus, S. M., King, G., and Porro, G. (2011). Multivariate matching methods that are monotonic imbalance bounding. *Journal of the American Statistical Association*, 106(493):345–361.

Iacus, S. M., King, G., and Porro, G. (2012). Causal inference without balance checking: Coarsened exact matching. *Political Analysis*, 20(1):1–24.

Imai, K., King, G., and Lau, O. (2008). Toward a common framework for statistical analysis and development. *Jounral of Computational and Graphical Statistics*, 17(4):1–22.

Imai, K. and van Dyk, D. (2021). MNP: Fitting the multinomial probit model. R package version 3.1-2. https://CRAN.R-project.org/package=MNP.

Jackson, C. (2016). flexsurv: A platform for parametric survival modeling in R. *Journal of Statistical Software*, 70(8):1–33.

James, G., Witten, D., Hastie, T., and Tibshirani, R. (2014). *An Introduction to Statistical Learning with Applications in R*. Springer Texts in Statistics. Springer, New York.

Jay, M. (2019). generalhoslem: Goodness of fit tests for logistic regression models. R package version 1.3.4. https://CRAN.R-project.org/package=generalhoslem.

Johnson, N. L., Kemp, A. W., and Samuel, K. (2005). *Univariate Discrete Distribution*. Wiley, New York.

Johnson, R. A. and Wichern, D. W. (1998). *Applied Multivariate Statistical Analysis*. Prentice Hall, Upper Saddle River, NJ, 4th edition.

King, G. and Nielsen, R. (2019). Why propensity scores should not be used for matching. *Political Analysis*, 27(4):435–454.

King, G., Tomz, M., and Wittenberg, J. (2000). Making the most of statistical analysis: Improving interpretation and presentation. *American Journal of Political Science*, 44(2):341–355.

Kleiber, C. and Achim, Z. (2016). Visualizing count data regression using rootograms. *The American Statistician*, 70(3):296–303.

Kleiber, C. and Zeileis, A. (2008). *Applied Econometrics with R*. Springer-Verlag, New York.

Kruschke, J. K. (2013). Bayesian estimation supersedes the *t* test. *Journal of Experimental Psychology: General*, 142(2):573–603.

Kruschke, J. K. (2015). *Doing Bayesian Data Analysis*. Academic Press, San Diego, CA, 2nd edition.

Kruschke, J. K. (2018). Rejecting or accepting parameter values in Bayesian estimation. *Advances in Methods and Practices in Psychological Science*, 1(2):270–280.

Kruschke, J. K. and Liddell, T. M. (2018). The Bayesian new statistics: Hypothesis testing, estimation, meta-analysis, and power analysis from a Bayesian perspective. *Psychonomic Bulletin & Review*, 25(1):178–206.

Kruschke, J. K. and Meredith, M. (2020). BEST: Bayesian estimation supersedes the *t* test. R package version 0.5.2. https://CRAN.R-project.org/package=BEST.

Kuhn, M. (2022). caret: Classification and regression training. R package version 6.0-91. https://CRAN.R-project.org/package=caret.

Larmarange, J. (2021). *labelled: Manipulating labelled data*. R package version 2.9.0. https://CRAN.R-project.org/package=labelled.

LeDell, E., Gill, N., Aiello, S., Fu, A., Candel, A., Click, C., Kraljevic, T., Nykodym, T., Aboyoun, P., Kurka, M., and Malohlava, M. (2022). h2o: R interface for the 'H2O' scalable machine learning platform. R package version 3.36.0.3. https://CRAN.R-project.org/package=h2o.

Leeper, T. J. (2018). margins: Marginal effects for model objects. R package version 0.3.23, https://cran.r-project.org/web/packages/margins/.

Legendre, A. (1805). Nouvelles Méthodes pour la Détermination des Orbites des Cométes (New Methods for the Determination of the Orbits of Comets). Firmin Didot, Paris, France.

Liao, T. F. (1994). *Interpretating Probability Models: Logit, Probit, and Other Generalized Linear Models*. SAGE Publications, Thousand Oaks.

Liao, T. F. (2002). *Statistical Group Comparison*. Wiley, New York.

Liaw, A. and Wiener, M. (2002). Classification and regression by randomForest. *R News*, 2(3):18–22.

Likert, R. (1932). A technique for the measurement of attitudes. *Archives of Psychology*, 22(140):5–55.

Linden, A. and Samuels, S. J. (2013). Using balance statistics to determine the optimal number of controls in matching studies. *Journal of Evaluation in Clinical Practice*, 19:968–975.

Long, J. S. (1997). *Regression Models for Categorical and Limited Dependent Variables*. SAGE Publications, Thousand Oaks, CA.

Lüdecke, D., (2018). ggeffects: Tidy data frames of marginal effects from regression models. Journal of Open Source Software, 3(26):1–5.

Lüdecke, D., Ben-Shachar, M., Patil, I., Waggoner, P., and Makowski, D. (2021). performance: An R package for assessment, comparison and testing of statistical models. *Journal of Open Source Software*, 6(60), 3139.

Lynch, S. M. (2010). *Introduction to Applied Bayesian Statistics and Estimation for Social Scientists*. Springer, New York.

Maddala, G. S. (1983). *Limited-Dependent and Qualitative Variables in Econometrics*. Cambridge University Press, New York.

Mahalanobis, P. S. (1936). On the generalized distance in statistics. *Proceedings of the National Institute of Sciences of India*, 2(1):49–55.

Mantel, N. (1966). Evaluation of survival data and two new rank order statistics arising in its consideration. *Cancer Chemotherapy Reports*, 50(3):163–170.

Martin, A. D., Quinn, K. M., and Park, J. H. (2011). MCMCpack: Markov chain Monte Carlo in R. *Journal of Statistical Software*, 42(9):22.

Massey, F. J. (1951). The Kolmogorov-Smirnov test for goodness of fit. *Journal of the American Statistical Association*, 46(253):68–78.

McCullagh, P. (1980). Regression models for ordinal data. *Journal of the Royal Statistical Society. Series B (Methodological)*, 42(2):109–142.

McCullagh, P. and Nelder, J. (1989). *Generalized Linear Models*. Chapman & Hall, London, UK, 2nd edition.

McCullagh, P. and Nelder, J. A. (1983). *Generalized Linear Model*. Chapman & Hall, London, UK, 1st edition.

McFadden, D. (1968). The revealed preferences of a government bureaucracy: Empirical evidence. *The Bell Journal of Economics*, 7(1);55–72.

McFadden, D. L. (1973). Conditional logit analysis of qualitative choice behavior. In P. Zarembka (Ed.), *Frontiers in Econometrics*. Academic Press, New York.

McKelvey, R. D. and Zavoina, W. (1975). A statistical model for the analysis of ordinal level dependent variables. *The Journal of Mathematical Sociology*, 4(1):103–120.

McLachlan, G. J. and Peel, D. (2000). *Finite Mixture Models*. Wiley Series in Probability and Statistics. John Wiley & Sons, Inc., New York.

Meehl, P. (1997). *The Problem Is Epistemology, Not Statistics: Replace Significance Tests by Confidence Intervals and Quantify Accuracy of Risky Numerical Predictions*, pages 395–425. Mahwah, NJ.

Meehl, P. E. (1967). Theory-testing in psychology and physics: A methodological paradox. *Philosophy of Science*, 34(2):103–115.

Menard, S. (2000). Coefficients of determination for multiple logistic regression analysis. *The American Statistician*, 54(1):17–24.

Milborrow, S. (2021). rpart.plot: Plot 'rpart' models: An enhanced version of 'plot.rpart'. R package version 3.1.0. https://CRAN.R-project.org/package=rpart.plot.

Morgan, S. L. and Winship, C. (2015). *Counterfactuals and Causal Inference: Methods and Principles for Social Research*. Cambridge University Press, New York.

Muennig, P., Johnson, G., Kim, J., Smith, T. W., and Rosen, Z. (2011). The general social survey-national death index: An innovative new dataset for the social sciences. *BMC Research Notes*, 4:385.

Murphy, K. P. (2012). *Machine Learning: A Probabilistic Perspective*. The MIT Press, Cambridge, MA.

Neyman, J. (1937). Outline of a theory of statistical estimation based on the classical theory of probability. *Philosophical Transactions of the Royal Society of London, Series A, Mathematical and Physical Sciences*, 236(767):333–380.

Neyman, J. and Pearson, E. S. (1933). On the problem of the most efficient tests of statistical hypotheses, *Philosophical Transactions of the Royal Society of London. Series A*, 231:289–337.

Nuzzo, R. (2014). Scientific method: Statistical errors. *Nature*, 506(7487):150–152.

O'Connell, A. A. (2006). *Logistic Regression Models for Ordinal Response Variables*. SAGE Publications., Thousand Oaks, CA.

Pearl, R. and Reed, L. J. (1920). On the rate of growth of the population of the United States since 1870 and its mathematical representation. *Proceedings of the National Academy of Sciences*, 6:275–288.

Pearson, K. (1900). X. On the criterion that a given system of deviations from the probable in the case of a correlated system of variables is such that it can be reasonably supposed to have arisen from random sampling. *The London, Edinburgh, and Dublin Philosophical Magazine and Journal of Science*, 50(302):157–175.

Pearson, K. (1904). *Mathematical Contributions to the Theory of Evolution XIII: On the Theory of Contingency and Its Relations to Association and Normal Correlation*. Drapers' Company Research Memoirs, Biometric Series, No. 1. Dulau and Co., London.

Peters, A. and Hothorn, T. (2021). ipred: Improved predictors. R package version 0.9-12. https://CRAN.R-project.org/package=ipred.

Peterson, B. and Harrell, Frank E., J. (1990). Partial proportional odds models for ordinal response variables. *Journal of the Royal Statistical Society. Series C (Applied Statistics)*, 39(2):205–217.

Peto, R. and Peto, J. (1972). Asymptotically efficient rank invariant test procedures. *Journal of the Royal Statistical Society. Series A (General)*, 135(2):185–207.

Pinheiro, J., Bates, D., Debroy, S., Sarkar, D., and R Core Team (2020). nlme: Linear and nonlinear mixed effects models. R package version 3.1-149. https://CRAN.R-project.org/package=nlme.

Plackett, R. L. (1949). A historical note on the method of least squares. *Biometrika*, 36(3/4):458–460.

Poisson, S. D. (1837). *Recherches sur la Probabilité des jugements en matiére criminelle et en matiére civile [Research on the Probability of Judgments in Criminal and Civil Matters] (in French)*. Bachelier, Paris, France.

Powers, D. and Xie, Y. (2008). *Statistical Methods for Categorical Data Analysis*. Emerald, London, UK, 2nd edition.

Prasad, N., Sharples, K. J., Murdoch, D. R., and Crump, J. A. (2015). Community prevalence of fever and relationship with malaria among infants and children in low-resource areas. *The American Journal of Tropical Medicine and Hygiene*, 93(1):178–180.

R Core Team (2020). *R: A Language and Environment for Statistical Computing*. R Foundation for Statistical Computing, Vienna, Austria. https://www.R-project.org/.

Rabe-Hesketh, S. and Skrondal, A. (2012). *Multilevel and Longitudinal Modeling Using Stata Volume II: Categorical Responses, Counts, and Survival*. The Stata Press, College State, Texas, 3rd edition.

Raftery, A. E. (1995). Bayesian model selection in social research. *Sociological Methodology*, 25:111–163.

Raftery, A. E. (1999). Bayes factors and BIC: Comment on a critique of the Bayesian information criterion for model selection. *Sociological Methods & Research*, 27(3):411–427.

Rao, C. R. (1973). *Linear Statistical Inference and Its Applications*. John Wiley & Sons, New York, 2nd edition.

Rassam, P. R., Ellis, R. H., and Bennett, J. C. (1971). The n-dimensional logit model: Development and application. *Highway Research Recording*, 369:135–147.

Raudenbush, S. W. and Bryk, A. S. (2002). *Hierarchical Linear Models: Application and Data Analysis Methods*. SAGE Publications, Thousand Oaks, CA, 2nd edition.

Reid, N. (1994). A conversation with Sir David Cox. *Statistical Science*, 9(3):439–455.

Ripley, B. (2019). *tree: Classification and Regression Trees*. R package version 1.0-40. https://CRAN.R-project.org/package=tree.

Robert, C. P. and Casella, G. (2011). A short history of Markov chain Monte Carlo: Subjective recollections from incomplete data. *Statistical Science*, 26(1):102–115.

Robin, X., Turck, N., Hainard, A., Tiberti, N., Lisacek, F., Sanchez, J.-C., and Müller, M. (2011). pROC: An open-source package for R and S+ to analyze and compare ROC curves. *BMC Bioinformatics*, 12:77.

Rosenbaum, P. R. (2010). *Design of Observational Studies*. Springer, New York.

Rosenbaum, P. R. and Rubin, D. B. (1983). The central role of the propensity score in observational studies for causal effects. *Biometrika*, 70(1):41–55.

Ross, S. (1998). *A First Course in Probability*. Prentice Hall, Upper Saddle River, NJ, 5th edition.

Rossi, P. (2019). bayesm: Bayesian inference for marketing/micro econometrics. R package version 3.1-4. https://CRAN.R-project.org/package=bayesm.

Rubin, D. B. (1973). Matching to remove bias in observational studies. *Biometrics*, 29(1):159–183.

Rubin, D. B. (1980). Bias reduction using Mahalanobis-metric matching. *Biometrics*, 36(2):293–298.

Rubin, D. B. (2001). Using propensity scores to help design observational studies: Application to the tobacco litigation. *Health Services and Outcomes Research Methodology*, 2(3):169–188.

Samuel, A. L. (1959). Some studies in machine learning using the game of checkers. *IBM Journal of Research and Development*, 3(3):535–554.

Sarkar, D. (2008). *Lattice: Multivariate Data Visualization with R.* Springer, New York.

Schauberger, G. (2019). EffectStars: Visualization of categorical response models. R package version 1.9-1. https://CRAN.R-project.org/package=EffectStars.

Schlegel, B. (2019). glm.predict: Predicted values and discrete changes for GLM. R package version 4.1-0. https://CRAN.R-project.org/package=glm.predict.

Schlegel, B. and Steenbergen, M. (2020). brant: Test for parallel regression assumption. R package version 0.3-0. https://CRAN.R-project.org/package=brant.

Schwarz, G. (1978). Estimating the dimension of a model. *The Annals of Statistics*, 6(2):461–464.

Sekhon, J. S. (2011). Multivariate and propensity score matching software with automated balance optimization: The matching package for R. *Journal of Statistical Software*, 42(7):1–52.

Shannon, C. E. (1948). A mathematical theory of communication. *The Bell System Technical Journal*, 27(3):379–423.

Shannon, D. F. (1970). Conditioning of quasi-Newton methods for function minimization. *Mathematics of Computation*, 24(647-656).

Signorell, Andri et mult. al. (2020). DescTools: Tools for descriptive statistics. R package version 0.99.37. https://cran.r-project.org/package=DescTools.

Simon, G. (1974). Alternative analyses for the singly-ordered contingency table. *Journal of the American Statistical Association*, 69(348):971–976.

Sing, T., Sander, O., Beerenwinkel, N., and Lengauer, T. (2005). ROCR: Visualizing classifier performance in R. *Bioinformatics*, 21(20):7881.

Smith, T. W., Davern, M., Freese, J., and Morgan, S. L. (2019). General social surveys, 1972-2018. [machine-readable data file] /Principal Investigator, Smith, Tom W.; Co-Principal Investigators, Michael Davern, Jeremy Freese and Stephen L. Morgan; Sponsored by National Science Foundation. —NORC ed.— Chicago: NORC, 2019. 1 data file (64,814 logical records) + 1 codebook (3,758 pp.). — (National Data Program for the Social Sciences, no. 25).

Smithson, M. and Verkuilen, J. (2006). *Fuzzy Set Theory: Applications in the Social Sciences.* Quantitative Applications in the Social Sciences. SAGE Publications, Thousand Oaks, CA.

Snell, E. J. (1964). A scaling procedure for ordered categorical data. *Biometrics*, 20(3):592–607.

Splawa-Neyman, J., Dabrowska, D. M., and Speed, T. P. (1990). On the application of probability theory to agricultural experiments. Essay on principles. Section 9. *Statistical Science*, 5(4):465–472.

Stan Development Team (2018). Stan modeling language users guide and reference manual. https://mc-stan.org/users/documentation/.

Stan Development Team (2020). RStan: the R interface to Stan. R package version 2.21.2. https://mc-stan.org/.

Stigler, S. M. (1982). Poisson on the Poisson distribution. *Statistics & Probability Letter*, 1(1):33–35.

Stigler, S. M. (1986). Laplace's 1774 memoir on inverse probability. *Statistical Science*, 1(3):359–378.

Stone, J. V. (2013). *Bayes' Rule*. Sebtel Press, Sheffield, UK.

Stuart, E. A. (2010). Matching methods for causal inference: A review and a look forward. *Statistical Science*, 25(1):1–21.

Tanaka, H., Uejima, S., and Asai, K. (1982). Linear regression analysis with fuzzy model. *IEEE Transactions on Systems, Man and Cybernetics*, 12(6):903–907.

Theil, H. (1969). A multinomial extension of the linear logit model. *International Economic Review*, 10(3):251–259.

Theil, H. (1971). *Principles of Econometrics*. John Wiley & Sons, New York.

Therneau, T. M. and Grambsch, P. M. (2000). *Modeling Survival Data: Extending the Cox Model*. Springer, New York.

Therneau, T. M. and Atkinson, B. (2019). rpart: Recursive partitioning and regression trees. R package version 4.1-15. https://CRAN.R-project.org/package=rpart.

Therneau, T. M. (2021). *A package for survival analysis in R*. R package version 3.2-13. https://CRAN.R-project.org/package=survival.

Therneau, T. M. and Atkinson, E. J. (2022). An introduction to recursive partitioning using the RPART routines. The Comprehensive R Archive Network. https://cran.r-project.org/web/packages/rpart/vignettes/longintro.pdf.

Tjur, T. (2009). Coefficients of determination in logistic regression models a new proposal: The coefficient of discrimination. *The American Statistician*, 63(4):366–372.

Train, K. (2009). *Discrete Choice Methods with Simulation*. Cambridge University Press, Cambridge, UK.

Tukey, J. W. (1972). *Some Graphic and Semigraphic Displays*. In T. A. Bancroft (Ed.), *Statistical Papers in Honor of George W. Snedecor*, pp. 293–316. Iowa State University Press, Ames, IA.

Van Noorden, R., Maher, B., and Nuzzo, R. (2014). The top 100 papers. *Nature*, 514:550–553.

van Ravenzwaaij, D., Cassey, P. and Brown, S.D. (2018). A simple introduction to Markov chain Monte Carlo sampling. *Psychonomic Bulletin & Review*, 25:143–154.

Vehtari, A., Gelman, A., and Gabry, J. (2017). Practical Bayesian model evaluation using leave-one-out cross-validation and WAIC. *Statistics and Computing*, 27(5):1413–1432.

Venables, W. and Ripley, B. (2002). *Modern Applied Statistics with S*. Statistics and Computing. Springer, New York, 4th edition.

Verhulst, P.-F. (1838). Notice sur la loi que la population suit dans son accroissement. *Correspondance Mathematique et Physique*, 10:113–121.

von Bortkiewicz, L. (1898). *Das Gesetz der Kleinen Zahlen*. B. G. Teubner, Leipzig, Germany.

Wackerly, D. D., Mendenhall, W., III., and Scheaffer, R. L. (2002). *Mathematicsl Statistics with Applications*. Duxbury Advanced Series. Duxbury Thomson Learning, Pacific Grove, CA, 6th edition.

Wald, A. (1943). Tests of statistical hypotheses concerning several parameters when the number of observations is large. *Transactions of the American Mathematical Society*, 54(3):426–482.

Wang, X., Chen, M.-H., Wang, W., and Yan, J. (2017). dynsurv: Dynamic models for survival data. R package version 0.3-6. https://cran.r-project.org/web/packages/dynsurv/index.html.

Watanabe, S. (2010). Asymptotic equivalence of Bayes cross validation and widely applicable information criterion in singular learning theory. *Journal of Machine Learning Research*, 11:3571–3594.

Wedderburn, R. W. M. (1974). Quasi-likelihood functions, generalized linear models, and the Gauss-Newton method. *Biometrika*, 61(3):439–447.

White, J. M. and Jacobs, A. (2021). log4r: A fast and lightweight logging system for R, based on 'log4j'. R package version 0.4.2. https://CRAN.R-project.org/package=log4r.

Wickham, H. (2016). *ggplot2: Elegant Graphics for Data Analysis*. Springer-Verlag, New York.

Wickham, H., Averick, M., Bryan, J., Chang, W., McGowan, L. D., François, R., Grolemund, G., Hayes, A., Henry, L., Hester, J., Kuhn, M., Pedersen, T. L., Miller, E., Bache, S. M., Müller, K., Ooms, J., Robinson, D., Seidel, D. P., Spinu, V., Takahashi, K., Vaughan, D., Wilke, C., Woo, K., and Yutani, H. (2019). Welcome to the tidyverse. *Journal of Open Source Software*, 4(43):1686.

Wickham, H. and Miller, E. (2021). haven: Import and export SPSS, Stata and SAS Files. R package version 2.4.3. https://CRAN.R-project.org/package=haven.

Wilcox, R. R. (2003). *Applying Contemporary Statistical Techniques*. Academic Press, London, UK.

Williams, R. (2006). Generalized ordered logit/partial proportional odds models for ordinal dependent variables. *The Stata Journal*, 6(1):58–82.

Williams, R. (2009). Using heterogeneous choice models to compare logit and probit coefficients across groups. *Sociological Methods & Research*, 37(4):531–559.

Wooldridge, J. M. (2010). *Econometric Analysis of Cross Section and Panel Data*. The MIT Press, Cambridge, MA, 2nd edition.

Xie, Y. (2015). *Dynamic Documents with R and knitr*. Chapman and Hall/CRC, Boca Raton, Florida, 2nd edition.

Xu, J., Bauldry, S. G., and Fullerton, A. S. (2019). Bayesian approaches to assessing the parallel lines assumption in cumulative ordered logit models. *Sociological Methods & Research*, 51(2):667–698.

Xu, J. and Fullerton, A. S. (2013). Comparing, confounding, or clarifying? Alternative measures of statistical group comparisons in binary regression models. *Chinese Sociological Review*, 46(2):91–119.

Xu, J. and Long, J. S. (2005a). Confidence intervals for predicted outcomes in regression models for categorical outcomes. *The Stata Journal*, 5(4):537–559.

Xu, J. and Long, J. S. (2005b). Using delta method to construct confidence intervals for predicted probabilities, rates, and discrete changes. Unpublished manuscript.

Yee, T. W. (2015). *Vector Generalized Linear and Additive Models: With an Implementation in R*. Springer, New York.

Yule, G. U. (1900). On the association of attributes in statistics. *Philosophical Transactions of the Royal Society Series A: Mathematical, Physical & Engineering Sciences*, 194:257–319.

Zeidler, E. (1996). *Oxford User's Guide to Mathematics*. Oxford University Press, Oxford, UK.

Zeileis, A. A., Kleiber, C., and Jackman, S. (2008). Regression models for count data in R. *Journal of Statistical Software*, 27(8):1–25.

Zelner, B. A. (2009). Using simulation to interpret results from logit, probit, and other nonlinear models. *Strategic Management Journal*, 30(12):1335–1348.

Zens, G., Frühwirth-Schnatter, S., and Wagner, H. (2020). UPG: Efficient Bayesian models for binary and categorical data. arXiv Preprint.

Zhou, H., Hanson, T., and Zhang, J. (2020). spBayesSurv: Fitting Bayesian spatial survival models using R. *Journal of Statistical Software*, 92(9):1–33.

Zou, H. and Hastie, T. (2020). *elasticnet: Elastic net for sparse estimation and sparse PCA*. R package version 1.3. https://cran.r-project.org/web/packages/elasticnet/.

Index

Note: Locators in *italics* represent figures and **bold** indicate tables in the text.

A

Accelerated failure time (AFT), 180–181
    exponential AFT regression, 181–183
    Weibull AFT regression, 183–186
Adjacent category regression, 111–113
ADRC, *see* Average discrete rates of changes
AFT, *see* Accelerated failure time
AIC, *see* Akaike information criterion
Akaike information criterion (AIC), 56–57
anova function, 53–54, 103
Area under ROC curve (AUC), 59
Arrow's theorem, 127
Artificial intelligence, 4
as.matrix function, 201
as.numeric function, 16
aspect argument, 43
Asymmetric error distribution, 91
Asymptotic normality, 48
Asymptotic variance-covariance matrix, 64
ATE, *see* Average treatment effects
Auto-correlation, 81
Average discrete rates of changes (ADRC), 70
Average marginal effect (AME), 72
Average treatment effects (ATE), 217–218

B

Bagging and random forests, 250–252
BASH, 12
Bayes, Thomas, 5
Bayes' formula, 6
bayesglm function, 76
Bayesian analyses, 8, 31
Bayesian approaches
    linear regression, 32–36
    regression modeling, 4–6
    survival regression, 194–195
        Bayesian estimation of survival
          models using spBayesSurv,
          202–204
        Bayesian estimation of Weibull PH
          Model using rstan, 195–202

Bayesian binary regression, 74–75
    assessment of null values, 84–85
    estimation of, 78–82
    post-estimation analysis, 82–84
    priors, 75–78
Bayesian Cox PH model, 203
Bayesian estimation, 76, 79, 130–136
    binary regression models, 77
    negative binomial regression, 162–167
    survival models Using spBayesSurv,
        202–204
    Weibull PH Model using rstan, 195–202
    zero-inflated Poisson regression,
        167–168
Bayesian inference
    framework, 62
    using Gibbs Sampling (BUGS), 31
Bayesian information criteria (BIC), 56–57
Bayesian LOO-CV, 83
Bayesian multilevel regression, 214–216
Bayesian negative binomial regression
    model (NB2), 162
Bayesian non-parallel cumulative ordered
    regression, 132–134
Bayesian notation, 5
Bayesian parallel cumulative ordered
    regression, 131–132
Bayesian polytomous regression
    Bayesian estimation, 130–136
    Bayesian non-parallel cumulative
        ordered regression, 132–134
    Bayesian parallel cumulative ordered
        regression, 131–132
    Bayesian stereotype logit model,
        134–136
Bayesian statistics, 3–5
Bayesian stereotype logit model, 134–136
Bayes' theorem, 5–6
Behavioral Risk Factor Surveillance System
    (BRFSS), 206

Bennett, J. C., 40
Berkson, Joseph, 38
Berndt-Hall-Hall-Hausman (BHHH)
      algorithm, 48
Bernoulli distribution, 38
Bernoulli random variables, 137
Best linear unbiased estimator (BLUE), 22,
      25
Beta distribution, 6
Binary coding, 37
Binary logit models, 75, 91
Binary regression, 37
      Bayesian binary regression, 74–75
            Bayesian assessment of null values,
                  84–85
            Bayesian estimation of, 78–82
            Bayesian post-estimation analysis,
                  82–84
            priors, 75–78
      history, 37–38
      hypothesis testing, 49–51
            Hosmer-Lemeshow test, 61
            likelihood ratio (LR), 51–56
            null hypothesis significance testing
                  (NHST), limitations, 61–62
            ROC Curve, 59–60
            scalar measures, 56–58
            score tests, 51–56
            Wald, 51–56
      interpretation of results, 62–63
            delta method, 64
            discrete rates of change (DRC),
                  69–71
            effects, based on, 68–73
            end-point transformation, 64
            group comparisons, 73–74
            marginal effects, 71–73
            odds ratios, 68–69
            precision estimates, 63–65
            predictions, based on, 65–68
            re-sampling methods, 64–65
      linear probability regression, 38–40
      maximum likelihood estimation (MLE),
            40–41, 45–46
            examples, 41–45
            nonlinear probability, 49
            normality, consistency, and efficiency,
                  48–49
            numerical methods, 46–48
Binary regression models (BRM), 3, 37–38,
      57, 59

Binary response variable, 1, 3
Binomial distribution, 137, 161
Bivariate Cox PH regression, 193
Bivariate matrix plot, *19*
BLUE, *see* Best linear unbiased estimator
Body mass index (BMI), 87
boot function, 68, 97, 149
Bootstrap method, 97
Bortkiewicz's Prussian Horse-Kick Data,
      138
brant function, 101
Brant test, 100–101
Broyden-Fletcher-Goldfarb-Shanno (BFGS)
      algorithm, 48
BUGS software, 32

C
CART, *see* Classification and regression tree
Categorical and limited response variables
      (CLRV), 1–2
      history, 2–3
      overview, 3
Categorical variable, 1
Causal inference, 3, 216–217
      average treatment effects, 217–218
      average treatment effects for treated,
            218
      coarsened (exact) matching, 226–228
      genetic matching, 224–226
      ignorability of treatment assumption,
            218
      Mahalanobis distance matching,
            223–224
      propensity score analysis, 219
      propensity score matching, 219–223
Censoring and truncation, 170–172
Central limit theorem, 4, 49
c function, 10
Chow's test, 73
Classical ordered regression models
      heterogeneous choice models, 119–121
      inflated ordered regression, 115–119
      model selection, 121–122
Classification and regression tree (CART),
      253
Classification trees, 243, 252–255
clm function, 102
Coarsened (exact) matching, 226–228
Coding schema, 37
Common-sense criterion, 48
confint function, 69, 145

Conjugate priors, 6–7, 75

Constrained partial cumulative model, 104

Contagion, heterogeneity, and over-dispersion, 150–152

Contingency coefficient, 2

Contingency table, 2

Continuation ratio regression, 106–111

Continuation/stopping ratio (CR), 106

Continuity thesis, 2

`convert.factor` option, 12

Correlation coefficient in linear regression, 2

Correlation of association, 2

Count, 87

Count regression
 basic count regression models, 139–140
  contagion, heterogeneity, and over-dispersion, 150–152
  count response variable, 140–141
  negative Binomial regression, 152–157
  plot observed *vs.* predicted count proportions, 141–142
  Poisson regression, 142–150
  quasi-Poisson regression, 152
 Bayesian estimation
  negative binomial regression, 162–167
  zero-inflated Poisson regression, 167–168
 Poisson distribution, 137–139
 zero-modified count regression, 157–158
  hurdle models, 159–160
  zero-inflated models, 160–162
  zero-truncated models, 158–159

Count response variable, 1, 140–141

Cox proportional hazard regression model, 107

Cox regression, 190–192

`cox.zph` function, 194

`createDataPartition` function, 235

Credible intervals of predicted probability, *135*

Cumulative density function, 187

Cumulative logit, 89

Cumulative model, 88

Cumulative regression, 89–98
 hypothesis testing and model comparison, 93–95
 interpretation, 95–98
 model setup and estimation, 89–93

`curve function`, 41, 43

Cut-point equations, 89, 96, 99–102, 114, 115, 122

Cut-points, 89

D

`data.frame` function, 29, 97, 148, 201

Data management, 25

Data sparsity, 76

Davidon-Fletcher-Powell (DFP) algorithm, 48

Decision/regression trees, 243–250

Degrees of freedom (df), 3, 52–54, 61, 75, 93, 104, 127

Delta method, 64

`deltaMethod function`, 66

Descriptive survival analysis
 Kaplan-Meier estimator, 174–177
 log-rank test, 177–180

Differential equation, 38

Dirichlet distribution, 134

Discrete probability distribution, 137

Discrete rates of change (DRC), 69–71

`dispersiontest` function, 151

χ^2 distribution, **61, 104**

Dosage effect of nicotine, 37

`dotchart` function, 15

DRC, *see* Discrete rates of change

`dydx` function, 70

E

End-of-study censoring., 171

End-point transformation, 64

Equal-tailed credible interval (ETI), 84

Error correlation and ZiOP, 118

Exponential AFT regression, 181–183

Exponential PH regression, 187–188

Extensions
 causal inference, 216–217
  average treatment effects, 217–218
  coarsened (exact) matching, 226–228
  genetic matching, 224–226
  ignorability of treatment assumption, 218
  Mahalanobis distance matching, 223–224
  propensity score analysis, 219
  propensity score matching, 219–223
 machine learning, 228–229
  bagging and random forests, 250–252
  classification trees, 252–255

decision/regression trees, 243–250
penalized binary logit, 241–243
regression and classification, 230
regularization, 230–241
and statistical learning, 229–230
supervised learning, 230
supervised learning and unsupervised
learning, 230
training, validation, and test, 230
multilevel regression, 205–206
Bayesian multilevel regression,
214–216
multilevel count regression, 211–214
multilevel logit regression, 206–211
Extremum estimators, 46

F

Factor change coefficient, 144
factor function, 18
Factorial design, 3
Failure time variables, 169
First-order Taylor series, 64
Fisher, Ronald A., 3–4, 40, 49
Fisher's exact test for 2 x 2 table,
3
Fisher's test of significance, 50
Flexible parameterization, 31
foreign package, 12, 26
Frequentist approach to regression
analysis, 4
function function, 41, 43
Fuzzy regression framework, 2

G

Gauss, Carl Friedrich, 22, 40
Gauss-Markov theorem, 24
Generalized linear mixed models (GLMMs),
205
Generalized linear models (GLM), 229
General Social Survey (GSS), 38
Genetic matching, 224–226
Gibbs sampler, 8
Gini index, 252–253
glmer function, 207–208
glm function, 46, 76, 131, 143–144, 207
glmmPQL function, 210
glm.nb function, 143
glmnet function, 232
Gosset's Student *t* distribution, 3
Graunt, John, 169
GUI (graphical user interface), 10

H

Hamiltonian Monte Carlo algorithm, 8, 32
Hazard function, 173–174
Hessian matrix, 47–48, 56, 117
Heterogeneous choice models, 119–121
Higher-order polynomial, 24
Highest density intervals (HDIs), 61, 84,
85–86, 165
Homoscedasticity, 24, 29
Hosmer-Lemeshow test, 61
test in binary regression model,
94
hurdle function, 159
Hurdle models, 159–160
Hybrid Monte Carlo, *see* Hamiltonian
Monte Carlo algorithm
Hyper-parameter cumulative logit model,
122
Hyper-priors, 75
Hypothesis testing, 49–51, *50*
Hosmer-Lemeshow test, 61
likelihood ratio (LR), 51–56
null hypothesis significance testing
(NHST), limitations, 61–62
ROC Curve, 59–60
scalar measures, 56–58
score tests, 51–56
Wald, 51–56

I

Identifiability, 24–25
Ignorability of treatment assumption, 218
Independence of irrelevant alternatives
(IIA), 88, 127
indeptCoxph function, 203
Inflated ordered regression, 115–119
Informative priors, 7, 75
Integrated development environment (IDE),
9
Intelligence quotient (IQ), 42
Intent-to-treat effect (ITTE), 218
Intra-class correlation (ICC), 208
Iterative dichotomizer 3 (ID3), 253

J

JAGS (just another Gibbs sampler), 31

K

Kaplan–Meier estimator, 174–177
Kaplan–Meier plot, 177, *178, 180*
Kolmogorov-Smirnov test, 223

L

`label` function, 17
Lagrange multiplier (LM) test, 54
Lasso coefficients path graph, 238–239
Lasso regression, 238–239
Latent variable approach, *45*
Latin Square, 3
Law of parsimony, 100
Least squares estimator, 23
Leave-one-out (LOO) cross-validation, 236
Legendre, Adrien-Marie, 22
Likelihood function, 6, 44, 75, 136
Likelihood ratio (LR) test, 51–56, 101, 103
Likert, Rennis, 87
Likert scale, 87
Limited variable, 1
Linear function of covariates, 114
`linearHypothesis` function, 51–52
Linearity, 24
Linear predictor, 45, 205
Linear probability regression (LPR) models, 38
 health on education with jittering, *40*
Linear regression diagnostics plot, *30*
Linear regression models
 Bayesian approach to linear regression, 32–36
 ordinary least squares (OLS) regression, 21–22
 assumptions of, 23
 BUGS-like Software, 31–32
 estimation and interpretation, 25–31
 identifiability, 24–25
 nonstochastic covariates, 25
 normality, 25
 OLS estimator and variance-covariance matrix, 23
 results, 22–23
 spherical disturbance, 24
 Stan, 31–32
 zero conditional mean and linearity, 24
Linear transformation matrix, 51
Link function, 205
lm function, 25–26, 131
Local average treatment effect (LATE), 218
Logistic function, 38
`logitgof` function, 61, 95
Logit model, 3, 111–112
Log-rank test, 177–180
`loo_compare` function, 84

`loo` function, 133
LOO (leave-one-out), 133
LR test, *see* Likelihood ratio test

M

Machine learning, 228–229
 bagging and random forests, 250–252
 classification trees, 252–255
 decision/regression trees, 243–250
 penalized binary logit, 241–243
 regression and classification, 230
 regularization, 230–241
 and statistical learning, 229–230
 supervised learning, 230
 supervised learning and unsupervised learning, 230
 training, validation, and test, 230
Machine/statistical learning, 4
Mahalanobis, P. S., 223
Mahalanobis distance matching, 223–224
`margeff` function, 109–110
Marginal effects at means (MEM), 72
`margins` function, 70, 73
Markov chain, 8
Markov Chain Monte Carlo (MCMC), 8, 31, 81, *83*
`MASS` package, 92
`MatchBalance` function, 222
Maximization, 46
Maximum likelihood estimation (MLE), 38, 40–41, 45–46
 examples, 41–45
 mean for binomial distribution, 42
 mean for normal distribution, *43*
 nonlinear probability, 49
 normality, consistency, and efficiency, 48–49
 numerical methods, 46–48
Maximum likelihood (ML), 3
McFadden's pseudo, 94
McFadden's R^2, 57–58
MCMC simulations, 32
MCME, *see* Markov Chain Monte Carlo
Mean absolute error (MAE), 237
Mean squared error (MSE), 233
M-estimation method, 63
Metropolis-Hasting algorithm, 8
Minimization, 46
Mixed models in (bio-)statistics, 205
MLE, *see* Maximum likelihood estimation
Model complexity, 114

Monte Carlo methods, 8
Multi-category variables, 2–3, 87–88
Multilevel count regression, 211–214
Multilevel logit regression, 206–211
Multilevel regression, 205–206
    Bayesian multilevel regression, 214–216
    multilevel count regression, 211–214
    multilevel logit regression, 206–211
`multinom` function, 123
Multinomial logit, 111–112
    effect star plot, *126*
    model, 3, 127
    regression, 122–129
Multinomial probit regression, 129–130
Multinomial regressions
    multinomial logit regression, 122–129
    multinomial probit regression, 129–130
Multivariate regressions, 140
mvord package, 92

N
Negative binomial regression, 152–157,
    162–167
Newton-Raphson method, 47, *47*, 48
Newton's method, 47
Neyman, Jerzy, 4
Non-informative priors, 7
Non-linear function, 64
Nonlinear transformation, 24
Nonstochastic covariates, 25
Normality, 25
Null hypothesis, 51, 53, 61–62
Null hypothesis significance testing (NHST),
    61, 100; *see also* Hypothesis testing
    limitations, 61–62

O
Object-oriented language, 9
Object-oriented programming, 10
Observational errors, 22
Occam's razor, 100
Odds ratio, 2
`oglmx` function, 120
OLS regression, *see* Ordinary least squares
    regression
OpenBUGS, 31
`ordered` function, 17
Ordered regression
    adjacent category regression, 111–113
    continuation ratio regression, 106–111
    cumulative regression, 89–98

    history, 89
    model, 89–92
    ordinal measures and regression
        models, 88–89
    partial, proportional constraint, and
        non-parallel models, 101–106
    proportional odds/parallel lines
        assumption, 98–101
    stereotype logit, 114–115
`ordinal` package, 92
Ordinal measures and regression models,
    88–89
Ordinal response variable, 1, 3
Ordinary least squares (OLS) regression, 4,
    21–22
    assumptions of, 23
    BUGS-like Software, 31–32
    estimation and interpretation, 25–31
    identifiability, 24–25
    nonstochastic covariates, 25
    normality, 25
    OLS estimator and variance-covariance
        matrix, 23
    results, 22–23
    spherical disturbance, 24
    Stan, 31–32
    zero conditional mean and linearity, 24

P
`pairs` function, 19
Parallel lines assumption, 98–101
Parametric proportional hazard regression,
    186–187
    exponential PH regression, 187–188
    Weibull PH regression, 188–190
Partial, proportional constraint, and
    non-parallel models, 101–106
Pearson, Egon, 4
Pearson, Karl, 1–3
Penalized binary logit, 241–243
`performance` function, 59
`phreg` function, 188–189
`plotAreaInROPE` function, 86
`plot` function, 18, 30, 145, 177
Plot observed *vs.* predicted count
    proportions, 141–142
`plotPost` function, 201
`png` function, 15
Poisson, Simon Denis, 137
Poisson distribution, 137–139, 141, 150
Poisson regression, 142–150

Poisson regression model, 154, 156
polr function, 92
Polytomous(multi-category) response
    variable, 1
Polytomous regression, 87–88
    Bayesian polytomous regression
        Bayesian estimation, 130–136
        Bayesian non-parallel cumulative
            ordered regression, 132–134
        Bayesian parallel cumulative ordered
            regression, 131–132
        Bayesian stereotype logit model,
            134–136
    classical ordered regression models
        heterogeneous choice models,
            119–121
        inflated ordered regression, 115–119
        model selection, 121–122
    multinomial regression
        multinomial logit regression, 122–129
        multinomial probit regression,
            129–130
    ordered regression
        adjacent category regression, 111–113
        continuation ratio regression,
            106–111
        cumulative regression, 89–98
        history, 89
        ordinal measures and regression
            models, 88–89
        partial, proportional constraint, and
            non-parallel models, 101–106
        proportional odds/parallel lines
            assumption, 98–101
        stereotype logit, 114–115
Polytomous regression models, 111, 116, 119
Posterior distribution plot, *36*
Posterior distributions of hazard ratios,
    201
Post-estimation analysis, 82
Precision estimates, 63–65
predict function, 65, 96–97, 155, 161
predict.multinom function, 125
predictorEffects function, 98, 145
predict (predict.glm), 59
predict.vglm function, 109
predprob function, 148–149, 155
Price, Richard, 5
print function, 165
prior argument, 131
Priors, 75

conjugate priors, 6–7
informative, non-informative, and other
    priors, 7
Probabilistic programming language, 8
Probability density function (PDF), 6, 71,
    173
Probability distribution, 45
Probability mass function (PMF), 142
Programming indices, 9
Propensity score
    analysis, 219
    matching, 219–223
Proportional hazard (PH) assumption,
    192–194
Proportional odds (PO) assumption, 98–101
PseudoR2 function, 58
p-value, 50, 62, 124

Q
Quantile-quantile (Q-Q) plot, 29
Quasi-Poisson regression, 152

R
R, 8–9, 76
    add comments, 15
    advantage of, 9
    argument, 97
    base package, 9
    as calculator, 10–-11
    check transformation, 18–19
    close log, 20
    codes, 11
    data, examine, 13–14
    drop missing cases, 19
    dummy variables and check
        transformation, 16–17
    function, 156
    ggplot2 package, 9
    graph matrix, 19–20
    individual variables, examine, 14–15
    label values, 17–18
    label variables, 17
    load data, 12–13
    open log file, 12
    ordinal variables, 18
    Rcmdr package, 10
    RStudio, 9–10
    save data, 20
    save graphs, 15
    set up working directory, 11
    source codes, 20–21

subset data, 13
    `tidyverse` package, 9
    user-written packages, 31
R^2, 57–58
Random-intercept model, 207
`read.dta` function, 26
Receiver operating characteristic (ROC)
    curve, 59–60
Referencing process, 10
Region of practical equivalence (ROPE), 61,
    84, *85–86,* 165
Regression analysis, 4, 89
    Bayesian approach, 4–5
        COVID-19, 5–6
    Frequentist approach, 4
    Markov Chain Monte Carlo (MCMC),
        8
    priors
        conjugate priors, 6–7
        informative, non-informative, and
            other priors, 7
Regression and classification, 230
Regression models, 114
    for ordinal responses, 89
Regularization, 230–241
    parameter profile for ridge regression,
        238
Re-sampling methods, 64–65
Ridge regression, 231
ROC curve, *see* Receiver operating
    characteristic (ROC) curve
Root mean square error (RMSE), 237
`rootogram` function, 156
Rootograms of negative binomial *vs.*
    Poisson regression, *157*
ROPE, *see* Region of practical equivalence
`rrvglm` function, 102, 114, 115
`RSiteSearch` function, 11
R-squareds, 237
`rstan` package, 33

S
`sample` function, 232
Schwarz information criterion, 56–57
Score tests, 51–56, 101
Self-rated health (SRH), 87, 241
Sensitivity, 59
Separation, 76
`setx` function, 145
`shell` function, 12
Significance level, 50

Significance test, 50
`sim` function, 145
Single-coefficient LR test, 54
Single-equation regressions, 4
Single-parameter hypothesis, 51
`sink` function, 12
S language, 8
Slope-as-outcome equation, 210
Spherical disturbance, 24
Stable unit treatment value assumption
    (SUTVA), 218
Stan, 8, 31–32, 78, 132, 194
Standardized mean difference (SMD), 222
`stan` function, 165
Stan software, 33
`stan_polr` function, 131
Stan's `neg_binomial_2_lpmf` distributional
    function, 163
Statistical learning, 229–230
Stereotype logit, 114–115
Stigler's law of eponymy, 137
`Subset` function, 26
`summary` function, 123–124, 131, 165
Supervised learning, 230
    and unsupervised learning, 230
`Surv` function, 177
`survfit` function, 177
Survival analysis, 171
Survival function, 176
Survival regression, 169–170
    accelerated failure time model, 180–181
        exponential AFT regression, 181–183
        Weibull AFT regression, 183–186
    Bayesian approaches to survival
        regression, 194–195
        Bayesian estimation of survival
            models Using spBayesSurv,
            202–204
        Bayesian estimation of Weibull PH
            Model using rstan, 195–202
    censoring and truncation, 170–172
    Cox regression, 190–192
    descriptive survival analysis
        Kaplan-Meier estimator, 174–177
        log-rank test, 177–180
    hazard function, 173–174
    parametric proportional hazard
        regression, 186–187
        exponential PH regression, 187–188
        Weibull PH regression, 188–190

proportional hazard (PH) assumption, 192–194

time and survival function, 172–173

Survival time (time-to-event) response variable, 1

T

Taylor series, 64

χ^2 test, **2–3, 178**

Thresholds, 89

tidyverse package, 12

Time and survival function, 172–173

Time-to-event response variables, 3, 87

Type I errors, 51

Type II errors, 51

U

Uniform priors, 7

update function, 93

User-written packages in R, 31

Utility maximization, 123

V

var function, 11

Variance-covariance matrix, 23

Variance partition coefficient (VPC), 208

Verhulst, Pierre-Francois, 38

VGAM, 92

vglm function, 108, 112

W

waic function, 133

WAIC (widely applicable information criterion), 133

Wald test *(W)*, 51–56, 93, 101

Watanabe-Akaike information criterion (WAIC), 82

Weibull AFT

models, 180

regression, 183–186

Weibull PH regression, 188–190

WinBUGS, 31

Windows CMD, 12

wireframe function, 43

writeLines function, 80

Y

Yule, Udny, 1–3

Z

Zelig package, 68

ZeligChoice, 68

zelig function, 145

Zero conditional mean and linearity, 24

Zero-inflated ordered probit model (ZiOP), 117–119

Zero-inflated Poisson regression, 167–168

zeroinfl function, 161

Zero-modified count regression, 157–158

hurdle models, 159–160

zero-inflated models, 160–162

zero-truncated models, 158–159

zerotrunc function, 158

ZiOP, *see* Zero-inflated ordered probit model

ZIOPC function, 118